GPRS
Demystified

John Hoffman, Editor

McGraw-Hill
New York   Chicago   San Francisco   Lisbon
London   Madrid   Mexico City   Milan   New Delhi
San Juan   Seoul   Singapore   Sydney   Toronto

The McGraw·Hill Companies

Cataloging-in-Publication Data is on file with the Library of Congress

1 2 3 4 5 6 7 8 9 0   DOC/DOC   0 9 8 7 6 5 4 3 2

ISBN 0-07-138553-3

The sponsoring editor for this book was Marjorie Spencer and the production supervisor was Sherri Souffrance. It was set in New Century Schoolbook by Patricia Wallenburg.

Printed and bound by R. R. Donnelley & Sons Company.

McGraw-Hill books are available at special quantity discounts to use as premiums and sales promotions, or for use in corporate training programs. For more information, please write to the Director of Special Sales, Professional Publishing, McGraw-Hill, Two Penn Plaza, New York, NY 10121-2298. Or contact your local bookstore.

This book is printed on recycled, acid-free paper containing a minimum of 50 percent recycled, de-inked fiber.

This book is dedicated to the men and women who have toiled to make GSM a global success, from which to implement this "thing" we call GPRS; to the subject matter experts who have so kindly given of their time and talent by contributing to this book; and to Kirsten and Remy who are just beginning their quest for knowledge...

CONTENTS

PREFACE

Can GPRS be Demystified?

When general packet radio service (GPRS) was first conceived, it was fuzzy in terms of hard details, even for those who envisioned it as the beginning of the GSM evolution to an all-IP packet data service. Over time, GPRS became clearer in the minds of the "enlightened," who heralded it as a major first step in the move from circuit-switched wireless services to an always-on service with bits of data sent to and from devices at lightning speeds.

As the world began buying into the idea of the "Mobile Internet" (read, enabled by GPRS) accessible via inexpensive handheld wireless devices, something happened. Nothing. Real or apparent momentum slowed to a crawl and the GPRS we envisioned didn't happen. Everything got fuzzy again.

OK, GPRS was heading in an unforeseen direction. But change at this point proved to be a good thing, bringing with it evolution, movement, and the kind of advancement that only comes from a step back. If GPRS wasn't a bullet train to the future, we needed to demystify some of our own assumptions and expectations. Today no one is prepared to claim that she or he really knows what is going on in the world of GPRS, but here you can find perspective on problems and opportunities. If you read this compilation of some thirty chapters by practicing GPRS subject matter experts, all contributing their thoughts to the GPRS puzzle, you will have learned a lot about GPRS in particular and something significant about technology in general.

What do we know now about how things will play out tomorrow? Well, it will rain when we think it will be clear, be calm when we have a kite to fly, hot when we dress for cool weather and stressful just when we need a break. GPRS will never be what was first envisioned, but it's undeniable in the way a hit song is undeniable. It's easy to glom and hard to forget. It means many things to many people. It's an efficient, even an elegant platform for the multitude of applications we conjure up when we try to explain how wireless mobility will change the way we live. In some respects, the link between GPRS and tomorrow is stronger today than it ever was in the past.

There is this "thing" someone named GPRS or GSM Packet Radio Service, and I can only hope that whoever came up with the name was not a marketing genius gone mad, but rather some techie preparing a paper on packet services—because GPRS is indeed an innovator's innovation.

We undertook this book in the hope that it will give you ideas. I, for one, am still dumbfounded by the ability to transmit 1's and 0's over the airwaves such that folks like you and me can use wireless devices to make our lives more productive, or less stressful, or whatever "better" means in your vocabulary. It could be blinding fast downloads, total uninterruptible access to messages that matter, or freedom from tethers and wires. Or it could be the ability to send and receive important information wherever we may be without having to worry about whether it works or how it works. It sounds like magic, and is sometimes described that way, but actually today it's GPRS—at least for the world of GSM users.

So here is my advice:

Don't Read This Book.

At least in the typical sense of reading a book because this really isn't a book. It is more of a reference manual. A grouping of essays, vignettes, snippets of mature thoughts on specific topics, but not really a narrative. I don't say this just because the editor is not an expert on any particular subject, or because I tend to write in a stream of consciousness barely bounded by capital letters, commas, periods and conjunctions. But this format really is peculiarly useful for recording technology lifecycles. Let me explain.

Maybe by now you have bought into my theory that no one really knows GPRS. This theory is based upon my belief that the moment when something comes clear usually ushers in the moment when everything changes. According to this theory, which I really do espouse, the times in between when much is fuzzy are really the productive times, the good times when everyone is energized as a technical community. Some aspects of GPRS have or will soon come into focus and stay that way. Other aspects of GPRS will move in and out of focus, perhaps repeatedly. This "book" tries to identify *those* aspects of GPRS and outline their importance. This means that one reader may believe the author has presented a sound and plausible story about GPRS while another reader will vehemently disagree.

Why should you read the "book" which isn't really a book? Because if the "we" who put fingers to keyboard have successfully done our job, our words will evoke debate or at least contemplation. Because GPRS is really a quantum leap forward, measured as both a technology and a way of communicating. If GPRS really rewrites our interaction with information, much as the PC did ten years ago, I won't be surprised. If I fail to evoke such thoughts from you, I won't be communicating.

So for the three or four readers who have gotten to this point, I both congratulate and worry about you both in the same breath. You are probably one of the targeted readers, those who are between the ages of birth and death, male or female, interested in better understanding what makes GPRS go and even a little bit about how it works. You may be in the "wireless industry," but you may not be technically literate. Although this is a work about technology, it isn't a technical work. This is good as this isn't intended to be a technical book. And if you are not in the wireless industry, all the better, as you won't be bringing hearsay and hype along with you.

This is not to say that some parts are not technical; they are. Certain subjects that make up the GPRS story must be founded on technology or nothing works. We try to highlight them without getting bogged down in the nuts and bolts and jargon. Some of it (maybe more than is necessary) may get boring. When that happens, skip it and move on. Or better yet, pick up *GPRS Demystified* and read a few pages of a chapter that holds some interest to you. Then put it down and let it lie for a while—enough time for the thoughts to bounce around your brain or be filed away for future reference. Then pick it up again...

Inside you will find information about the three components which have to be present to make GPRS work: the infrastructure or underlying technology, the terminal devices to access the technology, and most important, the services or applications which run over the technology. Taken together, these three components define a commercial GPRS service offering. Without all three, the others are useless. Without the fourth "component," a well-conceived business proposition, commercial success is unlikely. The overall business proposition of GPRS has to begin, as maybe it always did, with the concept of an end-to-end service constructed for optimal simplicity. Therein lies the real challenge for GPRS success.

I conclude with a special thanks to you, the reader. You have taken a leap of faith in picking up this book, hopefully as part of a rational act. You are to be commended for your actions, or perhaps cautioned for your lack of sound judgment. And you should know one more thing: your purchase will never earn myself or the authors who contributed their time and talents to the writing of *GPRS Demystified*, one euro, dollar, shekel, or kroner. All proceeds that may come our way will be donated to charity. We didn't do this for the money, for the glory, or because we wanted to write a book. We did it to help advance a new and critically important part of the GSM family of technologies. GPRS. Which, if we as a GSM community get it wrong, will relegate a perfectly good technology like GSM to the status of a has-been. As a collective body, we don't believe such will be the case. Thus, everyone agreed to contribute their knowledge to make this project a reality. And much like GPRS itself, *GPRS Demystified* has been fraught with life's ups and downs, but the dream never died, maybe because nothing good in our world is ever easy.

John Hoffman
July 2002

CONTRIBUTORS

Laurent Bernard
France Telecom

Charles Brookson
CEng FIEE AFRIN

Clif Campbell
Cingular Wireless

Simon Cavenett
Mondo Techo LLC

Robert Conway
GSM Association

Carolyn Davies
Baskerville: Part of the Informa Telecoms Group

Axel Doerner
Vodafone

Scott Fox
Wireless Facilities, Inc.

Kim Fullbrook
O2

David Gordon
Orange Israel

Conchi Gutiérrez
Telefónica Móviles

Ray Haughey
GSM Association

Gerhard Heinzel
Swisscom Mobile

John Hoffman
GSM Association

Babak Jafarian
Wireless Facilities, Inc.

Stephan Keuneke
T-Mobile International

Joerg Kramer
Vodafone

Rainer Lischetzki
Motorola

Philippe Lucas
Orange France

Jarnail Malra
O2

Yves Martin
Orange France

R. Clark Misul
Detecon

Lauro Ortigoza Guerrero
Wireless Facilities, Inc.

Carsten Otto
T-Mobile

Stella Penso
Turkcell

Carol Politi
Megisto Systems

Tage Rasmussen
End2End

Jessica Roberts
Nokia

Jack Rowley
GSM Association

Rafael Ruiz de Valbuena Bueno
Telefónica Móviles

Richard Schwartz
SoloMio

Mark Smith
GSM Association

Darren Thompson
VoiceStream

Colin Watts
Lucent Technologies

Randy Wohlert
SBC Communications

Graham Wright
Lucent Technologies

1

Introduction to GSM Networking

You May Know GSM, but Do You Know How It Came About and Where It Is Going? (...and Should You Care?)*

It is often easier to understand the present from a historic perspective that adds context and background to what can otherwise be a knowledge black hole. History is even more important when future gazing with technology. GPRS has little track record and few people have first-hand knowledge of what it is and where it is headed. However, by looking back, we can begin to cement a foundation on which to build the GPRS story line.

GPRS is not an autonomous entity. It cannot operate by itself and relies on a subsystem for its very existence. That subsystem is GSM, Global System for Mobile communications or Groupe Spécial Mobile, as it was first called by the Conférence des Administrations Européenes des Postes et Télécommunications (CEPT). Groupe Spécial Mobile was re-branded Global System for Mobile communications in 1992 when it became clear that GSM was not going to be just a pan-European wireless telecommunication system, as first envisioned, but rather a global technology embraced by regulators and operators alike for providing standards-based communications services.

GSM is a wireless technology for core switching that's based upon the Mobile Application Protocol (MAP) and an air interface using time division multiple access (TDMA). One of its key strengths is the fact that it adheres to open standards, which are agreed on and published for use by both manufacturers and network service providers. The protocol work, like the technology, began as regional initiatives and spread globally in increments. GSM Core Specifications, as they are known, were passed from the GSM Permanent Nucleus, originally set up by CEPT in 1986, to the European Telecommunications Standards Institute (ETSI), which took over the responsibility for their development in 1989. The Phase I GSM Core Specifications were frozen by 1997, when they filled 130 volumes and numbered over 16,000 pages of text. But that was not the end of the story, as the specifications for GSM-based mobile communications continued to develop. Today the Core Specifications are managed by the 3rd Generation Partnership Project (3GPP), which is

---

*Portions of this chapter are reproduced with permission from GSM Association documents.

responsible for the ongoing evolution of standards for the GSM family of technologies.

But that is getting ahead of ourselves.... Before we look at the GSM family of technologies, which, by the way, includes GPRS, it will be helpful to move quickly through the development and launch of GSM.

GSM was conceived by CEPT in 1982 when it became clear that the various forms of analog wireless communications networks used throughout Europe would not withstand the pressure of continued growth or the lack of cross-border compatibility. After the technology was endorsed by the European Commission in 1984, countries throughout Europe set aside spectrum for GSM in the 900 MHz band. A meeting of the Heads of Member States in December 1986 recommended a Directive prescribing that GSM should be introduced across Europe, beginning with the launch of limited services in 1991, then coverage of major cities by 1993, and finally, the linkage of service areas by 1995.

On September 7, 1987, 15 network operators from 13 European countries signed the GSM Memorandum of Understanding (MoU) committing to deploy GSM service by July 1, 1991. The GSM MoU, which became a Swiss registered corporation in 1995 and was later re-branded the GSM Association (GSMA), is an organisation open to licensed mobile network operators committed to building and implementing GSM-based systems and to government regulators or administrators which issue wireless communications licenses. Recently, third-generation mobile operators have been added to the GSM 900, 1800, and 1900 MHz families that the GSM Association represents. Unlike 3GPP, which is responsible for the development of official standards and specifications, the GSMA cannot draft technical specifications for adoption. Rather, it represents its members' interests to the various working groups and recommends requirements which are then translated into specifications by the 3GPP.

How GSM Grew

The GSM Association has defined a clear vision and mission: "To be the leading representative body in the wireless industry by orchestrating the continuous improvement and evolution of wireless communications for the overall benefit of system operators, consumers, and suppliers around the world."

The GSM Association works to achieve its mission in the following areas:

- **Development and Evolution**
  Aim: "To spearhead the continued, nonproprietary development and evolution of GSM to maintain its position as the optimal wireless solution now and in the future."
- **Roaming**
  Aim: "To lead the world-wide development, enhancement, and promotion of international roaming capabilities for the benefit of the Association's members and their customers."
- **New Services and Applications**
  Aim: "To lead the development and co-ordination of wireless data core services and applications."
- **Social Responsibility**
  Aim: "To support and encourage the safe and responsible use, as well as the availability, of wireless communications throughout the world."
- **Communications**
  Aim: "To provide effective operational and promotional communications."

Through the leadership of the GSMA and hard work of member operators and manufacturers, the first GSM networks were launched in the summer of 1992, following a year's delay. This was due to the lack of GSM terminal phones and instability in network developments. However, the one-year delay allowed networks to be enlarged and stabilized and enabled the introduction of GSM phones to commercial production. This is important when considering GPRS because it demonstrates that delays are part and parcel of the launch experience and do not have to impact on the long-term success of a technology.

About the same time as the first networks were launching commercial service, the first GSM roaming agreement was executed. Telecom Finland and Vodafone signed it on June 17, 1992 and, by year end, 13 GSM networks were on the air and offering GSM service in Europe. Although the principle of roaming had been established, real GSM roaming did not occur until 1993 when operators across Europe began to tie their GSM networks together and promote seamless automatic roaming. 1993 also saw GSM adopted for use in the 1800 MHz frequency band and commercially launched by Mercury One2One in the UK in September. In that same year, the number of GSM subscribers surpassed 1,000,000 and the GSM banner advanced beyond the continent when Telstra became the first non-European signatory to the GSM MoU. GSM had taken the first steps to becoming the global standard for mobile communications that it is today.

During these early years, the personal computer revolution was beginning to take shape worldwide and GSM was at the forefront of the rapid wireless growth in Europe. It was inevitable that attempts would be made to bring the two together. Initially, this was focused on GSM's inherent data capabilities of circuit-switched data communications at 9.6 kbps and the capacity to send and receive SMS text messages of up to 160 characters in length between GSM phones. ETSI therefore finalized the specifications for data services over GSM and operators began to work on delivering services. These were first based on short message services (SMS) and later on high speed circuit-switched data (HSCSD), which has a maximum throughput speed of 14.4 kbps and was launched in 1994 in limited service areas. These two services represented GSM's first tentative steps away from the voice sector and into the rapidly growing data arena.

By Telecom 95 in Geneva, the GSM landscape had indeed broadened to include wireless data services. Developers were even able to demonstrate slow scan video over GSM on the convention floor. However, the GSM industry couldn't or didn't keep pace with data-related enhancements, largely because it was so successful with voice. The ever-burgeoning expansion of GSM service around the world put the focus back on attempting to keep pace with the demands for wireless voice communications from the masses. At the same time, the personal computing industry leapt forward with faster and faster modem capability. GSM, with only its 14.4kbps HSCSD upgrade to boast, had dropped well off the pace in the data race.

However, GSM's growth continued. By 1995, it had spread through Russia, Asia, Africa, the Middle East, and on into the United States, where GSM 1900 was launched by American Personal Communications in Washington, D.C. in November. By year end, GSM was operating in the 900, 1800, and 1900 MHz frequency bands and in 69 countries with more than 12,000,000 customers.

GSM added its 100,000,000th customer in 1998; its 200,000,000th customer in 1999 and its 400,000,000th user was activated in 2000. At the end of 2001, there were almost 700,000,000 subscribers and the GSMA predicts GSM will surpass the 1 billionth user milestone early in 2003, or even earlier if growth continues at its previous rate.

While global expansion of the GSM technology continued, so did the concentration on voice services. A lone exception to this was the robust growth of SMS services, especially in Europe and Asia, where it seemed like every young adult could type text messages on their GSM phones faster than their parents could talk. SMS messaging accelerated at breakneck speeds with a billion SMS messages sent during the month of June 1999 and an estimated 15 billion SMS messages sent in December 2000. The GSMA forecasts that 400 billion SMS messages will be sent in 2002, continuing to build on the 250 billion sent in 2001. These are indeed spectacular growth patterns for what is a very rudimentary form of data communications.

Meanwhile, the global expansion continued with inroads made into Latin America with the launch of Entel PCS. A number of other Latin American countries adopted the GSM technology in Entel's wake but it was not until Brazil adopted the GSM spectrum band of 1800 MHz that it became firmly entrenched in Latin America.

By the end of 2001, GSMA had 598 members, including 491 operator members from 172 countries around the world. There were 646 million GSM customers talking on GSM networks around the world. While GSM was originally conceived as a pan-European mobile communications system, every continent now boasts GSM service, with the Asia/Pacific region now home to the largest GSM operator, China Mobile. GSM accounts for 71 percent of the world's digital mobile telephony customers and 67.7 percent of the world's wireless market and has become the de facto standard for mobile communications around the world.

The GPRS Mission

Enough of the history. This book is supposed to be about GPRS and so far the only commonality I've established between GPRS and GSM is they're both acronyms beginning with G. What gives? As you will read in later chapters, GSM does most of the heavy lifting for GPRS. The GSM backbone and transmission architectures are reused in GPRS and augmented by GPRS-specific add-ons such as routers, firewalls, GPRS service nodes, etc. But without the GSM network to piggyback on, GPRS would be a much more expensive and difficult technology to implement.

Furthermore, GPRS is part of a well-conceived "family of technologies consisting of GSM, GPRS, EDGE, and 3GSM (a.k.a. wideband CDMA, or WCDMA). Each is part of a standardised set of global wireless technology platforms, which itself is evolving through the ongoing co-operation between standards committees, operators and equipment manufacturers. Without GSM there would evidently be no GPRS. But does GSM have a future for growth into data without GPRS?

In the mid-1990s, data was a small part of the GSM service offering. Although the introduction of HSCSD in 1993 was meant to emulate the dialup modem service of fixed telephony, modem and broadband technology speeds raced forward while HSCSD remained modest at 9.6 or 14.4 kbps. This lag, coupled with the fact that HSCSD's hungry use of up to four time slots put it in competition with voice calls for network capacity, encouraged network operators to stay focused on their voice businesses. That focus was not just a form of risk aversion: there was plenty to do. Demand, driven by huge numbers of new subscribers, required radio engineers to find ways to increase capacity through frequency hopping, cell splitting and the acquisition of small slices of additional spectrum when analog service was pulled back.

At this point, the industry was poised for change. Most people believed that voice businesses would peak as penetration rates passed 70 percent in some countries. Data was seen as the next great frontier. But not in the circuit-switched sense...for the advent of the global Internet made it clear that packet-switched data, based on the Internet Protocol (IP), was the way of the future.

The adoption of the Internet by the consumer masses in the late 1990s cemented IP's position as the underlying protocol for data transmission. GSM was not mainstream in this regard, as both HSCSD and SMS employed circuit switching as the basis for delivering data traffic. This method was costly and slow due to spectrum capacity issues, technology hurdles, and the dominance of voice needs within the GSM systems. But if wireless data was to take hold in the GSM business, a transition to IP-based wireless data was required. IP is more efficient due to the ability to send and receive data in packets during slow periods of traffic, as well as the ability to mesh with the now pervasive global IP protocol of the Internet. By moving away from expensive, time-consuming, and inefficient circuit switching and toward the IP packet concept, GSM would try to move back into the mainstream of data.

Adapting IP technology to GSM was a radical objective that quickly became a necessity. ETSI responded by beginning work on the GPRS specifications in the mid 1990s, completing Phase I requirements in GSM Release 97 and further enhancing them with GSM Phase 2+ Release 99.

The ETSI GPRS specifications defined the manner in which customers could send and receive data in a packet protocol mode, without commandeering the many GSM network resources required for circuit-switched transmission. Packet transfer is well suited to "bursty" data transmissions, which consist either of frequent small volumes (less than 500 octets) or infrequent larger volumes in the several kilobyte range. But packet transfer doesn't guarantee quality of service and GPRS is designed as a "best-effort" technology. Error correction and speeds therefore vary according to the service level implemented, terminal device type, the applications to be run, and network availability. On the positive side, GPRS is also designed to be "always on," a capability which changes the bursty nature of data transmission from a hindrance to network efficiency into a benefit.

HSCSD circuit switching does not allow for this as it ties up a specified amount of GSM spectrum for the entire duration of a data session. With packet-based systems such as GPRS, data is broken down into small packets that are sent individually and reconstructed at the end destination. Therefore, they can be sent during slow periods of network usage and the efficiency of resources is increased. This creates cost savings for the operator, which can be passed on to the customer through lower tariffs than would be viable with an HSCSD service. The always-on ability of GPRS also removes the requirement for a dialup connection to be established before every transmission, which allows data such as email to be sent to the customer on a near real-time basis. In many respects therefore, GPRS can be seen as the wireless equivalent of Internet access over broadband connections such as DSL or cable connections—always on and available.

The ETSI specifications also define requirements for overlaying the GPRS packet-based service on top of the GSM network, utilizing key components of the latter to transmit and receive the GPRS packets. The addition of gateways, routers, and firewalls along with packet billing collectors (the data equivalent of call detail records which provide network usage information for the billing system) allows the GPRS network to run independently from the GSM network. However, it is still interconnected at key interfaces such as the GGSNs (gateway GPRS support nodes) and SGSNs (serving GPRS support nodes). Don't worry about these terms or what the components do...this is all spelled out in Chapter 2. Suffice it to say that by utilizing an overlay of the GSM system, GPRS takes advantage of the basic switching system used for GSM voice service and its underlying infrastructure.

Would this be a match made in heaven? Or a force fit of two distinct technologies merged together to the detriment of both? While the answer to these questions has yet to be written, it is a fact that GPRS service was officially launched by BT Cellnet (now mmO2) in a limited service area of England in June 2000. Only a few days later, T-Mobile launched GPRS service in Germany covering a larger geographic area. Over the remaining months of 2000, GPRS networks around the world became "commercial"—a term that can be a little misleading and is best defined in this context as friendly user trials with select or limited numbers of loyal customers. Why? Three reasons.

First, the technology was undergoing debugging. Although the specifications were detailed, launching a large-scale network in the real world resulted in numerous instances where paper and practice did not quite coincide. As anyone involved in the launch of the initial GSM service would testify, such occasions are both painful and absolutely vital to the long-term success of the service offering. Second, as a "bearer service," GPRS can only be judged properly when applications are running over it. Alone it is just a transport mechanism. Think of a computer's operating system on which programs are run for word processing, gaming, accounting, and other user-required services. The GPRS bearer is similar, in that on top of the base technology, applications and services that the customer wants in a wireless data world such as email, picture messaging, chat, gaming, access to corporate business networks and, of course, access to the Internet, will run. While services have been the last part of the end-to-end business model, they are in fact the most critical. Without services that meet customer needs, the GPRS technology will be nothing more than another technology breakthrough sitting idle, patiently awaiting its own demise through lack of interest. While no one thinks this will occur, unless the appropriate services and applications are developed to drive customer demand, GPRS may not thrive.

Last, GPRS launches were significantly hampered by the scarcity of GPRS terminal devices on more than one occasion. Few were available to early adopters and their software and functionality were at best untested, and at worst, still in beta forms. Because GPRS devices were not being manufactured in quantity were operators or manufacturers willing to risk recalls and field upgrades, thus GPRS service was initially limited to a few committed customers. And some would argue that in mid-2002 little has changed. But today the future does look brighter with new terminal devices being brought to market by MEN (Motorola, Ericsson and Nokia) along with new players such as Samsung, NEC, and another company you may have heard of, Microsoft. Microsoft has developed the Stinger platform that is being used by some device manufacturers as the base for their new GPRS devices that tie the Stinger platform to the Windows concept. Will this be a divine match or another failed crossover by a dominant player in one industry segment trying to leapfrog into another? The jury is still out, but Microsoft could become a "player" in the wireless GPRS segment in the future...exciting times!

This, however, is only the beginning of a new section of GSM history. Over time, it is expected that GPRS will become both a viable and a critical component of GSM service. With its ability to offer wireless IP services and always-on mobility, building upon the HSCSD and SMS data usage already in place, GPRS is poised to be the next great wireless success story.

CHAPTER 1

Global Domination of GSM

Rob Conway, CEO

GSM Association

From humble beginnings, GSM has risen through the ranks by making the wireless world "an offer it just couldn't refuse"—quality, simplicity, security, and seamless roaming. Today, the acknowledged "godfather" of modern mobile communications, it is a worldwide blockbuster that has spawned a family of related wireless platforms whose influence will be felt for generations to come. However, the amazing success story that has seen GSM spread to more than 180 countries, boast around three quarters of a billion subscribers and account for 71 percent of the digital wireless market owes its success to more than just technology.

GSM or, to give it its full name, Global Standard for Mobile, is just that, an end-to-end prescription for the delivery of digital mobile telephony services from radio access network to billing and handset to core network. It is this concentration on the entire package and not just the air interface that has set it apart from competitive technologies such as the time division multiple access (TDMA) and code division multiple access (CDMA) prevalent in the United States.

The advantages of GSM for operators come as much, if not more, from the economies of scale, interoperability, and roaming capabilities inherent in the technology as from its propagation characteristics. These benefits of a single, well-defined standard, developed in an atmosphere of stakeholder cooperation have been the basis of GSM's success and the foundation of its development into the next generation. But I'm getting ahead of myself. However inevitable the rapid growth of a single standard may seem now, it was anything but in the 1980s when the mobile communications industry embarked on a road to digital transmission with its second generation.

GSM began its development in the inauspicious back offices of Conférence Européenne des Administrations Postes et Télécommunications (CEPT), the European body then responsible for telecommunications. It was inspired by a European Union dictate for a single pan-European technology standard for 2G wireless telephony as part of a general deregulation policy for telecommunications. The aim was to break the monopoly positions of the old state-run telecoms service providers and open the market to competition. By defining the technology choice for 2G mobile services, the EU believed that it would open the market up to new entrants. The target was a highly competitive mobile communications market for Europe.

That, it can be said without hesitation, has been achieved. However, in doing so the EU, in the guise of CEPT, also guaranteed a pan-European market for GSM equipment and set GSM on its way to becoming the global force it is today: the assumption here being that the system worked. It is worth saying that while this is now a given, it took an amazing amount of time and effort by incredibly dedicated and talented people to make it work. Perhaps, most important, is to remember that GSM wasn't perfect from the start. It took time to get things right and it was no surprise for anyone involved in this development that the creation and deployment of GPRS followed a similar learning curve.

In hindsight, Utopia for 2G would have been for a single standard to be adopted worldwide, but as we all know hindsight is only possible after the event. In addition, no one then had any idea of how popular mobile telephony would become or how soon there would be a global demand. Therefore other key markets, most notably Japan and the United States, embarked on their own separate paths to digital. Japan decided on personal digital cellular (PDC), a new technology developed by Japanese operator NTT in cooperation with NEC, AT&T, Ericsson, and Motorola, while the US authorities decided against prescribing a single technology to be used on the frequencies it set aside for 2G. The United States had a single standard for analog and chose not to follow the same path for digital because, it was argued, doing so was anti-competitive. It may seem strange now, but the motivation behind the opposing regulatory positions taken in Europe and the United States was exactly the same—to drive competition and create better and cheaper services for users. The results, however, have been quite different.

GSM gained momentum fast and needed a champion, which came in 1987 when the GSM Association (formerly known as the GSM MoU) was founded. A powerful and focused force, this new organisation had a threefold mission—to drive GSM technical leadership; to handle regulatory and policy matters; and to promote and harness global cooperation. All of these were targeted at achieving a single goal—a seamless, limitless world of wireless communications.

GSM spread quickly across Western Europe. The interoperability of equipment meant operators had much greater choice than in the analog days and this freedom drove vendor competition. It also threw up the potential for cross-border communication that manifested itself in the first roaming agreement between the United Kingdom's Vodafone and Telecom Finland in 1992. The ultimate result was the widespread implementation of GSM roaming, providing users with what amounted to a plug and play mobile phone system operating in every GSM country.

The rest, as they say, is history. The role of GSM in Europe was secure and highly successful, but it was not yet a global force. 1992 saw the first non-European GSM network installed in Australia by Telstra, marking the beginning of the technology's intercontinental expansion.

This largely followed the pattern of traditional economic ties. In broad terms, the Indian subcontinent, North Africa, and the Arab States largely followed the European line, while South America and Southeast Asia were more in step with the United States where the TDMA, CDMA, GSM battle continued to rage. Therefore the inroads that GSM has since made into North and South America are particularly significant.

The development of GSM equipment specifically for the 1900MHz frequencies used by 2G communications in the Americas opened the way for GSM. Chilean operator Entel PCS, became the first South American convert in 1998. The most recent and major breakthrough came in 1999 with the decision by Brazilian authorities to adopt the GSM-friendly 1800 MHz band for its new round of licenses, following long and thorough campaigning by the GSM Association led by the former Chairman, Scott Fox. His arguments, and ultimately the reasons for the decision, boiled down to the facts that GSM is an established, proven technology provided by a highly competitive pool of vendors and that it facilitates roaming.

The 2G technology choice has become a "no brainer." Europe provided a critical mass for GSM, from which it has become a self-perpetuating success. The more networks there are, and the more countries they cross, the stronger GSM's benefits become. And for any operator, these far outweigh any possible advantages any competitor might have on pure technology grounds. It is now such a powerful argument that even in the CDMA heartland of the USA, there are some 35 GSM networks.

So we've come full circle. GSM has been a dramatic success on a global scale because it offers operators more than just technology. It's a community made up of operators, manufacturers, and now application developers, the survival and prosperity of which depends on the common platform of effective development and deployment of mobile technology. GSM has always meant more than radio technology and, with a clear evolution path to data with GPRS and 3GSM (or W-CDMA), it is set to do so for many years to come.

There is no going back now. The division between GSM, the mobile technology and GSM, the community, has blurred to the point of disappearance. They are one and the same. But the GSM Association is not complacent—it recognises that GSM has to evolve and grow. GSM can and will go the distance, not just in terms of geographical reach but also as a communications platform for the future.

3GSM (W-CDMA), a vital component of the family of GSM wireless platforms, is already receiving critical acclaim and support as the next-generation solution. In fact, the GSM Association estimates that some 90 percent of the world's wireless carriers, and *all* European network operators, have already chosen W-CDMA as the de facto global standard for delivery of 3G wireless services. Japan has adopted a W-CDMA-based technology. Korea, which was once a CDMA stronghold, has sanctioned the entry of 3GSM into its market, while TDMA operators in the United States such as AT&T, Cingular, and Dobson Wireless took the unprecedented step of changing their 2G infrastructure to GSM to leverage the 2G benefits. These companies now have an evolution to 3G through GPRS and EDGE.

What GPRS effectively gives us is time. One thing that the success of GSM has shown is that it's not a race to be first but rather a race to be right. What is important is getting the quality, the platforms, and the services right. Only then do you achieve market acceptance, adoption, investment return, and profit.

GPRS is not a shortcut but it does offer the market an easy-to-follow, comparatively quick-to-implement migration path to advanced service delivery. One that allows operators to maximize their existing infrastructure investment while at the same time exposing consumers to 3GSM-type services, enabling new multimedia messaging, location-based services, wireless-Internet, and M-Commerce applications.

The evolution of new platforms, GSM-based or not, brings many new challenges but few more pressing than second- and third-generation roaming. As I have highlighted, roaming played a seminal part in the success of GSM and it simply has to be supported for GPRS and 3GSM. Considering that over 66 operators are now licensed to provide 3GSM(W-CDMA) services, and more than 110 GPRS networks are either in commercial operation or testing, this is already a burning issue. And one that is being tackled head on by the Association today.

For GSM to move to GPRS and above, it must ensure that new data applications and services have the capability to roam. This is a crucial part of the GSM family's ongoing development and one that is actively being addressed today.

Seducing the Consumer

Aside from roaming, GSM success can be credited to its ability to bring a compelling service to consumers. While the market debate on the future of GSM, GPRS, and 3GSM services has largely involved the technology, the investment, and the killer applications, it is the customer who must be at the heart of any development. As the lessons from wireless application protocol (WAP) have taught us, the key to widespread adoption lies ultimately in the hands of the consumer, and seducing the consumer is the ultimate challenge.

With GSM and voice this was easy because the benefit was obvious and intuitive; with GPRS and data, the benefits come further removed from the technology. Therefore services must be designed, presented, and marketed in a compelling way.

Tomorrow's advanced multimedia services are not simply replacement technology. They must be positioned and embraced as new solutions, bringing new benefits and a new lifestyle opportunity. But, having just said that technology is not the answer, such functionality does impact on all areas of infrastructure, service, and handset technology. Technology should therefore been seen and used as an enabler of services, not an end in itself.

For service delivery, the most crucial element will be content flow. This means that seamless voice and data delivery, global roaming, and open platforms will be as vital to the future of GPRS and 3GSM as they have been to GSM's past. Next-generation handsets will also have to be developed. More emphasis on simplicity, usability, design, and appeal will stimulate more wireless communications devices. These will also become a dominant channel for retailers and service providers.

Without doubt consumer acceptance will drive every aspect of the new wireless world. GPRS and 3GSM must create a new promise to the consumer, and they must be able to keep and deliver that promise.

The GSM Association's membership has championed a new approach to help tackle and focus on consumer issues and requirements. For example, its Mobile Services Initiative took onboard the lessons of the past by bringing together the operator community—much as it came together to develop GSM in the early days—to provide clear guidance to handset manufacturers and software developers on the device needs of consumers of Mobile Internet services, today and in the future. A significant boost to the attraction of GPRS, Mobile Internet applications will refocus the market from the "speed" to the real, compelling wireless Web drivers—services and content.

Standing the Test of Time

The evolution of a next-generation wireless-enabled world should prove to be even more dynamic and prolific than the current one. It will have more players, delivering more services to more people and in more ways than ever before. While there are still many uncertainties as to the precise shape of the market or the technologies, it is certain the GSM family will remain an underlying force. The GSM dynasty will embrace the lessons learned, the platforms developed, and the infrastructure already in place, in order to remain a global platform that dominates the future wireless landscape with the same worldwide success as it has in the past.

CHAPTER 2

Roaming Across Borders

Mark Smith

GSM Association

GSM is the only digital wireless standard that provides coverage on all populated continents of the world. The fact that it's "everywhere," however, doesn't make it mobile. For that I need to be able to connect to the network wherever I am, not just in the vicinity of my provider's network. Much has been written about the inherent capability of GSM and its evolutionary family of technologies to offer international roaming for customers around the world. But what does "roaming" actually mean? Here's how the GSM Association, the world's leading wireless industry body, formally defines the term:

> *Roaming is defined as the ability for a subscriber to automatically make and receive voice calls, send and receive data, or access services when traveling outside the geographical coverage area of the home network, by means of using a visited network. Roaming is technically supported by mobility management, authentication, and billing procedures.*
>
> *Commercial terms between network operators are contained in Roaming Agreements. If the visited network is in the same country as the home network, then this is known as National Roaming. If the visited network is outside the home country, then this is known either as International Roaming, or Global Roaming. If the visited network operates on a different technical standard to the home network, service may still be offered via special equipment supporting Inter-standard roaming.*
>
> *GSM Roaming offers the convenience of a single number, a single bill, and a single phone with worldwide access. The convenience of GSM Roaming has been a key driver behind the global success of the GSM Platform.*

What on Earth Did We Do before GSM?

For the one in nine people who own a GSM mobile phone today, it's second nature to take their phones when they travel abroad on business or pleasure. They take it for granted that when they arrive in their destination country, within seconds of turning on their mobile phone, they will be authenticated by their home network operator, and connected to a visited network. They probably don't think of it in those terms, of course, but they expect to be able to make and receive calls and send text messages. This scenario is possible across some 180 countries where the GSM standard exists today only because GSM is the world's first and only global wireless network. Although customers in most of those countries still use it predominantly for voice calls and text messages, the arrival of General Packet Radio Services (GPRS), EDGE, and 3GSM (or W-CDMA) services opens up a whole new world of roaming with wireless data.

It is not hyperbole to say that GSM is a technological success story of our age that ranks alongside other wonders of our time such as mass air travel, the television, and the Internet. All these momentous developments are indelibly linked to one thing: humankind's natural instinct to explore and communicate beyond borders, beyond oceans, seas and mountains, to the far reaches of our world. The arrival of mass, affordable air travel in the late twentieth century brought with it a desire to stay in regular touch with family, friends, and colleagues at home. While fixed-line telecommunications served the purpose well for most of the century, the freedom to communicate whenever, and wherever we wanted—at home or across borders—was the ultimate goal of the early wireless visionaries.

In the Beginning...

The first types of mobile telephone systems were quite rudimentary but nevertheless were a significant first step to the mass-market mobile service we have today. In the United Kingdom the first system, called "radiophone" was introduced by the post office in the 1960s. Radiophone service used VHF frequencies to provide a limited service to a small number of customers. The customer equipment was so bulky—comprising cumbersome transmitting and receiving units—that it could not be easily carried around and was "mobile" only in the sense that it could be used in vehicles and installed car phones.

At that point coverage was available only in certain areas (e.g., London and Manchester, but not in rural areas or smaller towns and villages) and capacity maxed out at 14,000 customers. Radiophone was also expensive to use. Initially all calls were set up using an operator. Not surprisingly, mobile telephony was a premium product that only a privileged few used. Even with service improvements like direct dialing by the customer, capacity constraints prevented this service from being marketed to the general public.

Why was the technology used in the early mobile systems incapable of providing extensive coverage and the capacity to accommodate a mass-market service? This was because the radio frequencies allocated to the service were used very inefficiently. Dedicated radio channels had to be used over a wide area so the number of channels available limited the number of users in the coverage area. Since it was limited to a small number of users and couldn't leverage economies of scale, this type of system was expensive to provide.

First-Generation Cellular

These obstacles were overcome by advances in cellular radio techniques. The basic notion of cellular radio is that radio channels (or *frequencies*) can be used over and over again, thus allowing much greater capacity than the concept of dedicated radio channels, as used with the radiophone. To achieve capacity gains, the range of each channel must be limited so that it does not interfere with the same frequencies used in a nearby area. A coverage area is divided into cells, each with its own set of frequencies. Adjacent cells use different frequencies to avoid interference, and non-adjacent cells can reuse the same set of frequencies without constraint.

The advent of cellular radio techniques gave telephone companies an opportunity to provide service to a large number of users and led to the mass-market take-up that has characterized modern mobile communications. During the early 1980s, analog cellular telephone systems were experiencing rapid growth in Europe, particularly in Scandinavia and the United Kingdom, but also in France and Germany. Each country developed its own system, which was incompatible with everyone else's in equipment and operation. This was an undesirable situation. Not only was the mobile equipment limited in operation by national boundaries which, in a unified Europe, were increasingly unimportant, but markets for each type of equipment were likewise very limited. Again, economies of scale and the subsequent savings could not be realized.

Analog systems are known as the first generation of cellular radio systems. They usually used frequency modulation and interconnected with the public switched telephone networks (PSTNs). Each country had different frequency allocations and there was little industry collaboration on the development of these systems.

The system used in some of Europe was called total access communications system (TACS). It is based on, but is not compatible with, the United States system known as advanced mobile phone system (AMPS). AMPS was only "advanced" at the time, in that it heralded the first cellular-type radio system that could support a large number of users.

The chief problems of analog systems fell into three categories: Because of incompatible standards, they generally did not permit a subscriber to take his phone to another part of the world and use it successfully. Even where the standard was uniform, there were too few roaming possibilities. And, these systems primarily provided voice communications and could not easily accommodate the increasing requirement for data communications. As the limitations of the 1G became more obvious, they precipitated an effort to conceptualize a new cellular radio system. It was the proposal of the European operators that led to the development of the second-generation system known as GSM.

One Giant Leap to GSM

Today, most subscribers to mobile services in the world use the digital GSM system—some three quarters of a billion people in fact—with a few remaining pockets of analog users. In fact, many networks have already closed, or plan to close down their analog systems over the next few years. If the birth of the analog wireless phone in the early 1980s was one small step, the development of GSM was really the giant leap for mankind's ability to communicate across distance "wirelessly."

The objective was to develop a specification for a pan-European mobile communications network capable of supporting many millions of subscribers, handling the kind of escalating capacity that was putting immense pressure on the creaking analog networks. Via 1987's Memorandum of Understanding signed by 15 network operators from 13 different countries, work began to realize the vision of a flexible, reliable wireless system for cross-border communications. The vision of the early pioneers, however, reached further than that. Much further. In fact the vision quickly spread beyond Europe's horizon, reaching out to every populated continent on the planet during the next decade.

In 1989, GSM responsibility was transferred to ETSI, which published Phase I of the GSM specifications a year later. GSM was the first, and is still the only, mobile phone technology designed specifically to allow easy and secure roaming between different networks. The first commercial GSM networks went on air in 1991 (13 networks—in 7 countries), the first roaming agreements were signed in 1992, and roaming started soon after. It was at that moment that GSM began to deliver on its promise.

Roaming and Roaming Agreements

The GSM standard was designed to be very flexible, and today many services and features can be offered to both the network operator and the user. Because they work to a common standard, all GSM networks work roughly the same way and every network can offer service to users from any other network. As a result, GSM encourages a large amount of roaming traffic, and users quickly become used to receiving service when they travel abroad.

Traffic is also encouraged by the high level of security GSM provides. Powerful algorithms are used to authenticate users; the difficulty of cloning a user's identity has been proven out over ten years of GSM operation. Eavesdropping is deterred by encrypting the radio interface to a high degree.

Another key to GSM's mobility is the terminal's reliance on a removable subscriber identity module (SIM) card that holds the identity of the user. Because the SIM card can be plugged into any compatible mobile phone, the user can change handsets easily and take his identity with him. Also, a user's list of names and telephone numbers can be stored on the SIM card so that personal information can be transferred just as easily. The SIM card has been enhanced recently to add more processing power and memory for new types of operating systems and applications like m-commerce and banking.

Because key interfaces have been standardized, equipment from different manufacturers can work together in the same network. Because all equipment works to the same specification, the R&D costs can be shared, and equipment can be mass-produced for better affordability. This also gives GSM network providers a wide choice of equipment vendors as well as the ability to "mix and match" equipment from different vendors.

With all conditions in place for boundless roaming, what do operators have to commit to in order to make it happen? For two GSM network operators to agree to arrange roaming for customers in each other's country or region, they conclude what are known as *roaming agreements*. These are basically business agreements between two providers to transfer items such as call charges and subscription information back and forth, as their subscribers come in and out of each other's areas. When you return to Spain from Sweden, for example, your next phone bills will include the calls you made while in Sweden, because this billing information has been cooperatively transferred back to your home operator.

GSM Phase 2 Enhancements

Over the past three years, the GSM standard has been enhanced to meet the demand for better performance and new services. Enhancements are contained in a series of annual specification releases that build on Phases 1 and 2, called the "Phase 2+" set of capabilities.

They notably include better speech quality; new network services such as ring back when free; multiple subscriber profile; call transfer, calling name presentation, and pre-pay control; location-based services; and new dimensions for data transmission including HSCSD, EDGE, and GPRS—the latter being the main focus of this book.

The data standards represent the new promise in digital technology. High speed circuit switched data (HSCSD) can deliver data rates greater than 64kbit/sec, although 28.8 kbps is more usual at present. On behalf of the user, HSCSD reserves multiple time slots to provide the higher rates. A great advantage of HSCSD is that it can support "real-time" services like live video because each user has a dedicated resource. HSCSD service is already available today to more than 100 million customers across 27 countries around the world—in Europe, Asia Pacific, South Africa, and Israel. Some 40 GSM network operators have successfully introduced the service.

Enhanced data for GSM evolution (EDGE) is also a new modulation technique to expand the amount of data that each time slot on the GSM radio access can carry. Used in conjunction with HSCSD or GPRS, EDGE further boosts either one's rate improvement.

The leap to GPRS is a special case, and a revolutionary step for the GSM family, providing the technology for significantly higher data transmission rates.

Mobile terminals simplify roaming.   Mobile terminals have also undergone rapid development. Today, almost all new GSM terminals are dual-band (mostly 900 and 1800 MHz but some 900 and 1900 MHz) and many are tri-band (900, 1800, and 1900 MHz).

Tri-band phones, which cover all the frequencies on which GSM is currently used, can roam from country to country and region to region, ideally without the loss of any capabilities. While right now there may not appear to be a big market for this type of phone on the face of things, note that in 2000 the United States had 28 million travelers leave the country to visit locations with GSM service.

Terminals with GPRS capability are also making their appearance in ever-increasing numbers. Some offer features such as voice recognition, colour screens, and personal organisers. Terminal size has been reduced and battery life has been dramatically improved.

A Global Success

During a typical global month, GSM network operators will collectively handle more than two billion roamed calls (*source:* EDS). Roaming traffic is growing at a fast and furious pace. Just three years ago the level of roaming traffic was put at 750 million calls per month (August 1999, *source:* MACH). As roaming becomes more and more important both for business user and operator, quality of service and accurate billing will be the key to continued growth. To this end the International Roaming Expert Group (IREG) of the GSM Association is currently working on protocols and verification for operator databases, aspects of signaling, optimised routing, and end-to-end functional capability testing.

Roaming customers happen to be the most important customers in telecom today because, unlike local customers, they can choose to change networks whenever they want. They also constitute the customer base that expects and demands real mobility. The ultimate concern of that consumer is that her phone works seamlessly in another city or country. It matters to her that she only needs one number, virtually worldwide. And she cares that GSM offers meaningful variety in her choices of handset, including "loaded" models for affordable prices. The secret of the success of GSM is in its commitment to open standards and cooperation between operators and manufacturers—an approach that is being repeated with the ongoing wireless evolution that is GSM, GPRS, EDGE, and 3GSM.

CHAPTER 3

Short Message Service and GSM Circuit-Switched Data

David Gordon

Partner Communications Company, Israel

Just as traditional telegraphy pioneered data communications services, with telegraphs acting as one of the first "killer applications," text messaging pioneered wireless data services in the world of GSM. The short message service (SMS) has actually launched an evolution path, stemming from the exchange of simple text between terminals, down the line via transmission of parameters like location data, and on into application download. This lineage tree of data services continues over faster and more generic modes of data communications such as HSCSD and GPRS.

What Is It and How Does It Work?

In a sense, the short message service is where narrowband strikes back at the over-hyped vision of broadband. SMS is a feature made popular by GSM; although embedded within the 2G standard from its earliest days, it achieved its greatest popularity just as the world began to transition to the implementation of 2.5G and 3G. From the late nineties and even as these words are written, there is no sign of change in the rates of SMS growth. According to estimated figures published by the GSM Association, the number of SMS messages sent worldwide has grown from a billion messages per month in April 1999 to an average of more than a billion messages per day during the first quarter of 2002, with no sign of slowdown.

Strictly speaking, the SMS is not a circuit-switched feature. It is, however, transmitted over the signaling channel of the circuit-switched system. SMS transmits content composed by the user (originally text) in data packets together with all the unique signaling messages that support the traditional circuit-switched GSM voice network. The set of messages that control the routing and transmission of the SMS packets is part of the signaling communications layer known as the *mobile application part* (MAP). MAP includes messages that serve as instructions for crucial functions of GSM such as mobility management, call handling, activation of services per subscriber, and many more.

SMS implies the ability to send and receive text messages to and from a mobile terminal. By nature, it is not a real-time feature requiring an available channel for successful execution. Instead, the terminal, once associated with a network, is "always on" and ready to submit and receive messages, which may be held temporarily in the system if the phone is turned off or out of reach.

The SMSC

The SMS message is generated by terminal equipment, devices ranging from handsets to specialized application servers built specifically for creating messages. A crucial component of the system is the SMS Center (SMSC), a computer that acts as the "store and forward" module and is the "buffer," "amplifier," and "router" of the process. In other words, the SMSC performs the following functions: it *stores* the message until the recipient is ready to receive it and then *forwards* it; it will act as a buffer and avoid transmitting messages to an overloaded phone; it *amplifies* the message in the sense that it duplicates it and tries to contact the destination device time and again until a predetermined period of retry expires; and finally, it *routes* the SMS from its source to wherever in the world the recipient is residing.

Although the idea seems ridiculous in retrospect, early GSM phones could receive SMS messages, but not generate any. Only towards the later part of the 1990s was most of the installed base of phones able to send SMS messages. Two-way messaging was probably the first factor to ignite the explosion of the SMS business. Today the SMSC server is regarded as an essential part of any GSM system, and it is very rare to find a GSM system that did not launch from day one with full-fledged SMS services.

Message Structure

An SMS may include up to 160 characters of text, if you're writing in one of the core languages of the GSM community such as English, French, Spanish, or German. Unfortunately, scripts from countries outside the core community are often not included in the "default alphabet coding." In that case, they are supported by a different coding scheme, known as UCS2, which will enable only about half that amount of text per SMS—say, 70 characters.

This situation is improving. Some very recent work now enables SMS systems to embed left-to-right writing in a right-to-left stream of typed characters, as required by Arabic and Hebrew writers throughout the world, and other such efforts may increase script compatibility.

Keying Input

One of the major restrictions to SMS is the uncomfortable manner in which messages are keyed on the cramped 3×4 phone keyboard. That drawback hasn't precisely been erased, but today predictive text keying systems, the most well known of which is "T9" by Tegic, are good mitigators. They choose characters on the user's behalf by predicting the words he is keying, in accordance with general grammar and vocabulary knowledge and careful statistical analysis. In superior handsets these systems ease the keying of text; they make the phone keyboard more like a full-fledged computer keyboard by eliminating the need for three or four keystrokes to type a single letter. Predictive keying, along with pen-enabled smart phones and full keyboard ("QWERTY") phones, has boosted the propagation of SMS despite its restrictions.

What Is It Good For?

You'd be surprised how imaginative individuals and organizations can become with 160 characters of text. For starters, SMS administered the *coup de grace* to the paging business when it enabled two-way wireless messaging among individuals and organizations.

For person-to-person communications, especially among the growing computer-literate youth community, Internet-related features such as emoticons :-) and short cuts ("How R U? GR8!") began to drive the adoption of mobile SMS. It was only a matter of time, then, before the messages comfortably stored in the private memory of a SIM (subscriber identity module) would be linked to email systems. The first exciting applications to appear in this arena included mobile banking (accessing information on one's bank account and executing bank transactions), and basic telemetry services (transmitting urgent notice of computer failures to system managers).

The SMS is a winner at delivering interactive content, often on specific user demand. A wide array of content services is offered by network operators and third parties throughout the world; in a multitude of languages, services—news, sports statistics, travel information, weather forecasts, jokes, astrology—can be requested in two ways: by sending a service code or textual command via SMS to a value-added services platform, or by selecting from a menu planted by the operator in the SIM card. In the latter case, a technology known as the SIM Toolkit (STK) automatically generates the request and sends it via SMS.

The mass media has often made use of the SMS as a way to create interaction with audiences; we've seen applications from televoting to tele-poetry contests (*The Guardien*, April 2001). Another innovative example is Radio BU, a radio station in Israel where listeners communicating via the SMS of Partner Communications Company rate their favorite songs, receive notification of a title about to be aired, and send comments to popular DJs.

For the business user, the most valuable application so far is undoubtedly liaising with the desktop computer—to receive and send short emails, and to consult or update the desktop calendar.

Finally, the cost-conscious user, who sees a benefit in transmitting essential information for a fraction of the cost of a telephone call, is adopting the SMS.

Current Application Trends

Interpersonal communications were always a primary use for SMS, but as more and more young users joined the GSM user community, SMS fun products have become a high-consumption business. These members of the computer generation don't see the chore of keying text as an obstacle. Operators and content providers are jockeying to fill the youth agenda with appropriate content. Once supported by the terminal manufacturers community, applications like Ring Tones and Picture Messaging became so popular that they are now recognized as the current key driver of growth for SMS.

The instant messaging (IM) revolution started in the Internet, where it became an everyday activity for millions of users, and is steadily making its way to the top of the chart of SMS applications. IM is a fancy term for the computer "chat" that was always fashionable among closed user communities within computer networks. Today the typical mobile service provider offers its users "buddy lists" and "nickname" systems, by which they acquire the ability to "live" in virtual communities populated with mobile phone or desktop computer users who share common denominators. Look ahead for developments in this area pioneered by the likes of AOL's ICQ or Odigo.

In a more advanced technological mode, SMS may also serve as a carrier for code commands. GSM supports a feature knows as OTA (over the air) update or activation. A unique type of message is generated by a dedicated server and is programmed to update the content of files on the

SIM card, to install applications, or to initiate other processes on the mobile equipment.

Another example of the SMS bearing operational data and feeding external systems is applications in which SMS messages are employed for the transmission of location information in the framework of fleet management and emergency personnel dispatch systems.

SMS serves in various attempts to realize M-Commerce. Mobile payments are being authorized by a set of SMS acknowledgments, for example. Admittedly SMS has not yet managed to turn the mobile phone into an e-purse, and it seems this job is more likely to be assumed by more sophisticated platforms and more user-friendly interfaces.

SMS has turned out to be a valuable tool for road warriors who are heavy users of the international roaming feature among the GSM systems; they benefit from free voice mail notification and collect new messages that await them immediately after the plane touch down. People on the road are good candidates for the various SMS-based content services; lately, utilities like currency conversion or phrase translation services are provided as well. The latest concept in the arena of international roaming is that of the "Welcome SMS" (WSMS), a message sent out to all roamers upon arrival, notifying the visitor of the unique services of the hosting network.

The next frontiers for the mobile community are interworking SMS on the national level between networks, and interworking to non-GSM networks, e.g, across the Atlantic from the a GSM network in the United Kingdom to a CDMA network in the United States. In addition, we are lately witnessing the rapid growth of international SMS traffic. International SMS is a curious by-product of the establishment of international roaming. In enabling a GSM user from one country to be hosted by a GSM network in another country, the standard actually has created "open borders" on the signaling links which can be used for free SMS transmission.

It is now believed that linking competing GSM networks within the same country will accelerate the growth of national SMS traffic by as much as 50 percent. This traffic will no doubt be further enriched if and when SMS interworking is established with wireline terminals or other interactive devices, such as interactive TV set-top boxes. It is not far-fetched to assume that international SMS interworking will further boost the traffic volumes and potentially create a new and intense marketplace. In addition to the GSM operators, this marketplace is driven by globalization and the rising magnitude of travel, workforce mobilization, and cross-border relationships. This new marketplace is populated by new and innovative global SMS content and service providers; it also includes SMS aggregators that collect traffic, diverting it according to price and performance, and low-tier providers attempting to make a buck on route arbitrage, just as others have been doing for decades in the traditional international voice communication arena.

High-Speed Circuit-Switched Data

Although the inherent data transmission capabilities of GSM were for years the pride of the industry, we did not manage to leverage them to transform the world into a real wireless data community. The main reason for this omission was that throughout the nineties, GSM provided a rather pathetic data rate of 9,600 bits per second at a rather dear cost. By comparison to transmission rates that IT was getting over fixed lines, wireless data was unappealing. When awareness of this disparity became acute, it resulted in the two features known as high-speed circuit-switched data, or HSCSD, and GPRS. In technical terms, HSCSD means the co-allocation of multiple full-rate traffic channels.

As opposed to GPRS, HSCSD does not constitute a paradigm shift. The earliest versions of the GSM standard posited several variations of circuit-switched data, none exceeding the 9,600 bit barrier. Circuit switching was GSM's native approach to data. Just like basic telephony, a data session was allocated a full end-to-end channel of communications or "a circuit." All resources allocated were held for the full duration of the call before being returned to the network for use by other sessions.

Only after November 1997 did GSM include the concept of HSCSD. The idea is quite a naive one, in which circuit-switched resources are jointly allocated according to availability. The HSCSD-enabled terminal requests additional resources from the network. It is then up to a unit called a base station controller (BSC) to decide how much additional resourcing can be allocated to the terminal: it may be able to provide either one or two additional timeslots of radio resources, or none at all. The timeslots allocated are combined into a single, "wider," timeslot. GSM permits up to four transmission channels of the air interface to be joined into one, providing up to four times the standard 9,600 or 14,400 bits per second. This enables effective data rates of 43,200 bits per second, which is still among the highest rates the industry can provide commercially.

During the latter years of the last decade, before either HSCSD or GPRS were implemented, a fierce debate took place between the proponents of each. Without knowing much about GPRS, which we have yet to define in any detail, we can chart out some of those arguments here:

Feature	GPRS	HSCSD
Continuity of service	Always connected	Call setup necessary
Orientation	Packet-switched	Circuit-switched
Resources allocation	Spectrum efficiency	Spectrum inefficiency
Effective data rates, 2002	Up to 44–55 kbps per second	Up to 43.2 kbps per second
Maximum data rates	Up to 112 kbps per second	Up to 43.2 kbps per second
Terminal availability, 2002	Availability of commercial models from several vendors	Availability of commercial models from several vendors
Investment in infrastructure	High—new core network elements	Low—software upgrade of existing network elements
Future proof?	Core network is the framework for 3G services	No
Overall cost	Relatively high	Relatively low

The chart clearly indicates that while GPRS is geared towards the future, HSCSD constitutes a relatively narrow-scale upgrade on existing systems, with its main virtue being effective data rates of up to 43.2 kbps. When ease of implementation is taken into consideration, HSCSD, being mainly a software upgrade, does not require new network elements. The GSM operator not only avoids network redesign, but also may expect rapid implementation.

With those motivations, implementation of HSCSD effectively began around 1999. Even now, only a subset of the infrastructure vendors have implemented the feature, so worldwide coverage is limited by definition. Still, three years later, we know of 39 networks in 28 countries that have chosen to adopt the HSCSD feature, mostly in Europe, but also including South Africa, Hong Kong, Singapore, Kazakhstan, and Israel. These numbers translate into HSCSD availability for commercial use by almost 100 million subscribers worldwide.

These figures are significant and should not be overlooked by network planners. Yet, at this point in time, it seems that most of the GSM world has elected to skip HSCSD and implement GPRS as a path towards 3G services. Clearly, no network that has implemented HSCSD is considering skipping GPRS. These networks simply identified a window of opportunity or of necessity, and acted accordingly. Their payoff is the ability to provide their customers with actual data rates that are the highest in the industry at the present time.

Terminals and Applications

With about a dozen phones on the marketplace, it is clear that HSCSD did not engender a rich choice of terminal equipment. The first terminal was in a Nokia PC card modem format known as "Phone Card II." Its main disadvantage is that it cannot be used without the laptop computer for regular voice conversations, unless one must be willing to plug in an earpiece and dial a phone number via the software driver on one's PC.

A more advanced concept manifests itself in phones with an embedded HSCSD modem and infrared connectivity between phone and computer. HSCSD-enabled smart phones and personal digital assistants (PDAs) have a lot of potential, and the latest model of the Nokia communicator capitalizes heavily on the coupling of its color screen and the equipment's HSCSD capabilities.

Hypothetically speaking, endless applications are possible for HSCSD in wide-ranging mobile data applications, real-time transmission of images and video clips, and in conjunction with entrenched wireless office applications—email and Web browsing. Networks that have implemented the feature have reported its use in many embedded systems. It is still unclear if the HSCSD feature will have a contribution to make to Wireless Application Protocol (WAP), particularly as the first models did not support WAP over HSCSD.

The HSCSD feature is especially attractive to heavy business users, who are finally able to free themselves of the computer cord and its ever-present kit of conversion plugs. For them, HSCSD roaming, however expensive relative to home usage, is a viable alternative. Wherever HSCSD has been implemented by both the visited and the home network, roamers should be able to use it automatically, provided that the proper definitions are implemented on both sides. This "plug and play" feature establishment of HSCSD is in sharp contrast to the complexity of implementing GPRS roaming, as will be explained later. Recent work carried out by the GSM Association is further facilitating and accelerating the establishment of international HSCSD roaming.

HSCSD is the first technology that challenges the old concept of indifference of customer billing to quality of service (QoS). For HSCSD users, it is theoretically possible to request a certain level of service—defined as the number of channels that would be jointly allocated. The user could be charged dynamically, according to the number of channels actually allocated, for each interval where the QoS remains constant. As the quality of service might change during the course of the call, the total charge will be a sum of the charges for these intervals. QoS is a crucial issue, especially given that overloaded networks tend to allocate no more than one channel, thus downgrading HSCSD to the level of simple GSM circuit-switched data.

While HSCSD is not expected to be a flagship of wireless data technology, its proponents believe that for the time being it is the only viable wireless data infrastructure, and that we'd do well to exploit it as a test bed for next-generation services. The idea is that while network experts tune the network to achieve the promised data rates and functionality from GPRS, application experts will be able to shoot pictures, games, and music bits up and down their established legacy 2G networks at a fairly reasonable speed.

PART 2

A Perspective on GPRS Adoption Around the World

The first premise from which to open this topic is that the world is a pretty big place. As someone who travels around it far too frequently, I confirm the world is indeed large. And diverse, in geography, language, culture, economics, and a host of other measures. Such a premise is not controversial and you might well question what it has to do with this chapter, or even with the topic of GPRS. The answer is either a little or a lot...depending on how rapidly GPRS is adopted around the world.

Some Artifacts of History (Some Real, Some Myths, but All Written Here So They Must Be True)

GSM was first launched in July 1992. Today it is available in every continent around the world, which means that more than 170 countries currently offer GSM service. It took about five years, from the time the first operator began offering GSM in Europe, to the launch of GSM service in South America in 1997, for GSM to reach every continent. Five plus years for GSM to spread across the globe. Five years of licensing operators, hiring and training staff, designing networks, locating transmission sites, building masts, running cables, installing equipment, launching service, developing distribution channels, installing billing and customer care infrastructure, etc. Quite a feat if you think about it. And geographic, cultural, and economic aspects of the rollout raised the level of difficulty by a factor of X. Remember, GSM is founded on the premise of interoperability across networks and operations. Not only did GSM have to be installed and optimized in the countries currently offering service, it had to be the "same" in each to provide customers with automatic interoperability. Despite regional differences mandated by local regulatory requirements, the basic service works across borders and continents.

I mention the demands of the rollout in order to put GSM adoption into some kind of context before we discuss the rollout of GPRS. Some would say that GSM was a technology miracle. I put myself firmly in this camp. While it was happening, no one really stood back and realized just what a feat was being accomplished in such a short time span. But it truly was.

Just like GSM, GPRS began with a theory, the theory being that an IP packet data service was needed in order to provide *wireless data services with limited spectrum, bandwidth, and capital investment.* So someone, or more accurately a team of people, sat down and developed the specifications to allow such a service to be developed. That team came from ETSI, which developed and released the initial specifications for GPRS in October 1997 and called it Release 97 (clever naming convention—yes?) and followed up with Phase 2+ specifications in Release 99 in (you guessed it) March 1999. GPRS specifications detailing everything from how the technology works, to the protocols, interfaces, hardware, software, terminal devices, and everything else required to make GPRS a commercially viable product, had a blueprint to follow from the start. The steps from there? From there the infrastructure equipment had to be developed and commercially offered; terminal devices had to be designed and manufactured; regulatory bodies had to approve the technology; operators had to purchase and install the equipment, debug the service, and figure out how to offer and market the services; wholesale and retail billing programs had to be developed—this and more had to be done to bring GPRS to commercial reality.

On 22 June 2000, BT Cellnet (now o2) launched the first GPRS network in England. Within days, T-Mobile launched their GPRS network in Germany and soon others followed: Telstra in Australia, Mobikom in Austria, Telsim in Turkey, SmartTone in Hong Kong. These weren't real commercial launches in the truest definition; they were more like "friendly user trials" because the service wasn't stable and GPRS devices were about as rare as UFOs. Even so, in the span of just two and a half years, roughly half the time it took to realize GSM, the entire GPRS technology was developed and brought to market. Maybe this doesn't seem like a short time span anymore, but to those of us who participated, it is an unbelievable accomplishment.

So big deal...a couple of network operators were able to throw some money at a few vendors and push the service to market. No sweat as long as you throw enough money and push hard enough. But since GPRS' "commercial" launch in mid-2000, a wave of commercial launches has spread rapidly. In fact, GPRS service is now commercially available on over 150 networks in 50 countries around the world, with at least another 50 networks being implemented. If you do the math, almost one-third of the countries offering GSM services also offer GPRS, and it took them less than two years to do so. The adoption rate is much faster than GSM. Two years after GSM launched there were a mere 51 operators offering the service. And only in Europe. In somewhat less than the same time, GPRS was already commercial in Europe, Asia, North America, Middle East, Australia, South America, Africa, and Eurasia.

Why make a big deal out of GPRS service around the world? What does rate of adoption prove? Who really cares? All reasonable questions...true. Unless, that is, you are one of the ever-growing camp which believes wireless data services are the next McDonald's—serving millions and millions of [satisfied] customers around the world. If so, you care, and you need global GPRS service to attract your share of the multitudes. Or at least that is how some think the world will turn in years to come. (By the way, McDonald's operates in 120 countries as of year-end 2001, serves 45 million people every day, and began in 1955. Remember, GSM is available in over 170 countries, has over 700 million customers, and opened its doors in 1992.)

With GPRS service popping up around the world, when will GPRS roaming be implemented? The simple answer is that it has been, but only on a limited basis thus far. On 10 November 2000, PCCW/Hong Kong Telecom, SingTel Mobile, and MobileOne launched the first GPRS roaming service. KG Telecom was added eight days later. Leased interconnection lines were used to connect the GPRS networks. Since then, some GPRS operators have implemented GPRS roaming through the use of GRX (a data network built exclusively for GPRS backhaul) interconnection. In the spring of 2001, Telenor GRX and Sonera GRX became first to implement inter-GRX roaming through the Amsterdam GRX peering location.

Although GPRS service coverage continues to expand around the world, GPRS roaming remains on a slower track. Some consider this normal, since a critical mass of customers in the home market is required before GPRS roaming becomes an issue, whereas others believe that GPRS roaming availability will drive take-up in the home market. It's a classic "chicken and egg" question. Though no one knows for sure when GPRS roaming will be operational on the same scale as GSM roaming, 2002 will see great advances in the breadth and reach of GPRS roaming among the carriers. It is also expected that by late 2002 we'll see GPRS services expanded to over 100 countries around the world, with coverage of all continents and major countries by mid-2003. Not too bad for an upstart service. Nothing less will be tolerated by the buying public in today's world. Will GPRS become the wireless service staple of customers around the world, as GSM has been? Nothing less will be tolerated by the buying public in today's world. Global connections for an always-on, wireless data service are a must.

CHAPTER 4

GPRS Market Demand and Key Drivers*

Carolyn Davies
Baskerville:
Part of the Onforma Telecoms Group

As voice average revenue per user (ARPU) levels off and market growth slows, operators have been forced to look for new revenue streams and ways to differentiate their services. The principal method for diverting a rapid decline is widely seen to be wireless data services.

Data is not a new offering: it has previously failed on circuit-switched networks because of slow speeds, poor user interfaces, and the lack of useful applications. However, advances in network technology and terminals have created a more conducive situation where mobile data is once again viewed as an opportunity for mobile operators. GPRS presented the industry with the chance it had been waiting for to deliver convergence of mobile communications, the Internet, and a cost-effective and broad range of mobile data and multimedia services. We know that just as wireless voice services were not adopted by the mass market overnight, neither will wireless data services be. The pressure to deliver compelling applications today has been caused by the current market downturn, and spurred on by the huge investments made in next-generation networks and technology.

Many forecasts of the revenue that these services will create have been floating around the industry for some time. The ARC Group forecasts 120 million mobile data users by 2005; the UMTS Forum predicts that data will overtake voice and reach revenues of over $100 billion by 2005. Most operators rely heavily on their own, more meaningful predictions—namely, that they expect data ARPU to make up 25 percent of their total revenues by 2005.

Operators who are relying on data services to *retain* revenue, however, may be disappointed. There is already evidence that users expect certain services as givens, especially information services such as news and weather. Therefore these may have to be introduced just to stay level with other operators and the competition, and differentiation will lie in premium services such as games and location-based services.

This leads to the question of how so many leading players in an industry can arrive at their predictions when data services over circuit-switched networks have experienced slow adoption rates and there has been little proof that consumers require them. The answer lies in just two applications: SMS and I-Mode. These isolated success stories, coupled with the promise of greater bandwidth and technology from GPRS and subsequently 3G, has led the industry to believe that wireless data services will be the key revenue generators of the future.

Yet I-mode has also given the industry an important lesson to learn: in the rush to deliver new applications, the business case must not be forgotten. Revenue sharing models are essential for the success of all players. I-Mode also allows users access to third-party content, which is a key differentiator. It is unlikely for a single operator to provide all the required services over GPRS, and this will be the role of third parties for content providers.

Research findings from Baskerville's Executive Briefing Mobile Internet business models and revenue share trends (published October 2001) show that there is a long way to go in developing reliable and attractive revenue share models. At the moment, the operators offer a plethora of models for revenue sharing and the situation is very unclear to content providers. Operators must develop some standardisation before the content provision market can really take off.

For many, GPRS has been delayed by the limited offering of handsets currently available and also because of delays in ensuring the stability of the networks. But if consumers had a choice of handsets, which services would be available? It would appear that there are no services available today which can only be used with GPRS, and that fact will severely constrain the argument for upgrading existing WAP handsets. To counteract this, most operators have stated that, whereas GPRS does not offer any exclusive services, it does make existing data services such as WAP more convenient, and users can access WAP services at a much faster rate. Therefore, a user who spends approximately £100 for a new handset and subsequently pays around £5–10 per month for his subscription to access GPRS, will indeed only be able to access WAP services at a faster rate.

Key Data Services

Based on market research and the SMS and I-Mode experiences, the following areas are seen as key for wireless application development:

- Messaging
- Entertainment
- PIM
- Information

- Location-based services
- Basic M-Commerce
- Intranet/Internet

Important questions for GPRS application development include:

- Will the application be accessed via a mobile handset or a PDA?
- Will it work over differing operating systems?
- Is it secure?
- Will it be available to prepaid customers?
- How much bandwidth does it require?

As was the case for Internet services, where voicemail became popular both at home and in the office, there is a crossover to be anticipated between certain business and consumer applications. This is illustrated in a UMTS Forum diagram (Figure 4.1).

Figure 4.1

Consumer versus lifestyle usage and lifestyle segmentation.

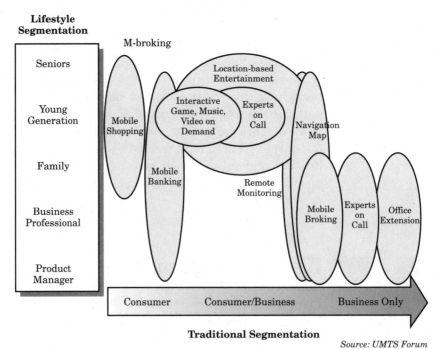

Source: UMTS Forum

The ARC Group conducted a Mobile Internet Survey in October 2000 and compiled the following results. They show the adoption rates of certain Mobile Internet applications from 2001 to 2006. By 2006, location based services will be the most popular, followed by PIM and entertainment (although currently it is the reverse, with PIM at the top followed by entertainment). Email, entertainment, and basic M-Commerce, in that order, have been the most popular applications on the I-Mode service.

TABLE 4.1

Mobile Users by Application, 2001–2006

Millions	2001	2002	2003	2004	2005	2006
PIM	275	467	700	922	1154	1445
M-commerce/retail	44	126	225	360	454	551
Financial services	149	265	410	598	771	935
Intranet	22	53	88	142	154	192
Internet browsing	94	154	250	428	654	924
Entertainment	174	292	434	631	836	999
Navigation/location	72	217	432	748	1181	1502

Source: ARC Group

The above table represents the regular usage of certain applications and is not intended to represent the total number of data users, as data users are likely to use more than one application.

Figure 4.2
Mobile users by application, 2001–2006.

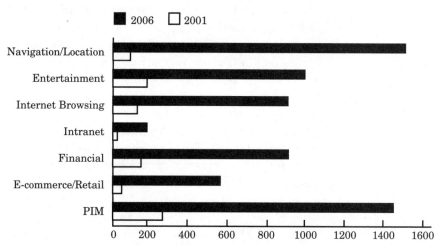

Note: Table 4.1 and Figure 4.2 above apply to all technologies and not exclusively to GPRS.
Source: ARC Group

Messaging

Messaging accounts for an average 10 percent of an operator's ARPU today, with the average user sending 12 SMS per month. For this reason alone, the industry predicts that messaging will be the largest revenue generator when data services take off. It is also seen as important not to cannibalise successful existing services such as SMS with the advent of GPRS. Why would a user still want to pay 8 cents per SMS when an email could cost 5 cents?

Early forms of messaging were SMS, voicemail, and fax services and have led to enhanced messaging services (EMS) such as picture messaging, ring tones, and animations.

Subsequently, EMS will evolve into multimedia messaging service (MMS) for 3G, which will involve rich media applications and more complex picture messaging.

Logica and CMG announced successful demonstrations of multimedia services in September 2001 over a live GPRS network. This event marked the realisation that the next generation of messaging, and also that the majority of multimedia services, need not wait for 3G bandwidth but can be readily delivered over GPRS networks. Current GPRS devices are mostly EMS enabled. According to Baskerville's Mobile Messaging Analyst, Sony Ericsson was until very recently the only vendor in the market with an MMS-capable handset with the T68i. Its major rival Nokia is slowly catching up, however. The vendor has announced three MMS handsets for the third quarter '02, with the 3510 already available. Other handset vendors are being much more cautious, with Motorola, Panasonic and Sendo so far announcing only one MMS handset each. In fact, Samsung has stated it will not have an MMS handset ready for the European market until the beginning of next year at the earliest. *It claims to be still testing the technology and says it will focus instead on color screens, which it sees as a more important functionality enhancement than MMS at this stage.* Logica says that this schedule translates to availability in quantity for the mass market at the end of 2002, with a push of multimedia-enabled handsets for Christmas 2002.

The focus on messaging applications for GPRS has been on email in particular. That's because email has seen an explosion in growth on the fixed Internet side, is extremely popular in both the corporate and consumer market, and is a low-bandwidth service. Email is also the most popular I-Mode service, followed by gaming and basic M-Commerce applications. Thirty-six percent of I-Mode airtime is taken up by email,

32 percent to other I-Mode sites, and 32 percent to third-party Web sites.

Specific to GPRS is the concept of chat services, enabled by the "always-on" connection GPRS provides. However, it has become clear that GPRS will not initially be able to support single- to multiuser connections.

Entertainment Services

Entertainment services over WAP have proved successful, in particular ring tones, horoscopes, and games such as those currently popular in the video games arena. These are seen as the future drivers of wireless entertainment services. The latter have been abetted by developments in the video games console market, which has seen the integration of email and contacts on the console as well as the introduction of wireless consoles. This growth in the games market is a key area for the mobile players, highlighted by the introduction of a mobile gaming device from Cybiko. Cybiko's product is a wireless gaming and instant messaging device with a large screen and keyboard. Users can download games when within range (typically around 100m) of a network transmitter, are alerted of any nearby users, and can chat via IM by "hopping" over nearby devices. Another example of the key role of gaming in the wireless market is Genie's WAP Fantasy football game, which attracted 5,000 players within three weeks of its launch, with no advertising.

The gaming market is highly appealing to operators who see it as a premium service that will aid in the fight against churn, as it enables the creation of communities. However, multiplayer games are needed to create communities and this service is still limited due to the constraints of current handsets. Revenue models are also still an open question, and one which needs to be solved before the true potential of the gaming market is realised. Different revenue models include sponsorship, advertising, airtime usage, and also commissions from the games developers themselves.

Another factor to consider is the life cycle of games. Many believe that, while multiplayer games will survive for some time as the concept of competition keeps them fresh, single-player games will have shorter life cycles and will demand a good deal of creativity.

Mobile game developers have been forced to reconsider their emphasis on rich graphics after evidence arose to suggest that GPRS networks are unreliable and are delivering speeds comparable with current circuit-switched performances. Mobile gaming services are expected to boom over the next five years as user numbers grow to around 850 million by 2006 from a modest 43 million at present, according to the ARC Group. However, high-bandwidth games are now being temporarily shelved because network speeds were overpromised.

CHAPTER 5

Market Positioning

(or Should You Just Give It Away and Make It Up in Volume?)

R. Clark Misul

Detecon

Great expectations have been created both in customers and in the stock market for emerging data services. The introduction of GPRS is generally perceived as the step needed to enable data services, attributing the vision of a new service paradigm to technological innovation. In the real world, the service needs more than that.

In the first round of service subscription offerings, the average user was not sufficiently empowered by useful applications and "background work" on new business models to actually open the market. Without these prerequisites, market positioning and focused advertising are impossible. As a result, wireless carriers are now under tremendous pressure from the financial markets.

Only recently, about a year after the first GPRS networks debuted, has the industry taken steps to influence market readiness through initiatives such as M-Services and new models for business partnerships in service provisioning. These efforts are ongoing. In the following pages, we'll investigate the prerequisites, identifying the key elements to make these business models viable.

Revolution and Evolution

Wireless data services will dramatically change the telecommunications business as we knew it up to now. In this section we'll examine some new architectural and business models that show the impact of such a revolution. It is worth noting that GPRS, while a revolutionary concept for the wireless carriers from a business point of view, is "just" an evolutionary step from a network point of view, as seen in previous chapters.

Contrariwise, as you'll see by reading forward, 3G will look much the opposite, a mere evolution of the same business models, but a revolution from the standpoint of networks and handsets. It is thus of paramount importance to the whole strategic path to "get all ducks in a row" right now from the business side.

GPRS Is "Just" a Bearer Service

In GSM terminology, this means that GPRS only transports information that higher layers of software, theoretically independent from the GPRS network, gather from servers and present to end users on their handset screen. It also, of course, transports information gathered from users and presented to server peers. The combination of GPRS and such additional application software enables the wireless data services, a brand new set of services that are structurally and functionally different from services hitherto. Effective implementation of GPRS-based services requires new types of terminals, new user interfaces, and new billing criteria. At the beginning, it may have only limited, or no, integration with the existing voice services.

Thus, GPRS data services can best be thought of as new product, following the typical new product life cycle, with an initial chasm like that depicted in Figure 5.1.

Figure 5.1
Wireless data services life (and market) cycle.

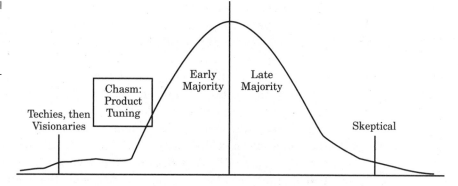

Figure 5.1 also shows a big spread in the real start of the sales ramp-up, a circumstance that's putting the whole industry under stress, reflected in the recent downturn of the stock valuation of most wireless companies. The way to reduce the initial chasm is by identifying the market segments most interested in the new services, and ensure that salespeople are armed with a "whole product" appealing to such segments. By "whole product" we mean the complete end-to-end set of features and functions that will give the end user the full service experience.

To be a bit more specific, the whole product for GPRS requires more than the key technology that bears the services, allows the appropriate speed, and supports usage-based billing. It also implies new billing features (such as event-based and transaction-based billing), appropriate terminals, personalized content from centralized resources (wherever they may be), seamless roaming also at the application level, and last but not least, easy-to-use software implementing specific services through interaction with complementary software at the other end of the network.

From what we've said, it appears that the main novelties in GPRS-based services are features of:

- The user interface
- Billing criteria and the related business models

The new user interfaces, which are being simplified further through the M-Services initiative and various applications in development, are described in detail elsewhere in this book. So are the new principles for billing. In this chapter we focus on the business models.

Reducing the initial product chasm requires real market focus. Services must be deployed with features adequate to the needs of the target market segment, such as quality of service, coverage in all appropriate areas, security, and content specialized by segment and presented in ways appealing to the segment.

As long as no single company has complete knowledge, control, and carrier class service management (provisioning and operations) of all elements that compose a whole product, all sorts of new business relationships have to be put in place. In fact, market focus implies a series of requirements strongly dependent on the specific segment, whose best implementation seldom resides in the realm of the carrier. Most likely, application service providers and vendors already specialized on the specific market are likely business partners for the carrier wanting to start a satisfactory relationship with the new customers.

Because of the novelty of wireless data services, the wireless operators are being challenged from many different sides. First, wireless operators who traditionally deal with circuit-switched networks and operations now find themselves immersed in an open packet-based environment where they are not the real experts. Other entities with which operators have no significant business contacts will play a major role in this transition, notably the IETF.

Second, operators who are used to very limited help (or interference) from outside companies for provisioning end-to-end services now need to make extensive use of partnerships. In fact, the full wireless data service is not going to be provided from a "stand alone" network owned in its majority by the provider, but will interact with a concurrent, parallel phenomenon called the Internet. This fact of modern life cannot easily be reflected in a partnership. This is mainly due to the fact that wireless carriers and Internet-based companies have different agendas. For example, the partnership utilized today with content providers for SMS value-added services is based on a new kind of content provision—one that didn't exist before carriers made access to their subscribers via SMS possible. The model was extremely simple, as in most cases the carrier was being paid on a per-message basis, irrespective of the content and the amount billed to the end-user. Because the Internet offers many alternative media by which customers can reach their content, of which wireless is only one, the content providers are much more independent and demanding as partners than those "created for the purpose" of serving wireless carriers.

So why are the carriers willing to pursue a new technology that opens the door to potential competitors like Internet content providers? The answer is in our inability to stop the slide of average revenue per user (ARPU) in basically all advanced wireless markets. In many instances, declining ARPU comes along with the first indications of market saturation from declining subscriber growth. But even where subscriber growth is still exponential, decreasing ARPU is an industry-wide reality. As an example, see Figures 5.2 and 5.3 below, which depict the United States market, one that's still far from saturation.

Figure 5.2
Estimated
subscribers.

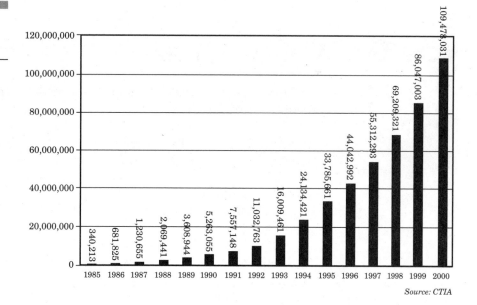

Source: CTIA

Shrinking ARPU creates a tremendous market push toward additional sources of profit on one side and effective churn-control tools on the other. Wireless data services have the potential to boost these key cash

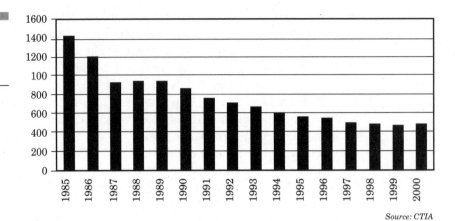

Figure 5.3
Total service
revenues per
subscriber.

Source: CTIA

flow enhancers, once new and re-imagined business models have been effectively implemented.

In the following discussion, we will try to outline the key principles for approaching the development of such new business models.

Who Are the Actors?

The following is a general list of interests that affect the full service offer either directly or indirectly:

1. *Network Operators.* They offer customers the means of access to a wireless network, by subscription.
2. *Network Vendors.* They implement the network elements that can support specified features of the service.
3. *Handset Vendors.* They implement radio interfaces on the network side according to specification to ensure interoperability with the base stations. On the user side, they implement the layers of software that facilitate information exchange between peers and present information to the end-users in ways humans can easily understand.
4. *Middleware and Mediation Device Providers.* They may offer nodes, software, or full-fledged services between operators or network nodes. Such products usually perform information pre-processing to ease exchange between peers (nodes, operators, end users, and network).

5. *Content Providers and Portals.* They maintain updated content and provide it on demand or according to personalized profiles.
6. *Application Service Providers.* They maintain dedicated servers running applications specific to a market segment and retrieving, where necessary, all information needed to provision a full service.
7. *Application Developers.* They represent to application service providers what network vendors are for the network operators.

That said, let's go on record that in real life more than one function may be performed by a single actor. It's not uncommon for a network vendor to also be a handset vendor, or a network operator to also be an application service provider. We will see in the next section how some (but not all) of these actors actually take part in service provision and its operational value chain.

A Reference Architecture

The following network service sectors constitute a minimum set of contributors to the overall service:

- Access network
- Core network
- IT and customer profiling
- Value-added service (VAS) servers
- Billing exchange

According to the model in Figure 5.4, many different service providers can participate concurrently in each full service to the user. They will be coordinated by the "billing" service provider, which is the one with a legally established retail billing relationship to the user.

To better qualify what these actors do, it is helpful to transpose them into a simplified network scenario (see Figure 5.5).

Each dotted square may be an instance of a family of providers of different kinds. In some cases, though, the service provider may incorporate more than one dotted square, in which case some of the dotted links to the billing clearing house (BCH) may not be there. Each provider shall be considered equipped with his own customer care and billing system (CCBS), with retail and/or wholesale functions.

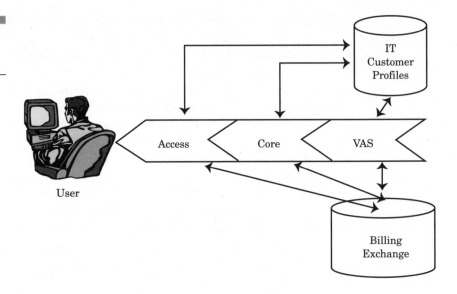

Figure 5.4
Service provision
modelization.

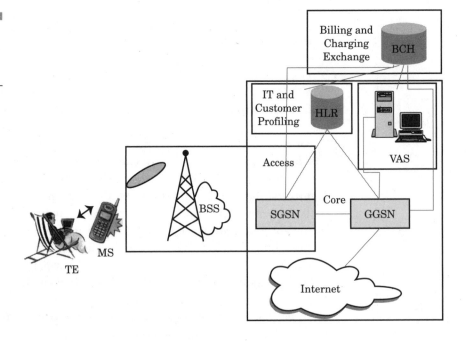

Figure 5.5
Network architecture
and service
providers.

Let's see some examples of each kind of service provider:

- **Access**—A current example is the visited network (VPLMN) where a user is roaming. A future example may be a new type of mobile virtual network operator (MVNO), specialized for coverage to multiple core network providers in areas where it is not feasible or convenient for core operators to build their own networks.

- **Core**—This is the current wireless carrier, when it elects to be only a "pipe" and let other carriers take advantage of its established switching/routing network.

- **IT and customer profiling**—A typical example is the current version of the MVNO (e.g. Virgin in the United Kingdom), where the HLR is the only part of the network the virtual operator owns (on top, obviously, of the CCBS). This type of service provider "owns" the end user and has the retail billing relationship with him. In the MVNO architectural model, a customer service-profiling database may not produce call data records (CDRs), but collects event detail from the various service element providers. Where the MVNO has its own direct commercial relations with other service element providers, a unidirectional link to the *clearing house* may be possible for collecting records.

  Another example is the security provider that hosts servers within an area protected by firewalls, within which exchanges are only allowed to and from specific sources according to agreed rules.

- **Billing clearing house**—This is an expanded version of what clearing houses already do for roaming charges. It will include new types of CDRs and TAP records, and it will add VAS service providers.

- **Value-added services**—Where the access and core networks converge with an Internet-based service, we find value-added services. They must be adequately customized for the wireless bearer's needs, able to access additional information such as end user's location, and personalized to give users quick access to relevant data. Current examples are the information and entertainment service providers operating over SMS.

Each separate entity contributes an element to the final end user service, but only one of them (or, in a less-than-ideal case, two of them) handles a direct relationship with the user.

A quite different consideration for positioning is the provisioning of various quality levels for each connection. Levels are usually distinguished according to various attributes defining throughput, error rates, delay, and reliability or transparency of the data connection. Many different models are being developed in 3GPP and IETF, but a generic model sees a *profiler* and a series of *actuators* in the provision of end-to-end quality controls in an Internet protocol (IP) environment. A profiler implements the control functions of the network and maps the provider's QoS policies to appropriate configuration parameters. An actuator implements the functions needed on the data session stream to apply the appropriate QoS class (Mapping, Classification, Resource Management, Traffic Conditioning). We can think of profiling as belonging to our IT service provider, with actuation functions divided among the access, core network, and VAS providers (see Figure 5.6).

Figure 5.6
The functions of IT
service provider.

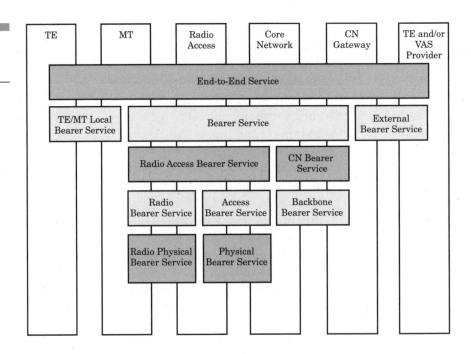

This structure implies a whole host of wholesale and partnership agreements that must be in place to allow proper coordination, provisioning, and revenue apportionment. Because the user only experiences and appreciates the QoS provided on an end-to-end service, all service providers on the path have to chip in to make such a feature possible.

Follow the Money!

Once we manage to take the architecture described above and accurately determine who owes money to whom for what, we've taken a key step in defining the new business models. Figure 5.7 is a "follow the money" diagram representing the various relationships of our model.

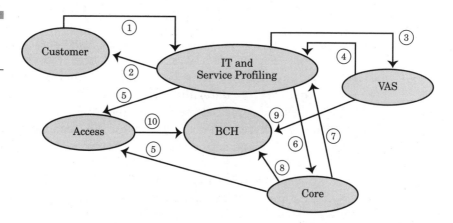

Figure 5.7
"Follow the money" diagram.

Please note that many other relationships are possible in the most generic theoretical diagram. In this one, we have only considered the relationships most likely to figure into a commercial service. While it's theoretically possible to create a commercial relationship and settlement charges between, say, an access network provider and a VAS provider, such a relationship would be dependent upon the execution of a transaction between the entity that owns the customer (i.e., the HLR) and the content-dependent VAS provider. This constraint makes it very unlikely, as it would permit revenue generation and customer interactions that bypass the service provider of record.

Let's add some detail to the kind of transaction represented by each arrow in Figure 5.7.

1. This is the normal transaction between carriers and customers, via the periodic bill. Billing may be based on multiple charging schemes, such as basic usage included in a fee with overage, subscriptions to specific push services, and per-usage pull ones. Each of these schemes may be offered pre- or post-paid.

2. This represents the possible rebates on usages, according to specific triggers (bundling of subscriptions, use of specific services, acceptance of targeted advertising...).

3. This represents payment of the specific VAS use, if the user pays for the service through her billing relationship with the IT service provider.

4. This represents payback from the VAS to the service provider for a particular use of network resources, advertising revenues, or payment of the agreed portion of revenues, in case of direct billing from the VAS provider to the end user.

5. This is the settlement from the core network to the access network provider for use of its resources. It is exactly the same kind of settlement that may occur between the IT service provider and the access network, as determined by the commercial agreements that have been executed between any two providers.

6 and 7. These turn on the commercial agreement set up between core network and service profiling. It may be payment for network usage, or rebates on specific events or amounts of use. Note that some events may be tied to specific VAS use; thus a virtual settlement between core network provider and VAS provider can also be envisioned.

8, 9, and 10. These represent the payment of the clearing house services (typically dependent upon amount of CDRs or a related metric).

As can be extrapolated from the features of the services provided by each actor, some are based on value added (high margin on revenues) and others on volume (low margin on revenues). Examples of the former are the IT and the VAS service providers whose margins allow a proliferation of competitors by geographic area and by market segment. An example of the latter is the BCH and the "purely" core network provider, whose basic need is to consolidate enough traffic and pull in enough additional revenue streams through value-added service offerings to ease their operations' burden. The latter thus deserve a bigger piece of the overall pie and a bigger market share.

Now, Who Teams with Whom?

From what we've seen in previous sections, the clearing house is not directly involved in actually providing the main services' features to the end user. Consequently, we will not further analyze the implications of each business model possible on the BCH. Business models emerge from, and are differentiated by, a number of variables and how they fall out. Relevant factors include:

- How the providers forge individual relationships
- Whether various providers are owned by a single company
- Whether providers allow complete open access to third parties for specific services, or allow only selected partners into the data services provision club (better known as a "walled garden")

In what follows we'll try to capture the main types of emergent business models. They will be defined from the perspective of the wireless network provider whose aim is to open up new services and forge new partnerships that enable such services.

Bit Pipe

Bit pipe is a typical model for an access network provider, or for a "full" carrier owning access, core, and IT networks. The tricky part of a bit pipe model is to ensure that enough volume of data transport is generated, but priced at the right level to avoid a *de facto* reduction of the carrier's bottom line. Remember that an equivalent minute of voice using the same resources (i.e., bandwidth) as a data service will produce higher revenues. This poses serious top-line limitations to computer-based connections, such as through GPRS PC cards. Such limitation, though, is found in virtually all new business models.

For a carrier, another significant limitation on this model is that proprietary information such as location or traffic profiling by type of service may need to be exported to VAS providers, that will use it for the full service provision. How to price and charge such information without having full visibility on the end service is truly daunting, which may end up inhibiting certain services.

On the other hand, for specific services such as vertical markets service integration for a carrier that's specialized in mass, horizontal marketing, this model may represent a way to collect revenues from segments that otherwise would get their services from other providers, while preserving marketing resources needed to cover non-core strategic markets. In fact, value added resellers (VARs) specialized in direct sales to specific vertical market segments may be able to exploit these markets better with appropriate custom packages.

Perhaps the most interesting new feature of the service provided under this business model will emanate from its ability, in the near future, to provide quality of service (QoS) classes. These classes can be compared to categories we all know today in the package delivery market: parcel rate, first class, and express mail. The customer pays a premium to be assured of delivery within a specific timeframe at a specific degree of reliability. In this latter case, a *variable premium* determined by the value of the content, as perceived by the customer, may apply.

Partnership

A partnership is a typical peer-to-peer commercial relationship on a service-by-service basis, where the driving force may be either the carrier (IT only, or IT, core, and access) or the VAS provider. Particularly interesting for the partnership model are creative solutions to the problem of negotiating a commercial agreement satisfactory to both sides in the hypothetical case of a VAS integrated with an IT (e.g., an IP-based MVNO with direct subscriptions). Such a service provider (think AOL or Yahoo!) may try to vie for superiority with the core network provider. The information each wants to control is the same proprietary data we considered in the previous case. The prize is who is perceived by the end-user as the actual service provider.

This business model will always be particularly affected, on a nation-by-nation basis, and in some cases region-by-region, by financial, regulatory, and licensing issues. Therefore, the most resilient way of partnering might be to specialize the functions of the end-to-end service provided by each of the partners into *commoditized* and *personalized* categories. Commoditized functions are those that are, or soon will be, sold retail "by the pound." That's happening now with voice minutes, and another example from history is generic Internet browsing. On the other hand, the end user presumably perceives the personalized functions as a distinctive feature worth a premium on top of the basic fees, such as the "infotainment" services offered for SMS. Feature distinctions afford each partner a level ground for negotiating the weight of each function in the context of a service package. In fact, such distinctions may be found in the entire end user service distribution chain (access, core, VAS, billing, customer care, and commercial distribution). Such distinction, though, will also depend on the level of competition in the specific market for each service distribution chain function.

Some examples may help clarify these concepts:

1. Access in urban or business areas tends to be commoditized because of the large number of competitors, but may be personalized in rural areas. In addition, access may have a personalized flavor because of added QoS features.
2. Core is typically commoditized but, again, specific kinds of connectivity or QoS features may favor differentiating categories of connection.
3. VAS is a personalized function in the end user service, amenable to a wide variety of market differentiations based on how it is segmented or specialized applications. Some VAS, being more common and widely accepted in the mass market, may end up quickly in commoditized status. At present there are no data on wireless data services use from which we can infer the end users' willingness to pay a premium, so it is difficult to forecast which applications will commoditize first. One likely example is plain mapping of specific addresses with point-to-point driving directions.
4. Billing and customer care can have both flavors, depending on the service offer and the value added. We'll expand on this in our discussion of how to measure a successful implementation of your business model.
5. Commercial distribution is probably the best example of the commoditized/personalized distinction. In fact, big distribution chains for the simplest packages in the mass market—say, convenience stores in the United States, or tobacco stores in Europe, and big electronic appliance retailers—are more likely to attain maximum commoditization. In this industry, direct sales stores from the carriers and service providers represent a first chapter of commoditization, with a residual level of personalization, while direct sales forces for vertical and enterprise customers are the best example of personalization.

The Walled Garden or the I-Mode Model

Remember the "club"? The carrier manages everything, including the service feature specifications to which the application developers (prospective VAS providers) have to conform if they want preferential access to its services. You will generally find this kind of organization in a walled garden. On the Internet, a walled garden is an environment that controls the user's access to Web content and services, such as WAP services or Research In Motion's Blackberry. In effect, the walled garden directs the user's navigation to particular areas where he can access preselected material and cannot access any other material. A service provider may or may not allow users to select some of the Web sites to be contained in or precluded from the garden. Although the walled garden does not actually prevent users from navigating outside its walls, it makes venturing outside more difficult for the user than staying within.

For a Web site soliciting visitors, the whole game becomes scaling those garden walls to reach members of a club who are otherwise unreachable. The approach, if properly managed in terms of available services and related features, has proved to be extremely successful. In fact wherever application developers and VAS providers mimic the "look and feel" of the access, core, and IT provider, thereby providing the end user with a homogeneous and predictable service experience, usage patterns generally improve as a consequence.

Other Models

This category comprises all the ways our previous models can be combined. It probably contains the most likely end scenario because rarefied business models are rarely tenable. When service adoption broadens to reach many different user segments, a variety of services tailored to specific customer needs and niches will be born on the VAS provider side. At that point, when services based on partnership or I-Mode models are no longer financially viable for a network provider, or when services fail to target a market segment the provider finds interesting, the pipe model can still be successfully utilized.

Wholesale versus Retail, Vertical versus Horizontal, Customer versus End User

Some guidelines are of paramount importance to a successful implementation of the business models just described between any two parties:

1. A clear understanding and acceptance of each provider's part for the others' goals and minimum needs

2. A clear understanding and acceptance on both sides of which is the served market. That understanding may include, if necessary, acceptable price points and price structures between the partners or to the end user.

3. Standardized interfaces (to the extent possible) for the exchange of billing and charging records. One easy benchmark is to ensure at least as many as you have for signaling and data.

4. Commercial distribution channels for the segments the partnership will serve must be in sync, motivated, and trained to smoothly promote the services strategic to the value chain.

5. Billing systems need to be:

   a. Flexible, to allow for multiple packages, differentiated by user, partner, service, market segment, related distribution channel and, obviously, applicable business model.

   b. Interoperable, in the sense of having enough common ground to interpret data in a concordant manner between peer billing entities.

The previous guidelines have to happen in each of a few broad cross-connected market categories to ensure early penetration and ramp-up of wireless data services. In those categories, requirements on billing systems, pricing, and distribution channels will change the type of agreement or the complexity of agreements depending on:

- Whether the relationship between the parts is wholesale or retail.
- Whether target market segments are vertical or mass market in nature.
- For vertical market and certain mass-market distribution strategies, whether or not they can accommodate n-tier customer models. These must be at least three tiers—consisting of an *account* (e.g., a large corporation) comprising many *subscribers* (e.g., regional offices) who support many *users* (e.g., the employees).

Some examples of the impact on billing requirements of the previous points are:

- Wholesale relationships do not need the bill rendering function (format and print) complexity that retail ones imply.
- Sales and distribution additional functions (such as point of sale—POS systems), to provision new customers, monitor sales, and evaluate quotas and commissions are substantially different in vertical or mass markets.

- Multitiered models may have different implications on discounting (for example a discount at the account cumulative level, but apportioned by subscription) and bill print-out (print details only at subscription level, totals by offices at the account level, with specifics dependent upon customer's request), possibly with cross-correlations between the two (you get a better discount for a simpler or standard billing structure).

In sum, the elements contributing to a profitable wireless data service offer based on GPRS can feel like a lot of puzzle pieces to fit together. Making some decisions will condition your options for others. It's hard to find a place to start. The prize for such an effort will be a completely renovated telecommunication business scenario, more able to take advantage of new opportunities and thus more solid in terms of steady and manageable profitability.

CHAPTER 6

The Enterprise Business Model

Stephan Keuneke

T-Mobile International

Product Offering

In general, the take-up of data services based on GSM has been rather slow. This was primarily due to the slow speeds of circuit switched data. Second, except for SMS, the terminals included no applications that made use of the network's data services. So, users had to connect and configure external devices like laptops, a task that could entertain an ambitious self-taught laptop-user for some sleepless nights. And finally, even if the connection to the laptop could be mastered, few off-the-shelf applications utilizing the GSM data services existed at all. Only a narrow segment of users had an absolutely pressing need for mobile data services that might effectively force them to cope with the obstacles just described. This segment consisted of businesses with a high percentage of mobile workers who rely on online connectivity even where no fixed line is available (e.g., sales personnel who would prefer not to ask a potential customer to use his phone).

With the introduction of GPRS, these things are about to change substantially: data rates are approaching the range of ISDN services. Devices like smart phones have data applications built in and work on open platforms capable of supporting client software for well-established enterprise software packages like MS-Exchange, Lotus Notes, SAP, and others. For business users, new GPRS-based product offerings have the potential to end the years of mobile data agony.

It must be understood however, that these product offerings will not always be available to customers in a shrink-wrapped, plug-and-play fashion. The key factor for a successful implementation of mobile data services in any enterprise is careful integration into the existing systems in use. Depending on the extent to which an enterprise has customized their mission-critical software packages, the mobile extension of those applications is more or less likely to require some special features that may not be available off the shelf. Therefore the enterprise markets will continue to be an arena for solution-driven, project-oriented business as well as for standard mobile data products. The following describes some product offerings we can expect to find in the future, as individual customer solutions and as standard products. Of course, the list is by no means exhaustive.

Mobile Intranet/Dedicated APN

The need for access to corporate information "anywhere, anytime" is the driver for most business customers to seek mobile data services. What the mobile worker wants is access to his email, his scheduling system, and many other kinds of databases he may need to tap regularly, if less frequently. Meeting these requirements, in its simplest form, means establishing a link between a mobile worker's laptop and the servers in his "home" intranet. But this, of course, has to be accomplished without turning the corporate intranet into a part of the public Internet, so control of access becomes a major issue for implementation.

Access control could be handled at the application layer: enterprises could opt for usage of the public Internet (the GPRS mobile Internet access offered to all customers) and rely on control mechanisms commonly used by commercial sites in the public Internet, enhanced perhaps by some protection against eavesdropping (probably SSL). However, most businesses will consider the level of security this method offers unsatisfactory. With dedicated access point names (APNs), operators are in a position to offer improved security for connectivity to corporate intranets. It works roughly as follows: only the subscribers belonging to company XYZ will be granted access to APN xyz.operatorMCC.operatorMNC.gprs. User-name/password authentication can be handled by the operator alone (that is, independent from the log-in procedures for the corporate intranet), or by means of a RADIUS-proxy that verifies authorization on behalf of a root-RADIUS within the intranet administered by the company's IT staff.

IP-VPN

For some enterprises, a dedicated APN may not be what they want, especially if mobile access is only one of several access methods they support. Businesses that have set up virtual private networks (VPNs) in the fixed world will want to maintain the same level of flexibility and security when extending their networks towards mobile access.

If strong security in the lower layers is required and public infrastructure is used, IP-SEC is often the method of choice. The technology involves setting up "tunnels" through a public network (the Internet), that isolate and protect the transferred data. To support or enable the end-point of these tunnels at the mobile end of a connection, specific functionality for addressing and other primary services must be introduced to the network infrastructure. Charging for this kind of service will most likely be based on monthly fees related to the number of users. The charges, of course, are in addition to charges for standard traffic handling.

Hosted Enterprise Portals

A data link over the air has some characteristics quite different from transmission over any kind of wire. Most notably the latency is far greater, and this subtle difference already creates problems when connecting to standard server software that assumes a client is connected via a fixed network. Furthermore, many client/server architectures utilize some "alive-checking," that is, they send some data every x seconds just to see if the client is still online, a practice that can be quite an expensive hobby if done (and paid for) via mobile networks that charge for volume. Finally, standard enterprise software usually provides info as it will be displayed on a standard (i.e., PC-sized) screen and does not take into account different devices that might be used for display.

These examples all point to the need for some kind of portal service dedicated to the mobile devices that want to connect to corporate information. It should also be acknowledged that the set of devices will change over time, so the task of providing adaptation for new types of terminals is an ongoing one. The mobile operator is concerned with these questions anyhow, so he is in a good position to offer portal services that solve them for corporate customers. These services can range from simply bringing the corporate phone-register onto some WAP pages all the way up to enabling full access to corporate information in a great variety of media types, including streaming video. Ideally the corporate customer won't need to make any changes to his intranet, as all transmission or media format problems for mobile devices are solved by the operator or service provider.

Products in this "Hosted Enterprise Portal" category will be combined with the access-related services described above, such as dedicated APN, that deliver secure data-transmission.

Telematic Solutions

Currently many fleet-tracking, machine-to-machine alert services, and other specialized mobile data applications are based on SMS, which is particularly suitable for the transaction-oriented nature of these applications. But there are limitations to this technology, most notably its maximum capacity of 160 bytes in a single message and the principal architecture of a store-and-forward service. The latter means no direct point-to-point transmission. Both limitations can be overcome with GPRS solutions. However, corporations have invested heavily in infrastructure to support their telematic solutions via SMS, and will therefore be reluctant to switch over to anything new before the old systems have amortized.

As a general trend, operators are attempting to move from solution-oriented business towards standard products that target smaller enterprises (see below). For telematics this can only be supported by offering customizable packages that are specific to an industry sector, since an "alarm-if-vending-machine-empty-application" and a fleet-tracking application do not have very many requirements in common.

Positioning

Within the business markets, GPRS addresses different segments and competes with other technologies. The following sections highlight the major issues.

Addressed Segments

The technical difficulties for mobile data described in the preceding section, along with the fact that smaller enterprises often have no dedicated IT department to tackle them, have led to a situation where mobile data is being used almost exclusively by large enterprises (defined as >500 employees). In most countries, however, the segment of small-to-medium enterprises (SME) is larger than the LE segment and has never been targeted; it therefore represents a good opportunity for operators and solution providers (see Figure 6.1 for the German market prediction).

Figure 6.1
Segments in the business market—example Germany.

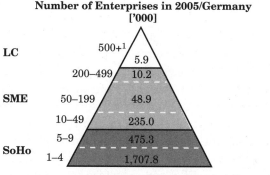

Number of Enterprises in 2005/Germany
['000]

LC	500+[1]	5.9
	200–499	10.2
SME	50–199	48.9
	10–49	235.0
	5–9	475.3
SoHo	1–4	1,707.8

[1]*Number of employees per enterprise.*

Should the industry succeed in creating plug-and-play products for mobile data, the small office/home office segment (SoHo) is also addressable. In this segment, customer care quality such as installation support on the phone will be the key differentiator.

Competing Technologies

GSM-CSD and GPRS are not the only options around when it comes to untethered data access. While past attempts with DECT and TETRA-based data solutions must be considered a failure in most countries, new and more promising technologies have started to hit the markets. Most notably, wireless local area networks (WLANs) based on standard IEEE 802.11 are gaining attention and are sometimes seen as the "UMTS-killer"—which obviously includes a major threat to GPRS as well. This threat is especially relevant for the business markets, as a large portion of users in this segment is already equipped with laptops that can easily be enhanced with PCMCIA cards for WLAN access.

Since WLAN operates in the license-free spectrum, any company willing to take the risk could roll out a "mobile" offering. However, the mobility of users will be limited by the fact that WLAN is intended for nomadic rather than mobile use; it does not work well if the terminal is actually in motion at faster than pedestrian speed. Furthermore, for a viable business model, operators need an infrastructure that incorporates elements for provisioning, charging, and support of customer care. This technical foundation is still missing for the most part. Some GSM equipment vendors have started to offer WLAN components that utilize the authorization and charging mechanisms already in place for GSM/GPRS, basing access to the WLAN service on a SIM card. Considering that the expertise and skills for running a public telco service is something that operators would claim to possess, this approach seems only logical. WLAN has far better prospects as a complement to GPRS/UMTS than as a competitor.

Another technology with the potential to change the landscape of mobile data is Bluetooth. Originally intended as a replacement for patch cables, Bluetooth has grown—at least in expectations—to become the platform for personal area networks, connecting and integrating a multitude of personal devices ranging from phones to PCs to cameras. Note, however, that the concept of a "personal" network implies a relatively small number of devices to be connected. A natural role of Bluetooth in conjunction with GPRS will therefore be as a means to connect peripheral devices to some kind of GPRS-connected hub, which may be a mobile phone.

The threat from Bluetooth today mainly resides in indoor coverage situations. Since Bluetooth does not support any handoff capability, it is not a significant threat to wide-area UMTS/GPRS revenues. The next version of the standard ("Bluetooth 2") is being designed to support a handoff capability and higher data rates (10 Mb/s gross bit rate), but Bluetooth 2 products are not expected to be available before 2005.

When positioning GPRS (and UMTS) -based products for enterprise markets, providers have to make decisions about their relation to the fixed world. A service provider has to go either for enhancing existing fixed networks with a mobile part, therefore focusing on systems integration, or for offering easy-to-use monolithic systems in the form of a greenfield installation or as replacement for legacy installations on customer premises. The two approaches require quite different competencies and processes, so operators face quite a challenge if they opt to go for both. For the systems integration approach, partnering with the big enterprise software vendors can be of great advantage. This is just as true for common technical development as for promotion and marketing efforts.

Pricing

With the introduction of GPRS as the first technology to enable volume-based pricing in the mobile domain, the almost religious debate about volume- vs. time-based charging in the fixed telco world moves into the mobile world as well. Similarly, the question of guaranteed service levels for the packet-switched service arises, and attempts to tackle this question so far resemble those in the fixed world.

Volume-based versus Time-based

Unlike the mass market, where there is at least some reasonable doubt whether John Doe understands what a kilobyte is, volume-based pricing has been the norm in business markets for quite some time. Prices for fixed connectivity for enterprises usually include a monthly fee for access plus charges for usage, the latter based on the volume of data transferred. Therefore the principal discussion about time- versus volume-based charging can, at least partly, be considered decided.

What complicates the issue for the mobile case, though, is the fact that existing mobile applications like WAP were tailored for the very limited bandwidth of CSD, whereas mobile intranet uses much higher bandwidth and volume. So, for operators who choose to charge by volume, the tricky issue is the relation of CSD tariffs (current as well as past) and volume-based charges. If the older narrowband/CSD tariffs determine a volume tariff, mobile intranet/mobile becomes prohibitively expensive. On the other hand, if mobile usage scenarios are used to calculate the volume price that users are willing to pay, the same price would make narrowband applications extremely cheap (which, for some reason, operators dislike). See Table 6.1 below for some rough calculations.

TABLE 6.1

Example Calculation "The Volume-Price Dilemma"

	Acceptable Volume Price Based on WAP Usage	Acceptable Volume Price Based on WWW Usage
Price base	0.3 euro per minute 9.6 (= 72 kb)	0.1 euro per page
Price for a WAP-deck (3kb)	≈0.013 euro	≈0.0013 euro
Price for a WWW-page (250kb)	> 1 euro	

The solution seems obvious from the marketing point of view: charge differently for different applications (e.g., WAP and WWW). From a technological point of view, unfortunately, this is not easy to achieve. Once an IP connection is established by means of a PDP context, there is no easy way to control what the user does. Technology for just that purpose is in development and is currently rather immature, one concept being service selection gateways that distinguish by the higher-layer protocols and ports used for the IP connection. Before this problem is solved, operators will naturally stick to the higher prices, enjoying yet another benefit: rather than just being the bit pipe, they can offer enterprises additional services or middleware to compress or reduce the amount of data transferred over the air.

Nontraffic-related Charges

Aside from traffic charges, network operators collect set-up and/or monthly fees. For intranet connectivity with a dedicated APN, fees may include a monthly "rent" for administering the APN, including authentication services based on remote access dial in user service (RADIUS). For solution-oriented businesses, a wealth of charging options will exist for hosting, application service provisioning, and other individually negotiated operator services. These are not specific to services based on GPRS, but they must be factored in.

Service Level Agreements

In the fixed telco world, service level agreements are commonplace for intranet connectivity services. Service levels (defined as levels of availability, bandwidth, error rates, etc.) are guaranteed in relation to specific service access points (SAPs). These can be regarded as the "wall socket" that the customer plugs into. What is important here is the fact that, until reaching this SAP, the operator is in full control of the environment, utilizing his own routers, switches, and any other hardware, including the wires used to transport the data. The mobile case is complicated by the fact that the "natural" SAP, the mobile terminal, is connected to the operator's network via an uncontrolled environment—over the air. The weather and the user's behavior, to name just two conditions, are beyond the control of the operator, who therefore can give no guarantee concerning the quality delivered at the terminal. On the other hand, guarantees related to any point or entity in the operators network (e.g., the GGSN, SGSN, WAP-GW) are not especially meaningful to the enterprise customer who primarily wants to ensure that his sales personnel can access corporate information at all times.

It seems clear that the business procedures of the fixed telco world for service level agreements cannot be copied as-is for the mobile world. Since the mobile data universe is just starting to grow, this problem still needs to be sorted out between the wishes of the customers and what the operators are actually capable of offering.

Promotion

As we mentioned at the very beginning of this chapter, mobile data solutions have in the past not been very successful offerings, at least when measured against the penetration of corporate contracts for voice services. Initially it seemed obvious for operators to use their corporate customer bases when promoting new GPRS-based products and solutions. In the enterprise, however, it is seldom the case that the IT department also manages the phone network, and often the contacts available to the provider's sales personnel are not quite the best addresses when it comes to mobile data solutions.

Depending on an operator's principal decision about positioning in the business markets—solution- versus product-oriented or both—different promotional strategies have to be pursued. For the solution-oriented business, the service provider must itself be perceived as capable of performing IT systems integration, or must communicate its closeness to system houses and/or software vendors. Many corporate customers have already instituted a close relationship with a system integrator—perhaps their enterprise resource planning (ERP) or sales force automation (SFA) systems vendor—and will be reluctant to let any other party mess with mission-critical infrastructure. Therefore, the way to the customer may often be traversable only for those providers who can partner with the original supplier of the enterprise software infrastructure. Common showcases at trade shows and other means of communication can support this approach.

When following a strategy to build a product-oriented business, a service provider must pay much closer attention to terminal equipment. Shrink-wrapped, ready-to-use products will in some cases require bundling with terminal equipment to be able to deliver the plug-and-play promise printed on the cover of the package.

In this chapter, obstacles as well as chances for GPRS in the business sector have been described. History has shown that early adopters of new technology most often are professional or business users. Looking at the capabilities of GPRS and its fit to the described needs of the mobile business users, we can expect this to be the case for this technology as well.

CHAPTER 7

GPRS in the Mass Market

Rafael Ruiz de Valbuena
and Conchi Gutiérrez

Telefónica Móviles

In its early stages, we saw GPRS marketed as a technology that enabled services mainly addressed to business customers, who would fully benefit from the higher data rates it provides. It is true that companies are usually early adopters of new technologies for data transmission at higher rates, which substantially increase efficiency at work. In addition, GPRS handsets were not economical from a mass market point of view at the very beginning and it is taking a while to see them in big volumes, just as we did in the case of WAP. None of this means that mass-market customers will not see the advantages of GPRS. Quite the opposite: from now on we expect to see terminals costing below 200 euros and lots of services targeted to nonbusiness customers that are possible now with GSM and WAP, but are not used or practically unknown because it takes GPRS to deliver those services in a fast, convenient, and affordable way. A fast, convenient, and affordable service is what the customer wants if his intention is to send a snapshot of his holiday back home. We will return to such services shortly.

Positioning

Once the advantages of GPRS become evident to everybody—and that will require wide availability of GPRS handsets at reasonable prices—GPRS will be seen as a way to access data more securely, quickly, efficiently, and cost-effectively than ever before. Therefore, GPRS will contribute significantly to the goal of accessing the right information, at the right time, from the right device.

At the moment, different technologies coexist in the wireless communications world for the transport of information over the air. Such technologies are independent of the nature of the information itself. One of them, GSM, is used for both voice and data wireless communications. However, recall that GSM is not the most appropriate technology for data transmission, since it uses a whole communication channel (a limited resource) during the entire duration of the call. Also, the speeds reached in data transmission with GSM never exceed 9.6 kbps. Channel limitations motivated development of a new technology for efficient data transmission in the mobile environment. In addition, GPRS data speeds clearly surpass those of GSM: with current GPRS handsets you can transmit up to 50 kbps depending on the terminal used and the network load. In other words, GPRS has multiplied the current speeds of GSM by at least a factor of five.

Voice will continue to use GSM technology, since a voice call (unlike data communications) requires the use of a communication channel in an exclusive and dedicated way during the whole time the conversation lasts. As a result, in the current mobile telephony environment, the most appropriate technology for voice transmission is GSM, just as GPRS is the most suitable technology option for data transmission. Until now we used GSM more crudely without regard to the nature of the information, voice, or data.

GPRS is also known as "2.5" generation system (2.5G), and it should be clear from that appellation that GPRS doesn't replace GSM. It is a complementary technology that coexists with GSM to substantially improve the data transmission capability in mobile telephony.

In order to be able to enjoy the advantages of GPRS, you must have a GPRS mobile phone, which is 100 percent compatible with the GSM technology. With a GPRS phone you will be able to benefit from better data services and, at the same time, from the same services provided by the GSM phones we know today (voice, mailbox, roaming, games, etc.). GPRS and GSM come "all in a box."

The following table compares GSM and GPRS from the perspective of the types of services they allow for:

TABLE 7.1

GPRS versus GSM

GPRS	GSM
Data transmission speed will increase, as a function of the terminal used and the state of the network, to a theoretical maximum of 115 kbps. (Due to terminal and network limitations, the current maximum is 50 kbps).	Maximum speed of 9.6 kbps.
GPRS will mean an important change in the way of *billing* the customer, because it opens the door to charges *per volume* of received/ transmitted information (euros per kbyte) and also *per event* (euros per image downloaded, for example).	Rates per call duration (cents per min).
Fast *connection establishment time* (practically immediate).	Call connection time of approximately 10 seconds.
GPRS will allow the user to be permanently connected (*always-on*) to the data services he or she is using.	A new call has to be set up every time you need to connect to the service.
GPRS represents an important step toward 3G or UMTS, and embraces the vision of the mobile device as a multifunctional device used for something other than speaking.	Second generation (2G). Particularly suited to voice services.

Product Offerings

Despite the fact that GPRS provides significant improvements to the way we experience data services today and should therefore be promoted as such, we can't forget that it is not a service in itself. Moreover, no service has been designed specifically for GPRS so far. The idea is that existing GSM mobile Internet services will now benefit from the advantages of GPRS and become more convenient to use (and pay for, as we will see).

Some service groupings that can be significantly improved with GPRS and can be very appealing to mass market customers are:

- **GPRS data transmission**—This includes the transmission of icons, photos, images in general, music, video...anything that can be packed into a data file. These files tend to be large (think about attachments to your e-mails) and therefore GPRS's higher transmission rates make sending them to a friend a faster and more enjoyable experience. Trying to do the same over slower GSM, at a maximum of 9.6 kbps, was more discouraging than fun.
- **Email**—Email is a particular case of data transmission worth mentioning in its own right. Two characteristics of GPRS make it especially attractive for email transmission: higher speed and the fact that users are "always on." (The moment an email arrives, you know it because your connection with your email is continuous.)
- **Video streaming**—Video streaming is another particular case of data transmission. Video streaming was almost impossible with GSM because of the low speed of transmission (video files are normally too big for a slow connection). Now we can see the first video clips moving over GPRS. Although there is room for improvement in transmission speed and data compression to make this service a better experience, it represents an important first step.
- **GPRS access to WAP services**—In the mobile Internet, WAP pages or WML[1] content will be accessed via GPRS. The protocol is used today for accessing content, but WAP over GPRS improves upon WAP over GSM in that data flowing to and from the wireless telephone will be stored and organized into efficiently transmitted data packets (that is the essence of GPRS). Contents do not change, only the way they're transported and the model on which they're priced. A GPRS user will be able to navigate freely through WAP pages, traverse menus which will guide him to a diversity of services offered by differ-

---

[1]Language in which WAP pages are written.

ent suppliers, personalize his own navigation menu and create his own WAP address book, utilize popular search engines, and more.

■ **GPRS Internet access**—GPRS lets anyone with the right handset and a portable PC or PDA surf the Internet freely, anywhere and anytime. The customer will not need to hire the services of an Internet service provider (ISP) because the operator will perform that function for customers of the GPRS service. At the same, time there will be no need for keys or passwords if you are content with anonymous access (the way you normally access the Internet from your PC).

With GPRS, users will initially navigate up to four or five times faster than with GSM. They won't have to worry about how long a connection lasts because GPRS supports tariffing per volume of data transmitted instead of per minute of use—a feature many consider essential for mass-market services. The new model of Internet access is much closer than the old to the traditional Internet access model, with the added advantage of mobility and minus the headaches of connecting operator and ISP.

Practicalities and FAQs

To conclude the chapter, we hope this section will help clarify and summarize the main ideas in play for mass-market GPRS services, while trying to answer important questions. Let's begin to sketch out some answers.

How Will I Perceive the Benefits of GPRS?

The big question: "What will be my first impression of GPRS access, whether through a wireless telephone, a PC, or a PDA, when downloading email or surfing the Internet?"

1. It considerably improves connection time in comparison to GSM or PSTN[2] (a *data session* is set up much quicker that a *data call*). Also, once the phone is on, it stays on; you do not need to connect again every time you require service.

---

[2]Public Switched Telephone Network (traditional fixed network).

2. Transmission/reception speed is much higher than in GSM. If you are a user of GSM data applications on the move, you can't help but perceive this change. People who use other means of remote connection (RTB, ISDN, ADSL) may not perceive a measurable improvement, but they will reap the added benefit of mobility.

3. Simultaneous voice and data calls. When the user of the service is connected by means of GPRS, he can receive voice calls on the same wireless telephone he is using for a data connection. And after answering the call, he can recapture the data session exactly where he left it. (With "class B" GPRS handsets, probably the most common GPRS handsets in the early market, this will generally be the user experience. More sophisticated handsets, "class A" GPRS, will handle voice and data calls in a truly simultaneous way.)

4. Always-on capability. The connection is permanent, and in this way a user's experience of applications like mail or the Web is much the same as in a fixed PC, where applications are always active and where, when necessary, information is transmitted or received automatically. The GPRS mail experience is quite close to instantaneous.

5. Billing. You won't have to think about how long it's taking to find your way around a WAP menu or remember to time your Internet sessions, because you won't be incurring charges by the minute. In general, this will mean lower bills for data services.

So What About the Prices?

Different operators have chosen many different ways to charge their customers in GPRS, from flat rates to tariffing schemes per volume and "bundled data" tariffs. You are most likely to see flat rates for WAP services such as information queries, in which the volume of information downloaded is not too big (maybe 2 kbytes per WAP page). For applications such as mobile office, requiring the movement of large volumes of data, operators like Telefónica Móviles are developing tariffs in the "bundled data" scheme: you pay a fixed fee provided that you use up to "X" Mbytes a month. Should you exceed X, an additional tariff of euros per kbyte will be charged for the overage only.

Here is an example bundled data chart from Telefónica Móviles:

- Tariff per use: 0,024 euros / Kbyte
- Bundled Data:

TABLE 7.2

Case Study:
Telefónica Móviles
GPRS Tariffs When
This Text Was
Written

"Monolínea" GPRS	Volume	Price	Extra Kbyte price
1	1 Mbyte	6 euros	0.012 euros/Kbyte
20	20 Mbytes	30 euros	0.006 euros/Kbyte
100	100 Mbytes	120 euros	0.003 euros / Kbyte

"Multilínea" GPRS	Volume	Price	Extra Kbyte price
200	200 Mbytes	240 euros	0.003 euro/Kbyte
1000	1.000 Mbytes	1.200 euros	0.002 euro/Kbyte

As you can see, the price goes down for a heavy data services user as mass-market services must.

How Different are GPRS and WAP?

The new wireless GPRS handsets will be WAP handsets at the same time, as long as WAP continues to be necessary for presenting Internet data on the screens of the GPRS handsets we know today.

WAP and GPRS are completely complementary in the same way that WAP and GSM are. The only change from one bearer to another is the way they transmit data from the Internet to the handset (previously WAP used GSM; in the future it will transmit data over GPRS). However, WAP continues to be necessary in both cases because the protocol and the language have not changed. We can express this relationship by saying that what GPRS changes in replacing GSM is the way in which content is transported, but not the way in which content is presented on the wireless phone.

Industry observers expect to see the first high-end GPRS handsets appear in 2002. They will cost more than $500 and display Internet pages in HTML format directly on a colour PDA-type screen. Once again, the transmission technology does not affect the format in which information is presented.

WAP–GSM will probably continue to be the most natural path for most residential customers to evolve from the traditional mobile telephony ("I only use the mobile to speak") to third-generation or UMTS services. As we mentioned earlier, WAP–GSM price ranges are not comparable to those of GPRS, at least for the time being.

What is the Difference between GPRS and UMTS?

Another common question is the role that GPRS plays in mobile telephony given that UMTS (or 3G in general) is about to arrive.

First of all, we believe that the commercialization of UMTS is not as imminent as it seems, because of the uncertain availability of 3G handsets in sufficient volumes and at a reasonable price in the short term. More realistic prognosticators speak of dates close to 2003 for a massive commercial launch of UMTS. For this reason, WAP over GSM can be considered a first step in which users become familiar with emerging mobile telephony models. The same applies to GPRS.

In UMTS, the customer will need a new terminal and a new SIM card (although he should be able to maintain his original number). Apart from this, it's good news for operators that UMTS will reuse most of the infrastructure deployed with GPRS, because data transmission in UMTS will use the same principles as GPRS. However, UMTS does not make any use of the GSM system. The latter will persist in the form of multistandard GSM–GPRS–UMTS handsets, and all three technologies will coexist for a long time.

We should be explicit, therefore, that the current *GSM network* is evolving toward an *enhanced GSM + GPRS network* that will be followed, in further evolution, by a *3G network* in which the core system resembles GPRS as far as data services are concerned.

CHAPTER 8

Portals

Tage Rasmussen

End2End

For several years the wireless arena has been full of claims about future data usage, data users, and future revenue from data usage. We have seen statements such as:

- By 2002, over half of all call minutes will be on mobile networks.
- By 2003, there will be more e-Commerce users on mobile devices than fixed PCs.
- By 2004, there will be more wireless connections to the Internet than fixed PCs.
- By 2006, 70 percent of mobile networks will be pure IP.

Today these claims are being revisited by a number of analysts as critical factors such as the limited success of WAP-based 2G networks, the delayed supply of GPRS devices, and uncertainty over the deployment of 3G networks have all adversely affected the likelihood of reaching the numbers cited. Despite all uncertainty, however, network operators and portal owners are all investing heavily in wireless data technology.

Wireless Portals

We find three main types of portals in the wireless space:

- Operator-owned or -affiliated portals (Vizzavi, BT Genie, Djuice, Orangeworld, T-Motion, Zed and Speedy Tomato)
- Independent wireless portals (I-Touch and Mviva)
- Fixed portals with wireless extension (AOL, Yahoo!, and MSN)

Club Nokia works a bit differently from any of these and probably should be mentioned as a fourth type of portal. We'll return to Club Nokia shortly.

Strong competition and large investments in the portals, particular in advertising and branding, are sustained by a conviction that the value chain will shift along with the technology. In the current situation, the network operators' "ownership" of the customers lets them claim the lion's share of revenue (Figure 8.1), but in a 2.5 or 3G scenario, players like content and application providers, wireless portals, handset manufacturers, and service providers will capture a share of the customer and thereby a significant piece of the revenue (Figure 8.2).

Figure 8.1
Current 2G value chain.

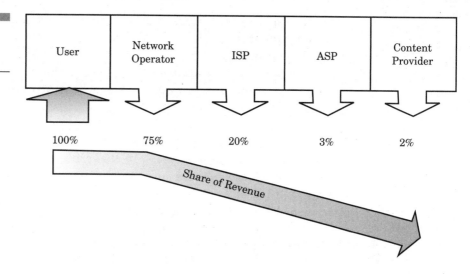

Figure 8.2
Expected 2.5G/3G value chain.

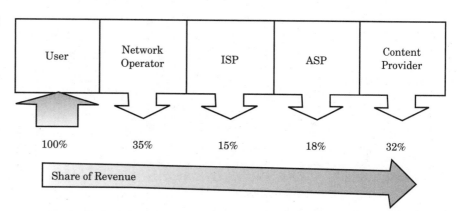

Whether this transition will really happen remains to be seen. Access to wireless data networks still relies on dialup to WAP gateways or to ISP access numbers, and thus is still under the network operator's control. The growth of 2.5G and 3G networks could obviously change this situation, while the intense competition will bring "airtime" revenue down.

Historically, the role of the portals was, as indicated by the name, to provide the users with access to Web sites. The early success of portals came about through excellent search engines that helped the users navigate among millions and millions of URLs. However, development in the fixed Internet space has clearly shown that the fixed portals are now under pressure from various sides:

- The experienced Internet user does not surf the Internet. Having found the desired sites, she will bookmark them and let her browser take her directly there.
- A number of dedicated portals, most notably e-Commerce portals such as Amazon.com and eBAY.com, are attracting an increasing number of direct users.

It appears likely that Internet users will increasingly bypass portals and go directly to desired sites. In this context it is difficult to imagine why users would become loyal to wireless portals in the first place, knowing that wireless Internet access will be more expensive than fixed access, which will persuade them to spend less time online.

What is it then that still causes major players like Vodafone and Vivendi to continue their huge spending on the Vizzavi portal, and causes medium players like Telenor and Telia to drive their Djuice and Speedy Tomato portals further and further from their home markets, where at least a degree of dominance has been achieved so far?

Well, it's a difficult question in all honesty. However, the logic could be that these companies are banking on their own strength to build brands, which will generate more "customer ownership" through email addresses, and on personalization of the portal to create stickiness. Both would be barriers to competition.

Fixed-Wireless Portals

In a highly competitive market, the major fixed portals such as T-Online, BT Internet, and Wanadoo are more likely to consolidate with their wireless "sisters," such as T-Motion, Genie, and Orangeworld, whereas other major portals like AOL, Tiscali, and Yahoo! may go out and find wireless alliances.

Club Nokia

Nokia is clearly the dominant player in the mobile handsets market and as such influences total market development greatly. The fact that Nokia did not launch a GPRS handset until 2002 is said to have deterred mobile operators from significant promotions of their GPRS networks, despite having invested heavily in those.

Nokia is also trying to move upwards in the value chain, or maybe it is fairer to say that Nokia is trying to play in various roles and thereby grab a larger piece of the total revenue. Over the years Nokia has gradually developed their Internet site, called Club Nokia into a real portal, despite having had to balance their promotion of Club Nokia against any appearance that they are in direct competition with their key customers.

Nokia has also launched serious development programs in the field of mobile Internet applications, such as Forum Nokia (http://www.forum.nokia.com/main.html). We have yet to see how those initiatives will be brought to market, whether directly promoted on Club Nokia or indirectly promoted via co-marketing on customer portals.

Service Providers

In the wireless space, the technical complexity of the system is far more formidable than in the fixed Internet space. Radio access networks are largely responsible for the complexity, which has given rise to a variety of providers specialized on simplifying one sector of the network or another for the user. Figure 8.3 identifies the main components of a wireless data network.

Figure 8.3
Wireless Internet access.

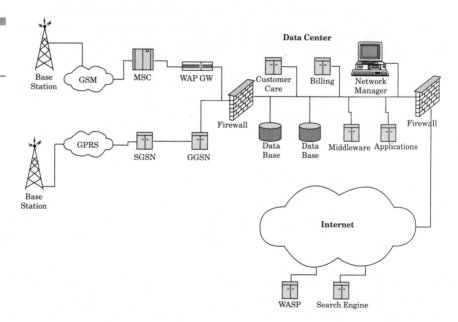

ISPs and MNOs

Unlike the fixed Internet, where the PC user usually selects an Internet service provider (ISP) if one is not selected by the browser, in the wireless space, mobile network operators (MNOs) are selected as the ISP in almost every case. These companies can and often do preload handsets with their own WAP gateways and access points.

The slow rollout of GPRS and the very meager success of WAP 1.1 initially left mobile operators and portals to cope with traffic figures trailing far behind the projected numbers. Despite the size and power of these companies, which would all like to see their customers using the portals as default access to the Internet, there is no evidence that they produced solid commercial successes.

In pursuit of customer loyalty, portals have to spend large amounts on promotion of their brands and give users access to the applications and content they perceive as useful and valuable. The pressing need for applications and content is creating a space for application service providers (ASPs) and even more so for wireless application service providers (WASPs).

ASPs and WASPs

The MNOs and wireless portals are aiming at seamless access for customers via PC or mobile device to a rich variety of services and applications including personal information management (PIM), email, news, sports, finance, information, games, and search engines. Unlike many other applications, SMS applications appear to have gained momentum and are attaining economy of scale. Entertainment applications are growing rapidly, particularly in Northern Europe and the United Kingdom, with Finland, Scandinavia, and Germany in the lead. In these markets we find mobile operators who generate as much as 10 percent of their revenue from value-added SMS messages that may or may not be premium rated.

Those services and applications are typically hosted and delivered by companies outside the MNOs and portal companies. This is the space of ASPs and WASPs. Among leading WASPs, we find companies such as:

- Digital Bridges
- Yomi Vision
- Materna

- Real Time
- Small Planet
- Add2Phone
- ZellSoft

In a WASP business model:

- The value proposition relies on application aggregation.
- The bottleneck has shifted from availability of applications to availability of the systems necessary for their delivery, namely customer provisioning, billing, and reporting.
- A short application life cycle and low entry barriers inhibit infrastructure investment.

This last characteristic means that WASPs do not provide turnkey infrastructure solutions, leaving the requirement for a reliable, carrier-grade infrastructure for application delivery still unaddressed. SMS-based applications (in particular) are being generated from hundreds of small companies, employing thousands of developers. Where can the portals get their hosting, managed services, SLAs, and customer support if not from the WASPs? We've already noted that providing those services in-house means assuming the onus of building physical connections to, and negotiating commercial agreements with, a large number of WASPs. Since the former is expensive and the latter implies revenue share schemes and license fees, business models would be hard pressed to accommodate an in-house "solution." This has generated a market space for a new breed of companies, which could be called wireless application infrastructure providers (WAIPs).

WAIPs

WAIPs address the needs of both portals and application developers. Their function is to:

- Extend resilient, carrier-grade platforms across geographies
- Bundle MNO and application developer needs to create scale economies
- Deliver applications to MNOs, and customers to application developers
- Provide dedicated hosting and managed services
- Provide integrated customer care and billing systems

Only a few companies, including Open Mobile, End2End, and M7 Networks, have established themselves in this space.

Figure 8.4 indicates the high-level network infrastructure of a WAIP.

Figure 8.4
WAIP network.

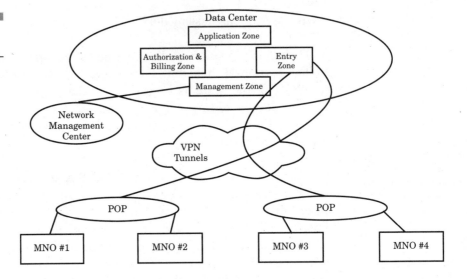

Content

As we've mentioned, the key to success for the portals is in their ability to provide applications and services perceived valuable and relevant by the customers. Such services must supply a rich variety of content from a number of sources.

Some of the major wireless portals have access to a very rich selection of content through joint ownership or alliances with major media companies. These include Vivendi (Vizzavi) and AOL Time Warner (AOL Mobile). However, the perceived value of the available content on a portal is highly dependent on the available technology, such as bandwidth and handsets. It should be expected that until a substantial rollout of GPRS, EDGE, and UMTS networks and handsets has been achieved, the majority of revenue will be generated from messaging, (email, instant messaging, and SMS) and relatively simple SMS-based entertainment applications.

Conclusion

Wireless portals may expect to find themselves in a situation similar to that of the fixed portals:

- The portals do not generate any access fees unless there is joint ownership with the ISP/MNO.
- Revenue sources for portals accordingly are limited to M-Commerce, advertising, and subscriptions for services and content.

There is a general perception that advertising is becoming a truly marginal source of revenue. M-Commerce is still very much in its infancy, but is widely perceived as a source of significant revenues through micropayment schemes and mobile banking.

Tables 8.1 and 8.2 below show the consensus of various analysts and research companies with respect to developing revenue sources among major mobile operators.

TABLE 8.1

Voice/Data Traffic Projection

	2000	2002	2005
Voice revenue	93%	74%	55%
Data revenue	7%	26%	45%

TABLE 8.2

Data Service Distribution Projection

	2000	2004
Voice traffic	90%	65%
SMS	7%	8%
Messaging	2%	7%
Data traffic	1%	12%
Content subscriptions	0%	4%
M-Commerce	0%	3%
Advertising	0%	1%

It remains to be seen whether this is an accurate forecast or, if so, if it can provide a sustainable business case for the wireless portals. Experience so far clearly indicates that until customers see services and content as relevant and valuable, they really won't have any interest in buying the technology. Portals will have to heavily promote not just their brands, but also their services and content. Players in this field should be prepared to be flexible and diverse. Bring a lot of services to the customers, and if the customers don't use them, remove them. Under no circumstances let the portals become flooded with unused services and content.

PART 3

GPRS for
Dummies
Like Me
(and Maybe You)

This chapter deals with the technical issues required to make GPRS a commercial reality. As such, it is usually considered the domain of techies, a domain in which most of us have no interest. Yet without this rather complex and detailed technology, GPRS would never become a viable business proposition. That said, why should you consider reading this chapter? The answer is straightforward: without a basic understanding of what goes into the technology, it is impossible to fully exploit its potential. For the technicians among you, this chapter will only brush the surface of the issues you'll face when implementing and optimizing a GPRS system. If you're a generalist, it won't make you into a technician, but it *will* give you a basic understanding of the planning and implementation process for a GPRS network. When crossing such a broad spectrum of expectations, we are sure to fail most of the time...but we still hope to provide at least one insight that will shine a light in the direction you are looking...so read on.

The Business Proposition

GPRS is an *overlay* technology. Translation: Everything to do with GPRS—messages, calls, hardware, software, signaling—is installed on or connected to the GSM network and transmitted over the air at the same time as GSM traffic. Generally speaking, the geographical coverage of GPRS approximates, or "overlays" the geographical coverage of the GSM service. GPRS is not a GSM service in itself because, although both GSM and GPRS inhabit the same geographical area and connect at the core network interface with the SGSN, they work independently of each other. The overlay concept allows GPRS to be fully integrated into the GSM core network, taking advantage of GSM's transmission network to transmit and receive GPRS packets. The piggyback design lowers up-front cost for GPRS considerably, as parts of the GSM core and the entire base station subsystem (BSS) are reused by GPRS traffic.

Real-World Economics

Raise your right eyebrow if you really care about such technical nonsense. Really, raise it if you think this overlay, piggyback stuff is important. Since most of you will be reading this in groups or at least in pairs, look at your colleague, significant other, or friend and see if their right eyebrow is raised. (It will be the one on his left as you sit facing him or his right one if you are looking at them in a mirror.) Everyone should now have his or her right eyebrow raised. Yes. Everyone reading these words should look kind of silly with their book in their lap and their right eyebrow raised contemplating why an overlay GPRS system is important to their life.

The reason is money. Dollars, pounds, baht, shekels, euros, pesos, you name it. Basically, you'll need less of it to implement and operate a GPRS system than for any comparable system. And if the system costs less to implement and operate, then it will naturally costs the customer less because, as we all know from economics 101: in a competitive environment, lower costs lead to lower prices over time. Sounds like a no-brainer, doesn't it? But before the government officials and regulators around the world start investigating GPRS prices, we must acknowledge what GPRS deployment means in the real world.

Picture a green field with no wireless service; no core networks; no signaling infrastructure; no billing systems; no customer care; no distribution network. No nothing. Now picture the work required to develop a new industry spanning the globe. Estimate how long it would take to build such a business. Add in the human toll of long hours, sleepless nights, and weekends of work. Consider the number of transmission locations to be constructed, the kilometers of cable to be strung between antennas and transmission equipment, and the fiber required to link these sites together into a cohesive network.

Now calculate the cost to develop, provision, install, and optimize such a network; to find and train the operating and customer care staff; to develop back office functions; and to rate and bill the services. Then estimate the timeframe for building a global business. Months, years, maybe dozens of years. It is easy to make a case that this undertaking would cost billions and take eons. That's *after* the specifications have been developed, published, and adopted. Since we all have been taught that time is money, add the cost of time to the cost of labor and the capital outlay and try to keep track of the zeros.

Why It's Ultimately Worth Doing

Someone has done the market research and the results reveal a global demand for packet data services, or else no one in their right mind would get involved. OK, maybe no one has actually done such a study, but the communications powers-that-be have made an educated guess that GPRS service is going to be needed and, most importantly, wanted, in the near future. Whatever your reason for believing in a GPRS market, the business case for building an overlay GPRS network and operating it in conjunction with a GSM business is compelling enough to take the risk. GPRS costs are reduced many times over by leveraging the existing global GSM business. As for the cost of time, it cost the GSM industry over ten years to build a global industry, and it could be argued that building the equivalent GPRS industry will take as long or longer. But it won't. By overlaying the GPRS infrastructure and business on top of the GSM industry, we get a global footprint, an operating structure, and a business environment in months, not years. GPRS saves time, it saves the operator money, and ultimately it saves the customer money.

Where to Begin

I'll leave the number crunching to the guys with green visors and move on to the basics of building and optimizing the GPRS business. Since we don't in fact have a greenfield situation, there are some givens to consider. The first is that if you start with a balky GSM system, you are guaranteed a balky GPRS system. Why? Because GPRS is always going to look a lot like the GSM system it is overlaying. Perhaps there aren't enough base stations, or they are not in the proper locations. Maybe they're not being monitored and maintained with state-of-the-art technology, or they don't have optimal interference suppression, or they don't hand off between sectors properly. In those scenarios, the problem isn't just that GPRS will exhibit the same flaws; rather, it will probably not work at all. In fact, it's possible that even if the underlying GSM network is properly designed, optimized, and operated, the GPRS system will still have to struggle to match customer service expectations. Why? Because opportunities for breakdowns within the packet system are going to be a lot more numerous than opportunities for breakdowns in the GSM network, GPRS operations have to contend with losses emanating from connection to corporate LANs, the Internet, firewalls, in-building use, highly complex applications, services that don't perform well in mobility situations....

Note that once again the payoff far outweighs the effort. With a relatively small investment in time and money, the transformation from circuit-switched GSM to the packet-switched future begins with the implementation of GPRS. The trajectory looks something like this: as the first step on the GSM evolutionary path, packet-based GPRS evolves to EDGE and from there into a fully integrated packet world for voice/video/data constituting what is commonly referred to as the Third Generation. This orderly progression leads to more service offerings, better use of spectrum, and the ability to deliver to many more customers than wireless digital technology can serve today. The vision includes no flash cuts, no service interruptions, no immediate migration of customer bases and best of all, an orderly progression from today's service offerings to those we haven't even thought up yet. The collective brains of the industry have developed a plan which we can all follow in the sure knowledge that the incremental steps lead to huge strides over time and not down a proprietary cul de sac of wasted investment.

Packets over Wireless

While the advent of GPRS brings the Internet packet world to wireless, not everything ports directly. Integration between the telephony world and the Internet world is difficult due to baseline differences for optimization and monitoring, routing and billing, numbering and addressing, naming and security. Of course many believe telephony solves all of these issues in a more efficient and much more logical manner than IP can—mostly through instituting detailed specifications adopted throughout the world. However, the Internet has been successful without adopting the same methods of doing business. It operates with fewer agreed standards and with a new naming structure but, as some contend, a less secure way of conducting business.

For telephony, numbering and addressing in both the wired and wireless worlds share the same structures: a unique numerical country identifier is followed by a further 6- to 9-digit unique numerical identifier for a specific telephone/terminal or device. Numerical reuse depends on the specific needs of a country, as defined by demand.

In the Internet model, however, unique addresses are mixed with reusable addresses that are dynamically assigned by service providers. A global authority, Internet Corporation for Assigned Names and Numbers (ICANN) allocates addresses on the basis of need and availability. When user addresses are assigned dynamically, it follows that there are more addresses to be assigned. Furthermore, because computers are not communicated "to" but rather communicated "from," few computers need to know their uniquely assigned address.

It sounds relatively simple, but it all has to change when wireless telephony with GPRS connects to the Internet. Because we can predict a near-term need to communicate and deliver information to GPRS devices, addressing differences between telephony and the Internet must be resolved. GSMA has taken the lead by settling on a way to provide GPRS addressing with ICANN. For the present, IPv4 addresses are statically assigned to GPRS nodes and dynamically assigned to GPRS devices. This combination enables GPRS roaming by fixing the addresses of nodes such as the GGSNs and SGSNs, but assigning addresses to devices only as they access the GPRS network. The compromise, for that is what this is, allows for integration of traditional IPv4 addressing to evolving GPRS technology. However, IPv6, which will dramatically increase the number of addresses available, is universally recognized as the way in which methods of addressing must evolve but the timing of such a transition is hard to predict. By agreeing to begin GPRS service with IPv4 and migrate to IPv6 over time, GPRS has been able to integrate services into the current model and at least get the ball rolling.

Similarly, GPRS access point naming (APN) had to be integrated into the Internet model. The Internet uses URLs to access specific Web sites throughout the world according to an agreed convention. GPRS again uses the same convention but has modified it to meet the needs of the wireless packet data world.

Evolution of Security

GSM was developed on the concept that, except for legally approved interception, security algorithms would be implemented as standard to prevent eavesdropping. Therefore, for GSM, even in the realm of global roaming, one network operator connects only to other "known" entities, i.e., other GSM operators or other fixed telephony networks. These entities have in common a certain regard for security that precludes unapproved use of eavesdropping devices by outsiders. In effect, the GSM security chain remains unbroken throughout the world.

The protection afforded by trust does not extend to GPRS, however. GPRS brings new and as yet not fully realized security threats to the "safe" home environment of GSM. When connecting the GSM network to a GPRS network through the SGSN, we're braving new connections to the less secure world outside. By this I mean the Internet. Although buffers such as firewalls and gateways are designed to protect the integrity of the service, by its very nature GPRS connects to the less secure world of the Internet. IP security is more vulnerable and less advanced; news articles about security breaches are published every day. Everyone recognizes that it's mandatory to keep these security threats outside the confines of the GPRS networks, but the task of ensuring protection remains daunting, especially if you toss in the possibility of wireless viruses and the ever-present threat of denial of service attacks (attacks that work by trying to overload the system resources).

The concept of end-to-end security takes on new meaning for the GSM community when its previously secure networks are connected to GPRS networks, which are connected in turn to the Internet, corporate LANs and WANs, Internet gateways, and other possible hacker avenues. Fortunately a lot of planning and investigation has gone into the GPRS technology to reduce and isolate potential security breach opportunities. Methods for detection, prevention, and cure are constantly being developed and improved. (Read more about this in Chapter 14.)

Conclusion

To review, this chapter exists to help readers understand something of how GPRS works so that they can use the technology in ways that maximise its potential and minimize its risks. The rest of the chapters in this section do the same in more detail. They will provide you with a high-level overview of the design, engineering, and operational issues which confront GPRS technology today. They are not intended to replace the GPRS specifications, nor replicate the technical training required to design, implement, and operate a GPRS network.

If you read these chapters you will walk away with enough material to become dangerous...dangerous enough to be helpful in a disconcerting way by challenging the status quo when something doesn't make logical sense.

Cool High-Speed Data Services

Randy Wohlert

SBC Communications

GPRS is sometimes confused with other things that have similar sounding names but aren't as advanced. The historical evolution of GPRS can be summarized as follows:

G   Pronounced "gee," this is an exclamation often uttered by GPRS fans.

GP   An exclamation uttered by GPRS fans who consume large quantities of beer and/or coffee.

GPS   Stands for global positioning system, a satellite based positioning system that has nothing to do with GPRS, but is often confused with it.

GPRS   General packet radio system, the most highly evolved member of the G family of acronyms, derived from Latin words meaning *cool high-speed data.*

GPRS Data Services

The wonderful thing about GPRS is that it enables your mobile phone to provide terrific data services. It does this by magically increasing the speed at which your handset can send and receive data, and by providing connections to the packet world (such as the Internet).

GPRS builds on GSM network infrastructure and capabilities to enable the transport of high-speed data. To accomplish its mission, GPRS uses the radio interface differently than GSM. It shapes data into packets, transports them parallel to the GSM network, and then interfaces with other data networks such as the Internet. In theory, GPRS provides data rates ranging from 9.4 kbps up to 172.4 kbps, depending on which coding schemes are employed and how the air interface is used. In real life maybe you'll get something between 14.4 and 115 kbps. Actual mileage will vary.

In addition to high-speed data, GPRS has very fast session setup times that give you an "always-on" experience. This enables efficient and satisfactory email handling and Web browsing. Life is good!

In this chapter we'll take a behind-the-scenes high-level look at how this magic is achieved.

GPRS Infrastructure and Components

Let's begin by considering what has to be added to the basic GSM network to provide the magic of GPRS. Where do we start? At the highest level, networks are viewed as being made up of two big black boxes. The black box at the heart of the network that transports speech and data is called the core network (CN). The black box that connects subscribers to the core betwork is called the access network (AN). Each black box is made of smaller boxes that are made of smaller boxes that are made of smaller boxes, ad infinitum. The black boxes are connected together by interfaces, which are usually identified by cryptic acronyms.

One of the best-kept secrets in the industry is that the magic in our black boxes is described in carefully crafted documents called standards—complex and convoluted though they may be. Standards spell out the requirements, architectures, and protocols to be used in building telecom networks. For GPRS, the relevant standards are developed by a group called Third Generation Partnership Project (3GPP). They hide their documents in plain sight (and that site is located at www.3gpp.org). With sufficient patience and the guidance provided in this chapter, you can go there and find out way more than you ever wanted to know about GPRS.

For example, a service description for GPRS (i.e., what it's supposed to do) is found in—surprise!—a service description specification identified by the cryptic label TS 22.060. If you go to the 3GPP Web site you can find it (maybe with some difficulty), download it, and be mesmerized by the magic within. It's a good starting point.

But rather than just diving into GPRS right away, let's initially establish a frame of reference by looking at the big picture. Although it can seem overwhelming at first, it isn't that hard to understand if you take each of its parts one at a time (and we'll do so—patience now). The big picture is going to show you the main black boxes of a GSM network, with some additional boxes added for GPRS, and the interfaces between them.

Look at Figure 9.1 and you will immediately detect that two new black boxes have been added to the network and now sit alongside the existing circuit switched (CS) network. They are GPRS support nodes (GSNs), which come in two flavors: the serving GSN (SGSN), and the gateway GSN (GGSN). These nodes perform packet-handling magic in the network, and we'll see how they do it later on.

OK, here's the big picture, straight out of the 3GPP specifications. Don't panic. We'll get through this together.

Figure 9.1
Basic configuration of a PLMN supporting CS and PS services and interfaces.

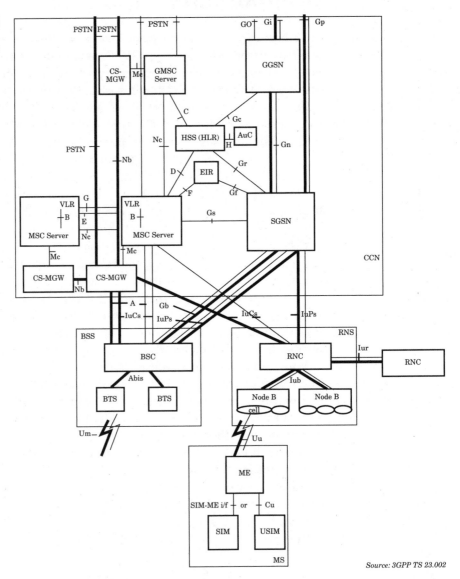

Source: 3GPP TS 23.002

I bet you've already found the SGSN and GGSN mentioned earlier (if not, look again). Each of the boxes in this picture represents a different network element, and the lines in between represent their interfaces. At a high level, this diagram shows that the GPRS stuff is put into the network alongside the existing stuff. The existing stuff consists of the network elements and interfaces that make up the GSM circuit-switched network. I'm making it sound simpler than it is because I haven't mentioned that most of the existing stuff will need upgrades. And you were worried that this might not be fun. Ha!

This diagram is useful to help keep things in perspective, so you might want to glance back at it every so often as you read (easier than memorizing it and then burning the original). Let's try to get comfortable with it. Starting at the bottom and working our way up (just like we do in our companies), you will find a box labeled "MS" with other boxes inside it. This is the *mobile station*, which is another name for cell phone. The cell phone connects to the network over the air interface. This diagram isn't just a GPRS diagram, but is also a generic network diagram, so it shows more than one air interface. GPRS networks built on GSM networks use the Um air interface to connect to the base station subsystem (BSS), not to be confused with that popular acronym BS, which stands for base station. Now you know wireless networks are full of BS.

The BSS connects to the SGSN, which connects in turn to the GGSN, which connects off the top of the diagram into deep space (also known as the Internet). Congratulations, you are now a GPRS architecture expert. The rest is just details.

Basically anyone who knows what's in the black boxes and the details of the interfaces can build a GPRS network in his own garage. That information is publicly and freely available in the 3GPP cryptograms. The big picture above was copied, with permission of course, from the 3GPP's generic architecture specification, which does a great job explaining network elements and their interfaces. Another diagram provides a slightly different view of all this, with more of a GPRS focus. It's found in the GPRS network architecture specification in 3GPP TS 23.060. Hmmmm, must be a pattern to all these specifications.

The GPRS network architecture specification not only provides detailed information about the GPRS architecture and the interfaces between network elements, but also explains what takes place between the various elements ("call flows") when GPRS sessions are set up, maintained, and torn down. So, for a really good time, browse through its nearly 200 pages, or some of the 70+ referenced specifications. Or just read the notes provided in the rest of this chapter. It's up to you. Since you bought, borrowed, or stole this book, I'm confident you will make the right choice.

Figure 9.2
Overview of the
Packet Domain
Logical Architecture

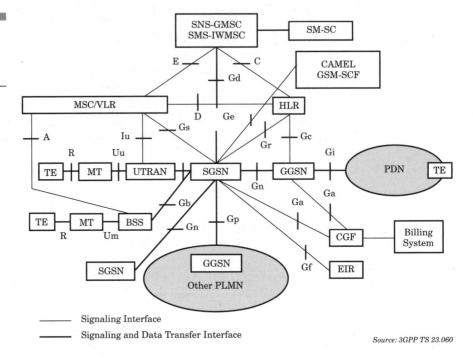

Source: 3GPP TS 23.060

OK, now that we have the big picture as a frame of reference,[1] let's look at some of the new and improved network components. Probably the most significant changes made to the GSM network in support of GPRS are how the air interface is used, the upgrade to the base station subsystem, and the addition of the GGSN and SGSN.

Air Interface Upgrade

GPRS would not be possible if some serious magic didn't take place at the air interface. With a traditional circuit-switched GSM air interface, you get one of eight possible radio timeslots for your use. Once you get a timeslot, it's reserved for your use only for as long as you want. That means you still have the timeslot even when you aren't talking. This allocation of resources may not make for the most efficient use of the air interface, but it does provide very high-quality speech. However, the amount of information that can go over the air in a timeslot is somewhat

---

[1]Help is at hand. Please turn to the appendix at the end of this chapter to consult an acronym list for network elements and interfaces.

limited. It's great for voice, but less than optimal for supporting bursty data applications.

Therefore GPRS takes a different approach. With GPRS, the cell phone dynamically grabs and releases one or more timeslots as it needs them to send and receive packets. New mechanisms are introduced to allow timeslots to be very rapidly seized and released, and then more new mechanisms are introduced to manage all this.

When communicating over a radio interface there are trade-offs to be made between speed and reliability. GPRS flexibly provides a range of data encoding options that uses different transmission speeds, offering different ranges of quality. Each of these different encoding options can be selected for use over different numbers of timeslots. Thus the throughput actually realized with GPRS depends on the coding scheme selected and the number of timeslots used, as well as other factors such as the amount of radio interference experienced, network lag, etc.

Base Station Subsystem Upgrade: Packet Control Unit

As we noted earlier, the GSM network was originally designed to transport voice using circuit-switched technology. This works well for providing high-quality voice connections, but not so well for providing bursty high-speed data. For the latter, the existing network must be upgraded in a way that enables optimal packet handling, but at the same time preserves cost efficiency by letting carriers re-use as much of their existing equipment as possible.

The biggest expense in a carrier's network is the base station subsystem. The BSS is the big black box in the access network, which the handset connects to. It consists of base stations, which are the radios in the access network and their controllers. A large carrier today will have thousands of base stations connecting happy subscribers to its network. It would be prohibitively expensive to replace all that equipment to handle GPRS data packets, so GPRS cleverly re-uses most of what is already there. It does this by structuring the GPRS data packets in such a way that existing BSS transport mechanisms can handle GPRS packets in a similar way to the circuit-switched packets.

GSM BSSs transport circuit-switched information in transcoding rate and adaptation (TRAU) frames. These are basically fixed-size data packets that come in different flavors depending on what kind of data is being transported (speech, data, extended data, management information, or idle). Therefore GPRS data is structured in packets the same size as TRAU frames. The GPRS version is called a PCU frame, a reference to the new and improved BSS black box that handles them, the packet control unit (PCU). In sum, the PCU is new for GPRS and enables the base station subsystem to handle GPRS packets in the same way as circuit-switched data.

All New: Serving GPRS Support Node (SGSN)

Data from the handset flies across the air interface, possibly using new encoding schemes and multiple dynamically assigned timeslots, to arrive in the access network's BSS. There it gets nicely bundled into PCU frames and shipped off to a brand new network element called the serving GPRS support node (SGSN).

The SGSN is essentially the part of the network that the handset talks to, by way of a connection over the air interface and the BSS. It functions as a router, either passing packets from the handset on toward their network destination, or receiving packets from sources within the network for forwarding to the handset.

A potential problem arises here. Handsets are mobile. Their location may change. So one of the SGSN's more important responsibilities is keeping track of where the handset is, even as subscribers scurry about. This function is known as *mobility management*, and it's challenging to do in real time as radio connections change from one base station to another (a phenomenon known as *handoff*). Mobility management is accomplished by defining "routing areas" for the SGSN to track. When a subscriber is moving around in any given routing area, the SGSN can find him and send him packets. Actually it's pretty precise. Routing areas are in turn made up of a number of different "cells," which are the areas that a single base station serves. When a handset is in what's known as a "ready" state, the SGSN can track the handset at the cell level. This helps a lot to make rapid packet delivery a reality.

In GPRS, the SGSN assumes responsibility for some of the functions that the BSS does in GSM, since it is now effectively the end point of communications with the handset. This means it is in charge of fun things like encryption and compression. It would be a mistake not to mention that the SGSN has responsibility for collecting charging information too. This information (a lot of it) gets bundled into call detail records, which are shipped off to a charging gateway that in turn interfaces with a billing center that sends you the neat monthly statement that, in the final analysis, makes all this possible.

All New: Gateway GPRS Support Node (GGSN)

We previously noted that the SGSN serves as a router. True and good. But when packets need to be sent or received from outside the carrier's network, they need some extra consideration. Special nodes called gateways provide the additional functionality for interfacing to the outside world, and for GPRS that's the gateway GPRS support node (GGSN).

In the world of GPRS, subscribers are free to roam around, and the SGSN can find them when it's time to deliver a packet. If their location changes beyond the routing area, they can reconnect to the local SGSN, but in the Internet this kind of mobility doesn't happen. Therefore exterior nodes that want to talk to the GPRS core network need a single point where all their communications take place (it can't move). The GGSN provides this point of contact, performing what is known as the "anchor function" in the world of GPRS, no doubt because it prevents packets from drifting about aimlessly in a sea of black boxes.

True Confessions

OK, as promised, we've revealed the high-level secrets of GPRS. However, as you probably suspect, there is more to it than we've covered so far. There are lots and lots of details that can make your life, hmmm, interesting. We didn't even begin to discuss protocols for example, or the various interfaces to existing GSM network nodes. And then we could talk about network access control functions, registration, authentication, authorization, message screening, routing, tunneling, compression, ciphering, logical link management, network management (especially radio resource management), compatibility issues, transmission details, information storage, how various identities are used, operational aspects, and interaction with other services. All pretty fascinating stuff.

You could petition the publisher to provide a more detailed sequel (*The Insider's Guide to GPRS*), which could take a while. However, you won't be completely abandoned at this point. For those who just can't get enough of this, we're including an appendix called The Insomniac's Guide to GPRS Specifications that lists some places where you can get all the gory details. Good luck.

Appendix

The Insomniacs Guide to GPRS Specifications

For those desiring additional or more technically-detailed information, the following list identifies GPRS-specific standards defined by 3GPP in Release 3 specifications. Note that earlier versions of these specifications exist in Release 99, and by the time you read this, subsequent versions will no doubt be available in new releases.

This brief list is essentially the tip of the iceberg. Additional references to related specifications are provided by each of the GPRS specifications listed below. TR 23.060 has a large list of referenced specifications, for example.

As of the time of this writing, these specifications are available (free) for downloading from the 3GPP website at http://www.3gpp.org.

3GPP GPRS-Related Specifications

1. TR 22.060 General Packet Radio Service (GPRS); Stage 1
2. TR 23.060 General Packet Radio Service (GPRS) Service description; Stage 2
3. TS 27.060 GPRS Mobile Stations supporting GPRS
4. TS 29.016 Serving GPRS Support Node SGSN—Visitors Location Register (VLR); Gs Interface Network Service Specification
5. TS 29.018 Serving GPRS Support Node SGSN—Visitors Location Register (VLR); Gs Interface Layer 3 Specification
6. TS 29.060 GPRS Tunneling protocol (GTP) across the Gn and Gp interface
7. TS 29.061 General Packet Radio Service (GPRS); Interworking between the Public Land Mobile Network (PLMN) supporting GPRS and Packet
8. TS 29.119 GPRS Tunneling Protocol (GTP) specification for Gateway Location Register (GLR)
9. TS 41.1061 General Packet Radio Service (GPRS); GPRS ciphering algorithm requirements
10. TS 43.064 Overall description of the GPRS radio interface; Stage 2
11. TS 44.060 General Packet Radio Service (GPRS); Mobile Station (MS)—Base Station System (BSS) interface; Radio Link Control/ Medium Access Control (RLC/MAC) protocol

12. TS 44.064 Mobile Station—Serving GPRS Support Node (MS-SGSN) Logical Link Control (LLC) Layer Specification
13. TS 44.065 Mobile Station (MS)—Serving GPRS Support Node (SGSN); Subnetwork Dependent Convergence Protocol (SNDCP)
14. TS 48.014 General Packet Radio Service (GPRS); Base Station System (BSS)—Serving GPRS Support Node (SGSN) interface; Gb Interface Layer 1
15. TS 48.016 General Packet Radio Service (GPRS); Base Station System (BSS)—Serving GPRS Support Node (SGSN) Interface; Network Service
16. TS 48.018 General Packet Radio Service (GPRS); Base Station System (BSS)—Serving GPRS Support Node (SGSN); BSS GPRS Protocol

CHAPTER 10

Planning and Dimensioning

Lauro Ortigoza-Guerrero,
Babak Jafarian, and Scott Fox

Wireless Facilities, Inc.

System planning for GPRS networks needs different types of expertise from those acquired in classical cellular systems. New coding schemes require cell and frequency planning to allow a specific average throughput in a cell. Also, new dimensioning problems arise as a result of the new packet data services that, until the deployment of GPRS, were not supported by the infrastructure. In the core network, new IP-based components and interfaces will pose a new challenge for dimensioning to determine the correct number of links necessary to carry the offered traffic produced at the radio network.

Frequency and Coverage Planning

As a kind of extension to GSM, GPRS is embedded in the physical channels (FDMA/TDMA) of GSM, but employs dedicated packet-based logical channels. In order to introduce GPRS to a GSM network, modifications are required. Some of the nodes already implemented in current GSM systems can be shared between GPRS and GSM, such as the base transceiver station (BTS), the base station controller (BSC), and the mobile switching center (MSC). Only two new node types, the serving GPRS support node (SGSN) and the gateway GPRS support node (GGSN) have to be introduced for the first time. In addition, this new technology requires the development of new mobile terminals.

GPRS can be implemented in existing GSM systems using the same cell structure. Depending on the coding scheme, new frequency planning may not even be necessary. If the operator does not want to use new coding schemes, modifications will be minor.

GPRS uses four different coding schemes: CS1 to CS4. They are differentiated by the number of redundant bits used to protect the information transmitted. CS1 has the most redundant information whereas CS4 adds no redundant information at all. For obvious reasons, CS1 produces the lowest throughput and CS4 produces the highest. All things being equal, everyone would want to use CS4 all the time. Unfortunately, the use of a particular coding scheme is dependent on the mobile station receiving enough signal power. Different power thresholds exist wherein the user maps its received power and selects the corresponding scheme. Table 10.1 shows the different coding parameters for the GPRS coding schemes.

TABLE 10.1

Coding Parameters for the GPRS Coding Schemes

Channel Name	Code Rate	Modulation	Radio Interface Rate per Time Slot (kbps)
CS-1	0.53	GMSK	8.8
CS-2	0.66	GMSK	11.2
CS-3	0.8	GMSK	13.6/14.8
CS-4	1	GMSK	17.6

Because different coding schemes have different carrier to noise ratio (C/N) requirements, the relative coverage area of each will also differ. Users with better signal quality (normally those close to the base station) will be able to transmit with CS4, while those with low signal quality will be able to transmit only with CS1, creating concentric cells of coverage as shown in Figure 10.1.

Figure 10.1

Concentric cells of coverage in GPRS due to the use of different coding schemes.

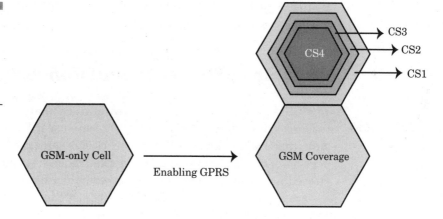

The physical channel structure of GSM systems is exactly the same as that specified for coding scheme CS1 in GPRS. Therefore, if the GPRS system is planned for use with CS1, no frequency planning is needed. The same system planning will work for both; specifically, the power level in 95 percent of the cell area will ensure a user at least CS1 in GPRS. If, however, the GPRS system planner decides he needs CS2 levels of throughput at a minimum, then the corresponding coverage area will shrink, as shown in Figure 10.1. At this point, the operator will need either new frequency planning or the implementation of new infrastructure (i.e., more cells). The same will happen if CS3 or CS4 is used. GPRS performance is dictated more by co-channel interference than by absolute sensitivity. In other words, performance is more often interference limited than noise limited.

Capacity and Growth Dimensioning

Capacity and growth dimensioning in a GPRS network can be divided into two distinct areas: dimensioning of the air interface and dimensioning of the core network.

Capacity Dimensioning for the Air Interface

Before we can discuss the GPRS dimensioning process, we need to describe some basic concepts for management of network resources. This section first examines the "capacity-on-demand" concept used in GPRS, and then explains the steps for dimensioning.

The Capacity-on-Demand Concept

In a GSM network, as in any other wireless cellular network, limited radio resources have to support a variety of uses—use for one purpose precludes simultaneous use for another. When GPRS networks are deployed as an overlay on the existing GSM networks, radio resources also have to be shared between circuit- and packet-switched data. Introducing packet services into an existing GSM network without allocating new spectrum can cause degradation of either voice quality or voice capacity. More specifically, it may increase blocking probability or lower voice quality and reduce the cell service area, defined as the area over which a specified outage probability limit is achieved. Conversely, if all radio resources were assigned to GSM services, GPRS users could experience a high blocking rate. But none of this is likely to happen. GPRS is designed to utilize *unused* radio resources to transmit short bursts of packet data.[1]

---

[1]When a GSM network is operating at a blocking probability of 2 percent, the average channel load is in the range of 60–80 percent, depending on the total number of channels used in the cell. Thus, there are 20–40 percent idle channels, on average, which may be used for data services of GPRS.

In order to understand how GPRS assigns resources to calls, let us look at some definitions. In GSM, a logical channel (or simply a channel) for voice or circuit-switched transmission is called a traffic channel (TCH). It's mapped into physical channels by the medium access control (MAC) layer. In GPRS, a logical channel for data traffic is called a packet data channel (PDCH). Any cell that supports both GSM and GPRS will have to assign resources to both, and therefore will use both TCH and PDCH. PDCHs are shared among different GPRS users in the cell.

Allocation of TCHs and PDCHs is done dynamically according to *capacity on demand*.[2,3] Within a cell, it is possible to allocate one or more PDCHs from a common pool of physical channels otherwise used as traffic channels. Physical resources are assigned to GPRS strictly according to the need for actual packet transmission. If there is no such need, then there is no allocation of physical resources. As many as eight PDCHs with different timeslots may be allocated to a mobile station (MS) at the same time within the same carrier.

This efficient use of scarce radio resources means that large numbers of GPRS users can potentially share the same bandwidth and be served by a single cell. The actual number of users supported depends on the application being used and how much data is being transferred. Because of GPRS' spectrum efficiency, there is less need to build in idle capacity only used in peak hours.

Network dimensioning.　In circuit-switched networks such as GSM, capacity dimensioning is a straightforward process. The total offered load, resulting from adding total speech traffic to total data traffic, is used as an input parameter in the famous Erlang B formula to calculate the number of required channels for a particular grade of service (GoS) (the blocking probability). The formula assumes that all blocked calls are immediately cleared and that the user population is much larger

---

[2]The capacity-on-demand concept states that load supervision should be done in the medium access control (MAC) layer to monitor the load on the PDCH(s), and the number of allocated PDCHs in a cell can be increased or decreased according to demand. However, the existence of PDCH(s) does not imply the existence of a packet common control channel (PCCCH). When no PCCCH is allocated in a cell, all GPRS-attached MSs automatically camp on the existing GSM CCCH as they do in the idle state. When a PCCCH is allocated in a cell, all GPRS attached MSs camp on it.

[3]GPRS also makes use of the master-slave concept which states that at least one PDCH (mapped on one physical time slot), acting as a master, accommodates packet common control channels (PCCCHs) which carry all necessary control signaling for initiating packet transfer as well as user data and dedicated signaling. The others, acting as slaves, are only used for user data transfer.

than the number of servers (there are tables available that show one variable as a function of the other two). This can help to determine the resources required per sector. The offered traffic in Erlangs per sector or per cell is traditionally obtained from the MSC and used in calculations. However, GPRS services add extra complexity to the dimensioning process: in the GPRS network, besides the traffic in circuit-switched data and speech, there is also packet-switched data traffic to consider.

Resource partitioning, which divides total available resources in a sector into a subset for voice plus circuit-switched data and another for packet-switched data, simplifies the dimensioning process. In this case, the resources required for CS data and voice services would be calculated using the Erlang B formula, as explained above. After this step, the resources required for PS data would be calculated using the ratio of the aggregate throughput in kbps expected in the sector and the maximum throughput in kbps a TCH can serve.

GPRS does not presently use static resource partitioning. Most GPRS systems use capacity on demand to allow all service types to use any idle resource. Although this concept is powerful in that it allows GPRS to manage idle resources efficiently, it complicates capacity dimensioning. The presence of three different kinds of services (voice, CS data and PS data) means ongoing competition for the same set of resources. It is only because GPRS manages resources dynamically, or on demand, that a single network can resource all service types.

There are three possible scenarios for network dimensioning in a GPRS network. These are defined by the relationship existing between voice, circuit-switched data, and packet-switched data traffic. Scenarios can be classified as follow:

1. The voice plus CS data traffic is larger[4] than the PS data traffic.
2. The voice plus CS data traffic is comparable[5] to the PS data traffic.
3. The voice plus CS data traffic is smaller[6] than the PS data traffic.

The first case will exist in the early stages of GPRS or in mature GPRS networks with a low penetration rate of packet services. The second case will exist in mature GPRS networks and/or initial GPRS net-

---

[4]The number of resources required for PS data is at least 10 times smaller than those required for CS data and voice.

[5]The number of radio resources to attend voice plus CS data traffic is almost equal to those required for PS data.

[6]The number of resources required for PS data is at least 10 times larger than those required for CS data and voice.

works that include areas with high penetration rates. Finally, the third case will exist when all data traffic is packet-oriented and some of the voice traffic has migrated to voice over IP. The three cases are briefly explained in the next sections.

Case 1

Here the dimensioning process follows the guidelines used for dimensioning a GSM network, as explained before. It takes into account only the voice and circuit-switched data dimensions of the RF network and ignores packet-switched data. A GoS of 2 percent (which means a blocking probability of 2 percent) is normally used for dimensioning purposes and the Erlang B formula is applied. Since packet-switched data traffic is relatively small, it can be supported by the idle resources of the GSM network with no adverse effect on circuit-switched traffic.

Case 2

The second scenario will require an in-depth analysis of the service mix in each cell. The Erlang B formula alone will not yield useful results because packet-switched data transmissions have been added to the equation. There is no exact general analytical solution for this case (unless all the service types can be adequately modeled in a multidimensional traffic analysis) that will permit us to calculate the required resources. Instead, alternative solutions have to be considered, some of which include approximated analytical solutions, computer simulations, or a combination of both.

An approximated analytical solution will result in overdimensioning or underdimensioning problems. Dimensioning solutions will differ from the exact solution as a function of the assumptions made to simplify the analytical method for modeling the traffic mixture. The most common method consists in separating circuit-switched voice and data traffic from packet-switched data and dimensioning them separately. Resources necessary to carry the packet-switched traffic are dimensioned first, followed by the circuit-switched data traffic. The steps are as follows.

Packet-switched data traffic dimensioning. Achieving the theoretical maximum GPRS data transmission speed of 171.2 kbps would require a single user to take over all eight timeslots (in GSM) without any error protection. It is obviously unlikely that a network operator would permit all network resources to be used by a single GPRS user. Additionally, the initial GPRS terminals are expected to support a limited number of timeslots (probably one to three). Therefore, available bandwidth for a GPRS user will be severely limited. Theoretical maximum GPRS speeds should be checked against the reality of constraints in the networks and terminals. Operators recognize that in reality mobile networks always have lower data transmission rates than fixed networks. Due to the constraints specified above, high mobile data speeds may not be available to individual mobile users until the implementation of enhanced data rates for GSM evolution (EDGE) or third-generation W-CDMA or cdma2000. In fact, GRPS throughput is normally between 25 and 45 kbps.

The following information must be considered in GPRS dimensioning for Case 2:

- The minimum CS to be used in the cell (i.e., CS1, CS2, etc.)
- The maximum data rate for the GPRS carrier. In theory, up to 8 timeslots per carrier can be assigned to a GPRS user, but in practice to the maximum is three. GPRS has four different coding schemes but initially only CS1 and CS2 will be used, for a minimum throughput of around 8.8 kbps and a maximum throughput of 33 kbps. GPRS networks are designed to provide the desirable data rate in 95 percent of the cell area.
- The aggregated throughput[7] produced by packet data services during the busy hour.[8] This is the sum of all the individual throughputs produced by each service offered in the network (email, FTP, Web browsing, etc.).

The required resources (number of time slots)[9] for packet data services is simply the ratio of the maximum traffic expected in kbps and the throughput a GPRS time slot can handle in kbps for the minimum CS to be provided.

Circuit-switched data traffic and voice dimensioning.   The steps to correct dimensioning are:

---

[7]A usage factor can be used, for instance 85 percent.

[8]The busy hour for PS data can be different for CS data or for speech. In this case, the hour in which the most PS data traffic (kbps) and the most CS data plus voice traffic (Erlangs) is used.

[9]Clearly, this dimensioning process will lead to overdimensioning the RF network because it uses the minimum throughput of a GPRS carrier rather than the average throughput.

- Calculate total voice traffic in Erlangs during the busy hour
- Calculate total circuit-switched data traffic in Erlangs during the busy hour
- Add voice and data traffic in Erlangs
- Calculate the number of timeslots required using the Erlang B formula for a GoS of 2 percent.

Aggregate traffic

- For the number of timeslots required, add the timeslots required for packet-switched traffic data to the timeslots required for circuit-switched data and voice. The average load in timeslots that carry circuit-switched traffic is in the range of 60 to 80 percent. The rest of the capacity is used as a margin to absorb the burstiness of packet-switched traffic.
- Find the total number of carriers required for the aggregate traffic by dividing the number of timeslots by eight (the number of timeslots in a GSM frame).

We know that computer simulations offer an exact dimensioning solution at the design level. Computer tools such as OPNET are commonly used to model GPRS networks. Essentially, a simulator models the network based on air interface parameters and the traffic produced by the users. Service types (voice, email, FTP, online banking, Internet browsing, etc.) and user mobility (if handoff is taken into account) will be based on the operator's demand. Computer simulations can be as complete as the operator wishes, but complexity increases as the level of detail rises.

Some network operators have shown interest in building a simulator but have been discouraged by the amount of work and level of expertise entailed. Others have decided to follow the method that offers approximated solutions at the expense of over- or underdimensioning.

Case 3

The solution presented in the first scenario cannot be applied here by inverting the roles of PS data and CS data plus voice. Instead, this case can be solved following any of the methods explained in Case 2.

Growth dimensioning. Traffic forecast in GPRS networks is tightly linked to morphologies and network coverage areas. Packet-oriented traffic will be the main reason for operators to deploy GPRS. Although most of the technologists and many of the business people focus on speed, GPRS' always-on feature will be a more important feature for users. Users will be able to pay by the bit instead of for their connection time. Therefore, even if it takes a little longer to download a file, it will be less expensive for customers to use GPRS. This appeal of cost savings will fuel up data and Internet usage in GPRS networks. For GPRS operators, predicting the growth of data usage in general, and packet-switched data in particular, will be the main consideration for designing and dimensioning the networks.

Impact on outage probability.   Providing capacity on demand means that GPRS data services will not have much impact on the capacity of GSM voice services. GPRS actually creates an additional capacity for data services. However, a slight system performance degradation is still to be expected. Outage probability increases and coverage area decreases due to additional interference produced by packet data transmissions.

When an operator plans a GSM system, the outage probability in a cell is normally designed to be within some limit for a certain traffic load. Co-channel interference is predictable even under low voice traffic loads. If the offered voice traffic reaches outage level, the overlaying packet data services will degrade either voice quality or voice capacity. When the outage probability is below the maximum acceptable value, additional capacity can be obtained by overlaying the GPRS packet data services on GSM voice services, but capacity extension is limited by the level of outage considered tolerable. This is because of the maximum acceptable value for outage.

Core Network Dimensioning

The separation of packet-switched and circuit-switched traffic happens in the base station controller (BSC). Traffic load in the interface between the base transceiver station (BTS) and BSC (Abis interface) will increase because of extra packet-switched traffic. When traffic is divided, packet data is sent to the MSC through the A interface, and circuit-switched traffic is sent to the SGSN through the Gb interface.

The new interfaces in the GPRS networks follow dimensioning rules for data networks. The Gb interface is Fast Ethernet based on Frame Relay. Where the overlay network is capacity limited, it will be designed for the estimated number of "active" and "standby" users within the coverage area of one BSS. Where the overlay network is coverage limited, SGSN capacity will be the main dimensioning parameter for the Gb interface. As you can see in Figure 10.2, there will be other entities added to the GSM network architecture such as the packet control unit, which is added to the system according to air interface demand. Billing systems will be added according to estimated data users or GGSN capacity. Finally, interfaces Gs and Gr will use same methodology as GSM networks.

Figure 10.2
New components in GPRS networks.

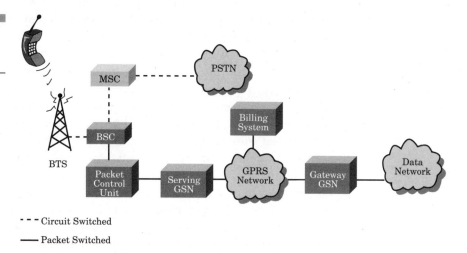

- - - Circuit Switched

—— Packet Switched

GPRS Network Optimisation

As networks are upgraded, mobile operators will begin to face challenges posed by the convergence of voice and data in GPRS deployments. Optimizing a GPRS network is completely different from optimizing a GSM network. Quality of service (QoS) can no longer be measured solely by voice quality. Instead, it will be driven by many factors, including throughput, delay, blocking, retransmission, or packet loss.

To maximize investments in infrastructure, each part of the network needs to be optimized. The optimization process extends to the radio interface and value-added applications, but also includes network integration as a whole.

In other words, the optimization of a GPRS network starts with the GSM network. (A well-designed GPRS network is always deployed on top of an optimized GSM network). Next, there is a process to ensure that the new IP services offered by GPRS work properly, and that all protocols involved in the data and signal transmission work as defined. Optimization concludes with a process to ensure efficient integration of the GPRS and GSM networks. These steps are elaborated in the following sections.

GSM Network Optimization

The optimization of a GPRS network starts with the optimization of a GSM network (if the GPRS network is to be deployed for the first time, then the GSM network needs to be optimized). Although the optimization of a GSM network implies not only the radio interface, the coverage, the core network, and the dimensioning issues, only the coverage and capacity (network dimensioning) are discussed here.

Good coverage is important because the average throughput of a GPRS session is directly dependent on the coding schemes (CS1 to CS4)[10] used to secure the information transmitted. In turn, the level of security is dependent on signal quality. The better the signal quality, the lower the level of protection and, therefore, the higher the throughput. When a mobile station initiates a GPRS session, it will always start transmission with CS1 irrespective of its position within a cell or its signal quality (i.e.,: its C/I ratio[11]). As the session progresses, measurement reports are exchanged with the base station to let each know the power levels received. If the C/I meets or exceeds the C/I threshold for the use of CS2, then the mobile station will automatically make use of CS2. Should a further increase in the power level be experienced, CS3 will replace CS2, eventually giving way to CS4.[12]

Note that when CS4 is used, redundant protection is eliminated and maximum throughput is attained. Spots with bad signal quality will force mobile stations carrying an active session to use a CS with more redundant information and the reduction in average throughput that accompanies it. Tools to optimize GSM coverage have been available for a while and are normally complemented by drive tests.

Optimizing the GSM network implies dimensioning as well, given the fact that voice and data users will coexist in the network and share its resources. Even though GPRS does dynamic allocation of resources between voice and data, the correct amount of resources must be main-

---

[10]CS1 is the coding scheme with maximum protection and minimum throughput. CS1 to CS4 coverage areas form concentric circles around the base station (considering omnidirectional antennas), in which CS4 and CS1 are respectively the smallest and the largest circles.

[11]Obviously, the C/I will have to be good enough to establish a call.

[12]CS3 and CS4 will not be implemented in the first phase of GPRS deployment.

tained by each cell if voice and data services are to meet QoS requirements for both circuit and packet transmissions. As previously noted,
due to the nature of the data traffic, Erlang tables are not enough to
dimension the network requiring computer simulations of different traffic mixes for proper observation of performance.

Value-Added Services

The second step in optimizing a GPRS network is to ensure that a packet
data session starts and continues according to the expected performance of
a similar session over IP. All major QoS parameters should be checked in
this phase. Usual practice calls for a data collection device to gather data
from the Abis and the Gb interfaces, and a post-processing tool[13] to corre-
late information from both the access network and the core network. (The
latter ensures a complete history of the data sessions analyzed.) The last
check at this stage is the protocol stack used for communications between
the BSS and the SGSN, including Frame Relay, BSSGP, SNDCP, GTP, etc.

Integration of Value-Added
Services in the GSM Network

The final phase of the optimization process integrates the value-added
services in the GSM network, and it only occurs when the first two
phases have been completed. At this point overall network performance
needs to be examined as a whole, with particular emphasis on the cooperation established between the GSM and the GPRS network. Also
check signaling protocols and data planes. The former gives all kinds of
information about network service, BSSGP, LLC, GMM, SM, SNDCP for
Gb, and RLC/MAC for the Abis interface. The latter is a window to data
transfers from or to the mobile stations, IP or X.25, respectively.

If all the steps previously indicated have been carefully followed, the
network is in proper condition to start operating. Additional steps are
necessary to keep proper functionality of the network, but they will be
discussed in future sections.

---

[13]Optimizing a GPRS network requires the use of tools that provide detailed reports and
statistics to instantly focus on the weakness of the system such as bad parameters, dimensioning, cell planning, frequency planning, access, or even some mobiles. These tools are
normally post-processing data tools (software) that analyze the data collected at several
points in the network (normally the Abis and the Gb interfaces).

CHAPTER 11

Implementation and Testing

Colin Watts and Graham Wright

Lucent Technologies

The GPRS standard has been specified by ETSI to provide enhanced data services for GSM subscribers worldwide. By implementing GPRS as an overlay to a GSM voice network, the operator gains control over substantial revenue-generation opportunities and subscribers gain access to a whole new dimension of connectivity and services.

Let's briefly review the technical basis for these gains. The arrival of GPRS introduces *packet-switching* technology into GSM-based mobile networks. Packet switching is the basic technology that has powered the tremendous growth of the Internet, and literally means that data travel over the network in the form of small bundles ("packets")—structures that can be transported quite efficiently. GPRS also has the ability to aggregate multiple timeslots for a single user. Together, these technologies mean that subscribers can use significantly higher data rates than are currently available with GSM, giving faster response times, wider choice of applications, and a much-enhanced end-user experience. Applications can, of course, include truly interactive services such as gaming, and simple multimedia capabilities that pave the way for 3G services to come. Whereas traditional GSM data services have used dedicated circuits, GPRS packet-switching technology allows *sharing* of circuits by multiple users, thus reducing the cost of delivering the service for the operator.

In this section, we will discuss some of the building blocks of a typical GPRS network installation, describe how they interwork, and explain some of the more important acronyms that bombard us every day.

GPRS Infrastructure and Components

One of the drawbacks of GPRS is that it introduces so many more acronyms into our working lives. And to make matters worse, they are not restricted to three-letter ones—there is at least one eleven-letter acronym, the base station subsystem general packet radio service protocol virtual connection. (Fortunately this is usually condensed to simply the "BVC.") You will be pleased to know that in this chapter we intend to introduce the minimum number of acronyms necessary to further understanding and help readers associate GPRS with what they already know about GSM.

In order to support the introduction of GPRS into a GSM network, an existing base station subsystem (BSS) must be upgraded. The upgrade adds the functionality to separate GSM voice traffic from GPRS data traffic. The GPRS data is then packetized for onward transmission over the GPRS network to fixed or mobile destinations or to other data networks—including the Internet. A similar process applies in reverse for incoming data where the BSS reformats the data and redirects it to the receiving mobile device.

A GPRS network comprises three main functional elements:

- The GPRS support nodes (GSN), which contain the major functionality required to support GPRS
- The Internet Protocol (IP) backbone core network, which links the GPRS support nodes
- The operations and management ventre for GPRS (OMC-G), which is the network element manager for GPRS nodes

Several ancillary elements are also needed to support a GPRS network. These include the charging gateway (CG); performance gateway (PG); border gateway (BG); and the network servers required for IP operation (including the domain name server (DNS); dynamic host configuration protocol (DHCP) server, and remote access dial in user service server (RADIUS). Yes, here come the acronyms!

GPRS Speeds and Handsets

Before diving into further detail about the implementation of GPRS, it is worth considering how GPRS offers greater data rates and therefore bandwidth to the subscribers. This is achieved in practice by the combination of two mechanisms:

1. The GSM standard air interface or radio channel frequency is divided into eight timeslots (TS). In circuit-switched GSM operation, each timeslot is capable of transporting 9.6 kbps of voice or data, and each call involves a single timeslot in each direction. Data use has so far been mainly for SMS messages, and some limited fax and other mobile data usage for which this speed sufficed. But it is not really fast enough for new data services like Web browsing, and results in a very poor end-user experience. (The industry has hopefully learned a lesson in trying to apply WAP as a "sticking plaster" to cover the wound, while awaiting the cure for the underlying ailment.) In order to increase user bandwidth for these services, GPRS

handsets are capable of combining timeslots in both the uplink and the downlink directions.

2. The second mechanism introduces new "coding schemes" that specify how much error protection is to be applied to the user data. This gives the operator or service provider the flexibility to offer higher bandwidth with less error protection, or lower bandwidth with greater protection.

For more detail on data rates, timeslots, and coding schemes, please turn to Chapter 10, "Planning and Dimensioning."

GPRS handsets are classified according to the number of timeslots in upward and downward directions, and the coding scheme(s) supported. Early handsets supported one TS in the uplink and two in the downlink. The latest ones support two TS in the uplink and four in the downlink with coding scheme CS2, giving a *theoretical maximum* of 53 kbps, less overheads.

In the first live GPRS networks, applications delivered at speeds of 20 to 30 kbps and have shown a marked improvement in usability over those achieved with circuit-switched data, and have been warmly received by users. This is the vital end-user experience that both justifies and drives the 2.5G and 3G programs. At the time of writing, most vendors have not yet implemented the higher rate coding schemes, CS3 and CS4. Obviously, increasing the data rate over the air interface has repercussions on the other links in the communications chain and may give rise to capacity and scaling issues "downstream" in poorly designed networks. The role of higher rate coding schemes, and also the role of enhanced data rates for GSM/global evolution (EDGE) is the subject of some debate in the industry, and will be heavily influenced by achieved 3G rollout timescales.

Base Station Subsystem (BSS)

The BSS, as referred to in Figure 11.1, connects the mobile phone subscribers to the GSM/GPRS core network. It comprises the base transceiver station (BTS) and the base station controller (BSC) and is linked to the operations and maintenance center (OMC-R). No new BSS network elements are required for access, as GSM and GPRS can both use the same BSS (see Chapter 9). However, the BSS needs upgrading to be able to handle data packets, via an entity known as the packet control unit (PCU). (This may be a software-only upgrade in well-designed

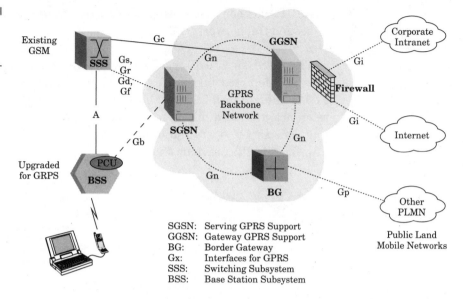

Figure 11.1
Simplified GSM/GPRS network.

SGSN: Serving GPRS Support
GGSN: Gateway GPRS Support
BG: Border Gateway
Gx: Interfaces for GPRS
SSS: Switching Subsystem
BSS: Base Station Subsystem

Some elements such as the charging gateway function, have not been shown to aid clarity.

equipment.) The PCU handles the data "packetizing" process and communicates with the GPRS support nodes over the Gb interface.

Switching Sub-System (SSS)

The switching subsystem for GSM (SSS), as referred to in Figure 11.1, includes the mobile switching center (MSC); the short message service center (SMSC); home location register (HLR); visitor location register (VLR); equipment identity register (EIR); and the operations management center for switching (OMC-S). No new network elements are required for an existing GSM SSS to support GPRS. A software upgrade to the HLR is required to add GPRS-specific subscriber data.

GPRS Backbone Network (GBN)

The GPRS backbone is an overlay network that supports the GPRS functionality and contains:

- **GPRS Support Nodes (GSN)**—The two new network elements for GPRS, these being the serving GPRS support node (SGSN) and gateway GPRS support node (GGSN).

- An IP backbone consisting of routers to transport the data packets around the network.

The ETSI specifications for GPRS cover a number of 'G' interfaces that are depicted in Figure 11.1, and are further described in Table 11.1—GPRS Interfaces—as shown below. These are introduced here for reference and each one is more than just a connection, having specific functions in the operation of GPRS networks. (Read more about the G interfaces in the previous chapter.)

TABLE 11.1

GPRS Interfaces

Interface Name	Description
Gb	The Gb interface connects the SGSN and the BSS. The interface is a Frame Relay bearer service over which radio-related protocols operate.
Gr	The Gr interface is the SS7 signaling interface between the SGSN and the HLR. It is used when the SGSN contacts the HLR to obtain GPRS-specific subscriber data.
Gd	The Gd interface provides the SS7 interface between the SGSN and the short message service center (SMS-C) to provide routing of SMSs over GPRS.
Gs	The Gs interface is implemented as SS7 between the SGSN and MSC/VLR.
Gf	The Gf interface links the SGSN to the EIR. Implemented as an SS7 link, it provides a MAP interface to check the authenticity of the MS in the EIR.
Gn	The Gn interface is the core of the GPRS IP backbone and interconnects the GSNs. It is used to transport user data using the GTP protocol.
Gc	The Gc interface links the GGSN to the HLR in the case of network-initiated PDP context activation. The GGSN uses it to query the HLR for subscriber data if it cannot identify the SGSN serving the addressed mobile. Gc is also an SS7 link.
Gp	The Gp interface connects the GSN to other GSNs in different PLMNs. This is provided at the border gateway and is used for roaming GPRS users.
Gi	The Gi interface provides the packet data transmission between the GGSN and other external packet data networks such as the Internet and corporate intranet.

SGSN

The serving GPRS support node is the core element of a GPRS network responsible for mobility management, session management, tunneling and routing, compression, authentication, encryption, and the generation of charging data records.

The SGSN has a foot in both GSM and IP camps, as it is responsible for enabling the connection of a mobile terminal to an IP network and also for controlling GPRS mobiles within GSM cells. The SGSN is the GPRS equivalent to the mobile switching center (MSC) in GSM, with which it has to interwork in order to control the mobiles. (At this point we must remind you that most GPRS mobile devices will continue to access voice services via the MSC.) Think of this in terms of two bosses, each with their own area of responsibility, who control the same pool of people for different tasks. Thankfully, the SGSN and MSC are much more efficient at sharing the responsibility of controlling the mobiles without the power games and politics of their human equivalents!

This means that a mobile may in fact be using the GPRS network to "surf the net" for example, while also using a standard circuit-switched GSM network in order to make and receive voice calls at the same time. (Humans are very good at multitasking, so it is only fair to provide matching mobile communications capabilities.) In practice, it is thought that many users will simply switch between GPRS and voice services, blissfully unaware that they could be making even better use of their time.

So how does a message actually get from sender to recipient? (Now for the onslaught of technology.) When a GPRS subscriber wants to send or receive data, the mobile performs an "attach" process to establish a "session" between the mobile and an SGSN. (This is like logging on to a PC network before requesting or sending any data.) Once the mobile is attached to the SGSN, it remains attached until instructed to the contrary (hence the "always-on" phrase used by vendors and by the press in mobile Internet articles). At this point the mobile can only send and receive SMS messages via GPRS, as it does not yet have a connection to a packet data network. In order to send or receive data other than SMS, the process of "packet data protocol (PDP) context activation" is carried out. (This is the equivalent of activating your modem and dialing up your Internet service provider.)

One of the more challenging aspects of GPRS is, in fact, the user's mobility. Nodes in other packet data networks see the mobile as just another packet data node that is accessible via a particular route—in this case via a GGSN. They do not know that in fact this end-user node is a mobile device and that it may be moving through a cellular environment.

The SGSN, which was introduced in some detail in Chapter 10, provides an essential network capability by controlling the packet routing and tunneling for the mobile from the GGSN to the BSS. The entry/exit point for the passage of packets is fixed at the GGSN at the time of PDP context activation. The SGSN handles the necessary changes in routing to accommodate the moving mobile, while the nodes in the packet data network remain oblivious to the "moving" IP address. This may sound overly complicated, but we must remember that we are communicating with the mobile in both the packet- and circuit-switched dimensions simultaneously and therefore have to work within both sets of rules.

Figure 11.2
SGSN routing
areas and tunnel
re-routing.

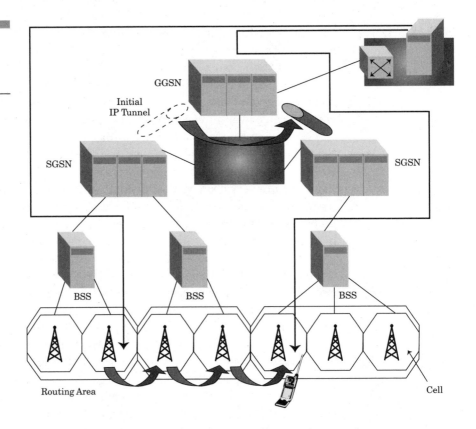

As we saw in Chapter 9's discussion of mobility management, the SGSN tracks the mobile within its routing areas, ensuring that the data connection is always maintained. The mobile must inform the SGSN when it crosses the boundary of a routing area. Furthermore, since each routing area is controlled by only one SGSN, the mobile must inform a new SGSN of its presence as soon as it moves into its area of coverage.

(An analogy would be presenting your passport for inspection by customs before entry into another country.) The new SGSN then communicates with the old SGSN and moves all the tunnels and routes over to maintain the data path for the mobile, as depicted in Figure 11.2.

GGSN

The gateway GPRS support node (GGSN), as its name suggests, is responsible for providing gateway access to other packet data networks (PDNs) from the wireless provider's GPRS infrastructure. The most recognizable examples of PDNs are the ubiquitous Internet or corporate intranets; and the former can provide access to the application servers deployed by operators for supporting new value-added services.

Basic Functionality

We've said that GPRS data travels over the backbone network in packets. In order to transport and protect the data between the GPRS support nodes, a protocol called the GPRS tunnel protocol (GTP) comes into play. The GGSN terminates the GTP tunnels and forwards the data packets to a destination network, such as your corporate intranet. To do so, the GGSN has to maintain an "associations table" that tells it how to deal with each tunnel, its end point, and the data packets it contains.

The GGSN can route the packet based on its IP address, or it can pass the packet directly into another tunnel connected to a corporate access point. (An access point is defined as the logical exit point in a GPRS network from which the packets are routed to other destinations.) Decisions on how to direct packets are made when the access point is created. In Figure 11.3, we can see a basic overview of the main functions involved.

The GGSN Major Interfaces

The Gn interface carries the mobile-terminated and -originated packets within the backbone network to and from the GGSN using the GPRS tunneling protocol (GTP). This protocol makes the underlying transport network "transparent" to the user data, making it possible to use an invisible addressing scheme to carry user data within the IP part of the GPRS network.

Figure 11.3
GGSN basic logical functions.

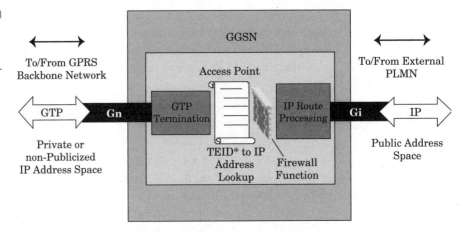

TEID = Tunnel End Point Identifier

In order for its data to reach the GGSN, an SGSN will need to know the IP address for the Gn interface. It finds out through a domain name server (DNS) lookup operation. The DNS function provides a direct translation between an IP address and its more commonly used name like www.microsoft.com. The GTP tunnel terminates at the access point where the user packet is passed through the gateway to the correct Gi interface based on IP address, or relayed through a VPN.

The GGSN also needs to map *incoming* packets to the correct GPRS mobile. As it has no direct visibility of the mobile (remember that device mobility within the GPRS network is handled by the SGSN), the GGSN uses a lookup table to match the packet's destination IP address (i.e., the mobile) to a particular tunnel endpoint identifier (TEID). This TEID is unique in describing a particular GTP tunnel into which the packet will be relayed. Once there, the packet is relayed to the mobile handset's SGSN.

The Gi interface can be said to effectively act as a proxy for the mobile to the external network. Looking at the Gi interface from an external router, it will appear that the mobile device is directly connected at this point because no further IP "hop" is required.

Security

The GGSN is a bidirectional service point, acting as both a data bearer and a network guardian protecting the GPRS network from unwanted intrusion. A good GGSN will include state-of-the-art firewall capabilities to protect the GPRS network from invalid data and irregular access attempts.

In addition to protecting the operator's GPRS bearer network, the GGSN should have the capability to protect the users' packets on the Gi side. The most commonly adopted method for IP packet protection will be IP security (IPsec) which is a protocol designed to protect data as it moves between different points.

IP Address Management

It is impossible to write about GPRS without at least mentioning some of the workings of IP. I have tried to keep this to a minimum in this chapter, but one more topic has to be introduced in order to comprehend IP *in the mobile domain.* May I suggest that you pour yourself a cup of coffee and relax as we briefly tackle the subject of IP address management? You will learn much more about the particulars of numbering and addressing in Chapter 12.

IP addresses can be static or dynamic. A *static IP address* is a fixed address in the network and usually refers to a network element (for example an SGSN)—externally, it will often have a user-friendly name associated with it like www.lucent.com. A *dynamic IP address* is assigned from a pool of IP addresses for the duration of a data call to or from a mobile user. It will route the call through the data network and be returned to the pool for reassignments once the data transaction is complete.

When dynamic addresses are needed to relay incoming data to the mobile handset or outgoing data to its destination, they are allocated by the GGSN. IP addresses are sourced in a number of ways:

1. From a dedicated pool of addresses that the GGSN manages—usually this would be per-access point.
2. From a dynamic host configuration protocol (DHCP) server, which manages IP address assignment
3. From a RADIUS server that authenticates and authorizes users dialing into the GPRS network, ISP customers for example, whereby an IP address is assigned after authentication has taken place.

Other essential acronyms and buzzwords.   As with any technology, a healthy, if superficial, understanding of all the associated "buzz words" is the key to the art of after-dinner conversation. I will now therefore do my bit for the human race by introducing you to yet some more terms that see daily use in the world of GPRS.

The GGSN needs to have the flexibility to handle *different types* of data packets, and be able to interact appropriately with the various protocols involved. In GPRS, the data packets sent from a mobile are known as packet data protocol (PDP) Contexts, as explained in the previous section on SGSN. However, two types of PDP context may exist between the mobile device and GGSN. The first of these is IP based and the second is based on the point-to-point protocol (PPP).

IP-based PDP contexts will generally be used in the case where the ciphering and authentication achieved at mobile registration is sufficient; in other words, where no *secondary* authentication is needed to service the connection beyond the GGSN to external networks. This is likely to be the case for the majority of public infrastructure users.

The PPP type of PDP context will prevail in the case of remote-access service provisioning to corporate intranets. This is used for fixed wire dial-up remote access to company networks today. In our case, the mobile passes PPP control packets during the establishment phase of the PDP context towards a RADIUS host, generally located at the corporate end. Generally these requests are forwarded on the Gi interface over Layer 2 tunnel protocol (L2TP) connections. It is also possible for the GGSN to terminate PPP and forward the request over normal IP-based virtual private networks (VPNs).

A future-looking GGSN should offer the capability to support not only both PDP types, but also additional PPP connectivity options, in a flexible manner.

Planning GPRS Configurations

GPRS support nodes from some vendors use the same hardware platform for the SGSN and GGSN. Within the same chassis, nodes can be configured in software for SGSN, GGSN, or combined functions. This common usage of products and cards helps to minimize installation time and the operator's need to hold spare equipment in readiness against the possibility of a failure. Costs are thus reduced.

In general, GPRS suppliers offer flexible and scalable support node architectures, which can be configured in different ways for GPRS solutions to meet operator requirements. There are several factors that need to be considered when planning a GPRS network, including (but not limited to):

- Will the GPRS network design be distributed or centralized? The availability and cost of links, density of subscriber base, and proximity to network management centres may determine this.
- Is redundancy required? By co-locating nodes, there can be some increased protection against equipment and network faults.
- What are the likely traffic volumes at start-up and in later projected phases? An "entry level" GSN node can be installed for GPRS introduction and new interface cards can be added as the traffic volumes increase.
- Does the network already contain IP or ATM equipment? Operators can install GPRS into an existing network, or set up GPRS as a VPN on a public data network, thus sharing the capital and operational costs.
- Is sufficient bandwidth available from the BSS to the SGSN Gb interface to support greater volumes of traffic? Likewise, is there sufficient capacity on the Gi interface from the GGSN out to the Internet and corporate intranets?

A fundamental requirement for the design of GSN equipment is *modularity*. This provides operators with the flexibility to accommodate expansion in their networks and increases in traffic and subscriber numbers, while protecting their investment in installed equipment.

As the GPRS market matures, many operators are discovering that they also have to deliver higher service requirements in some parts of the coverage areas rather than in others—where mobile devices are concentrated, for instance, or where wider bandwidth is required. In this situation, they can configure a variety of network architectures to accommodate higher requirements by using combinations of centralized and decentralized GSNs as follows:

1. **Centralized SGSN/GGSN**—This solution is suitable for systems where the traffic load is expected to be low in all areas, and for start-up situations.

   This configuration has the flexibility to evolve via software upgrade to a combined or distributed architecture as the volume of data traffic and number of data users increases.

2. **Centralized GGSN and distributed SGSN configuration**—This solution is suitable for systems where traffic load varies from area to area.

In this configuration, the operator can add additional GGSNs at a particular site and additional SGSNs in distributed sites. Additions result in a proportionate increase in the long-distance GPRS backbone traffic.

3. **Distributed SGSN/GGSN configuration**—This solution is suitable for a high traffic load in all areas.

A distributed SGSN/GGSN configuration is achieved by installing additional SGSNs and GGSNs in particular sites. As a result, the long-distance Frame Relay traffic is reduced, but there is a requirement for some long-distance GPRS backbone traffic, for the intra-PLMN (public land mobile networks) roaming between different GGSN areas.

GPRS IP Backbone (Core) Network

The GPRS IP backbone network is used to connect the GPRS support nodes and provide interfaces to other PLMNs. It is an IP, or IP-over-ATM network consisting of routers of various types and possibly core switches. The configuration the operators select will depend on several factors including the size of network, the volume and nature of traffic, whether the operator has existing facilities that can be used, and the operator's plans for future services.

The GPRS IP backbone operates as an IP domain with a domain name server (DNS) for logically mapping all connected elements to IP addresses for that domain. It provides a translation function between the given IP address and more user-friendly names like www.yahoo.com. The IP backbone also includes network services like:

- Dynamic host configuration protocol (DHCP) to allocate and manage the IP addresses
- RADIUS to validate and authorize users dialing into the GPRS network
- Firewalls required for connections to external networks to protect against unauthorized access

GPRS roaming is supported through a border gateway implemented by a router, which provides connectivity to other GPRS networks via PLMNs. Mobility management in the GPRS network is implemented in the SGSN, supported by the HLR.

Element and Network Management

The operations and maintenance center for the GBS (OMC-G) provides centralized management of the operations and maintenance functions of the GPRS support nodes (GSNs).

The OMC-G supports, at a minimum, the following OA&M functions:

- Fault Management
- Configuration Management
- Performance Management
- Systems Administration

Interoperability between Components and Manufacturers

One of the guiding principles of the ETSI specification for GPRS was to guarantee high standards of interoperability between equipment from different manufacturers. Interoperability testing (IOT) ensures compatibility between network elements from different suppliers, at a functional level on a given interface, in accordance with the relevant standards and specifications.

In order to further both the reach and the accuracy of interoperability testing, a Network Vendors IOT Forum has been formed. Members are major mobile infrastructure manufacturers with the prime objective of identifying and solving interoperability issues. It currently includes Alcatel, Ericsson, Lucent, Motorola, Nokia, Nortel, and Siemens.

Part of the work undertaken by the members of the Forum is to define test specifications for interoperability testing, and to assess the suitability and specification of test tools. The forum also facilitates agreements between members for the actual performance of such testing using these specifications and tools. IOT verification is achieved by executing a predetermined set of test cases between network elements from different suppliers, and is usually specific to one interface at a time. This testing should be performed one vendor at a time in a controlled environment. Note that IOT verification does not imply full specification conformance—this is clearly outside the scope of the IOT Forum. Such conformance is viewed as "internal testing," and as such is a prerequisite to IOT.

There are two main reasons why IOT is necessary in GPRS networks:

- GSM operators want to purchase from more than one supplier.
  Perhaps a primary vendor is too expensive or is not responding to changes in the market, or perhaps another vendor shows greater ability to evolve with the market. With the support of IOT, GSM operators can cherry pick best-in-class products using the latest technologies, and can position themselves to successfully open up new markets. *IOT is therefore an invaluable aid in giving operators the confidence to consider new suppliers on merit.* Incumbent vendors do not always view IOT as affectionately, however, as it encourages open competition and can also force prices down.
- GPRS subscribers *need* to roam.
  The success of GPRS depends on giving subscribers the freedom to roam to different countries and different operators' networks, as they do with other GSM voice and data services. Equally essential is support for roaming to another GPRS network where the SGSN is supplied by a vendor different from the one in the home network. In all these cases, interoperability testing needs to be carried out to ensure that products from different vendors interwork reliably to provide seamless services and a compelling end-user experience to subscribers.

Internet and Corporate LAN Connections

Gateways to external packet data networks can include the Internet, connection to Internet service providers (ISPs), and connection to corporate intranets. What is sometimes referred to as an *Internet exchange* (INX) connects the GPRS IP-based network to the Internet. In cases where there is more than one physical connection with either network, the INX will be called upon to provide routing capability. The type of connection from the INX to the GPRS network or the Internet depends upon the operator's type and number of physical links. Implement all connections with the Internet via a firewall in order to maintain security control over incoming traffic. In the event that direct connection to an ISP is not provided, it is recommended that operators implement a protocol such as IP security (IPSec) or Layer 2 tunneling protocol (L2TP) for an acceptable level of security.

The GPRS INX site typically comprises:

- IP router
- Firewall towards the Internet
- IPSec or L2TP support

Figure 11.4
Connections to external packet data networks.

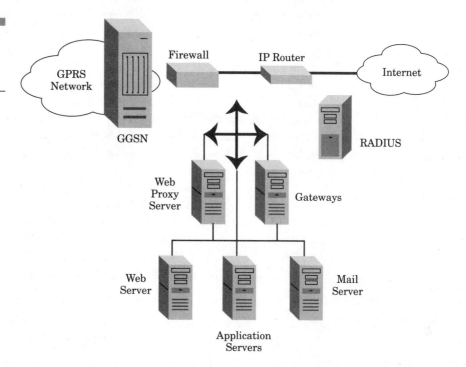

Security

While operators will have their own established security policies, a GPRS implementation is likely to require several levels of security, depending on its architecture and the applications offered. In a typical GPRS network, firewall protection is a must on the Gi interfaces out from the GGSN and the Gp interface at the border gateway. RADIUS will validate and authorize users dialing into the GPRS network, a procedure sometimes referred to as AAA (authorization, authentication, and accounting).

Billing System Connection

GPRS will have a *very significant impact* on customer care, billing, and service provisioning. Mediation and billing systems already exist to manage a wide range of services, and they are flexible enough to let operators compete effectively by offering new charging structures when necessary. It is obvious that operators need to verify that their proposed billing solution is compatible with the network implementation, and is flexible enough to absorb the addition of new applications and features.

A GPRS operator's billing system can be implemented as anything from a single box up to a complex multinode VPN setup. It is likely that the physical connection will be a LAN in the operator's corporate network, but it could be extended over the WAN, where charging data from distributed GSNs must be collected. However billing facilities are implemented, the security and integrity of the IP connection between the charging gateway and billing system is critical to protect revenues generated.

The charging gateway function (CGF) controls the collection and distribution of charging data records (CDRs) from its associated SGSNs and GGSNs. Charging records are then downloaded to the billing system automatically, or they can be polled by the billing system. Figure 11.5 shows the main elements of a typical GPRS billing system.

Figure 11.5
Main elements of a typical GPRS billing system.

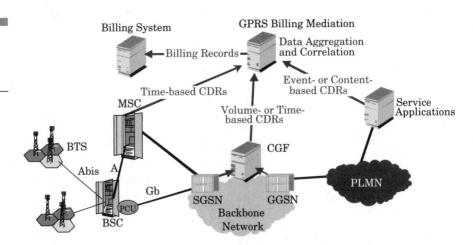

Within a GPRS environment, transaction and content-based billable events are likely to be generated by application servers, WAP gateways, and wireless portals. These applications may generate enhanced data records (EDRs) from premium services that need to be processed in conjunction with the CDRs.

You may have thought that the people that invent acronyms have no sense of humor. How then, did they come up with this next one? Such enhanced billing may well involve CAMEL (customer application mobile enhanced logic) functions (especially when the GPRS subscriber is roaming away from his home network). The billing system needs the capability to receive multiple flows of usage, and then to guide and rate them to the proper account. CAMEL is also required for prepaid services. Prepaid service has far exceeded initial expectations with voice traffic and it is expected to be extended to GPRS users soon. The recommendations in CAMEL Phase 3 outline mechanisms for prepaid service over packet - switched connections.

Despite the initial complexity of implementation, there is no doubt that GPRS introduces many more flexible and creative billing options than are currently available with GSM. These include:

- Flat-rate billing (Internet model)
- Per-packet billing
- Volume-based billing
- Value-based billing

Factors that may influence these rates include:

- Quality of service
- Time of day
- Content, call type, services used
- Source or destination

Billing products on the market today offer sophisticated and convergent data models for billing GPRS services, many include pre-integrated and sustainable order management, for customer care, provisioning, and the efficient delivery of services.

Many billing vendors will also require a mediation device to convert 2G billing into a 2.5G system (in other words to merge billing input from circuit-switched voice and SMS calls with packet-switched information) so that subscribers can have a single bill for both GSM voice and GPRS data calls.

Installation and Acceptance Testing

Comprehensive deployment, installation, and network verification services are crucial in providing the right combination of hardware, software, and support systems for the desired functionality at the lowest cost. Warehousing, transportation, erection, aligning, and mounting of wireless network equipment can be included in installation services.

Installation is one area where it is possible to work with time-saving processes like prefabricated cabling and preinstallation configuration of equipment. Such processes benefit operators' fast deployment schedules. A highly effective vendor strategy is to pre-build equipment to various key stages; this effectively "modularizes" the installation, minimizes the deployment timeframe, and minimizes impact on the operational GSM network.

Test and acceptance activities are performed based on the GPRS operators' requirements, but are often documented in a customized site manual that contains detailed building and equipment locations and traffic engineering information. Key variables are also documented, including cable runs and IP addressing plans. The site manual is handed over as part of the customer acceptance procedure. Testing with laptops containing CD-ROM, automatic test procedures, and documentation on CD-ROM helps the fast deployment process.

GPRS requires a *different suite of tools and skills* from normal GSM networks. For example, technicians have to become familiar with IP and perform IP traces. Test tools may be needed to support GPRS interfaces—especially Gb and Gr. Traditionally, installation staff work with operations people, but in GPRS environments they'll have to work closely with the IT department as well. Virtually all early movers in the GPRS operator arena were surprised by the level of "alien" (IT) knowledge required to install and commission their first networks.

Network acceptance is usually the final phase of a deployment project. It is performed before the network is ready to accept subscribers, and it gives assurance that the network will deliver the quality of service needed to satisfy subscribers. The network verification and acceptance procedures benefit the GPRS operator in numerous ways, including:

- **Reduced costs**—Systems and equipment verification will greatly reduce network repair and maintenance requirements that would add to costs and could cause service interruptions.
- **Faster revenue generation**—By following through with logical, well-documented installation and testing procedures, operators can achieve faster in-service dates, providing earlier returns on investment.

■ **Increased customer satisfaction**—Network verification services help to deliver a more sophisticated and reliable network that contributes greatly to increased customer satisfaction. This is particularly critical when launching new services, such as GPRS-based data applications, where customer acquisition and retention is of paramount importance.

System Optimization

GPRS network optimization can be a complex undertaking because the interaction of so many variables can influence data throughput rates.
GPRS performance is usually quantified using three major factors:

■ **Peak throughput rate**—The peak rate at which data is successfully transferred and defined by the parameters maximum bit error and mean bit error.
■ **Latency**—The time taken for data packets to pass through the GPRS bearer.
■ And of course, **reliability**.

Optimizing your GPRS network is essential for seamless support of the different types of traffic generated by *applications* with such a range of disparate characteristics. For example:

■ Intermittent, bursty data transmissions, where the time between successive transmissions greatly exceeds the average delay
■ Frequent transmission of small volumes of data (i.e., machine-to-machine applications such as telematics)
■ Infrequent transmission of larger volumes of data, for example file transfer/download.

In addition, the network components themselves within a GPRS network may have different characteristics:

■ GPRS links will have low throughput and high latency.
■ Transmission delays and throughput of GPRS links will be highly variable.
■ GPRS links may be subject to deterioration if the mobile moves out of the coverage area or radio conditions change.

We must not forget the extra factors resulting from the essential *mobility* of a GPRS-enabled device. Bandwidth, latency, delay, error rate, interference, and the all-important end-user experience may all vary as a GPRS user moves within and between areas of coverage. In order to provide optimal service, the network is required to adjust to these changes in a transparent and integrated fashion.

Thus, a good strategy is one that takes a step-by-step approach, both in the laboratory and in field environments, to tune existing parameters within the system, optimize various RF conditions, and evaluate the effects of overload handling mechanisms. In addition, expect the interaction of end-user behavior and applications to further affect overall performance.

Standard performance "benchmark" tests are used to validate and quantify the operation of the equipment at each phase of the GPRS call. The same test should be used to verify the operation of the system after any software or configuration change to any element of the system. (Consult Chapter 12 for step-by-step guidance.) Key parameters to be measured are:

- **Latency**—This will be measured using a ping test to determine round-trip time from a GPRS mobile (or PC attached to a mobile) to the SGSN.
- **Throughput using TCP/IP**—This is measured during file transfer using file transfer protocol (FTP). The performance is recorded in both the uplink and downlink directions. File sizes of 5 K and 10 K bytes are frequently used.
- **Peak throughput using UDP/IP**—This is usually measured in the downlink direction only, by sending a stream of 100 packets of 512 bytes apiece. The interval between packets is varied to change the overall throughput rate.

Each test is carried out several times and the minimum/mean/maximum performance carefully recorded in each case. These tests have been chosen to provide a set of *representative* results that are reasonably easy to reproduce. Benefits in terms of service quality can also be obtained by integrating IP QoS measures and ATM QoS with transport mechanisms like class of service (CoS) and flow control.

CHAPTER 12

IP Addressing

Kim Fullbrook and Jarnail Malra

02

IP networking is at the heart of GPRS. A GPRS system must be integrated with the IP networks of *all* its customers—end users, corporate intranets, and the Internet included. Internet protocol (IP) addressing is probably *the* most important subject in today's IP world. An understanding of addressing issues is vital for anyone planning to implement or operate a GPRS network, especially since the number of mobile devices now in use has tremendously increased demand for IP addresses. Various network elements in the GPRS network infrastructure have IP addresses, as does every GPRS mobile device it supports. There are several IP addressing schemes and choosing the right one for your deployment is key, as it can influence network design and the type of services that network can support.

The importance of IP in GPRS is underlined by the fact that, although GPRS was originally designed to support X.25 connectivity as well, mobile network operators have never exploited this capability and there are no X.25 GPRS products on the market. Some of the main factors to consider when selecting an IP addressing scheme for a GPRS network will be explored in this chapter.

Types of IP Address

In the Beginning...

The first popular version of the Internet protocol—better known as IP—was actually the fourth major version and is usually called IPv4. Your IP address pinpoints the location of a device on a network (your PC or mobile), and works much like a street address to identify where you are and how to deliver information to you.

When IPv4 was standardized in the early 1980s, every single device connecting to the Internet needed a unique IP address. This requirement seemed reasonable at the time, because the address structure uses a 32-bit number that can theoretically provide up to nearly 4.3 billion (2^{32}) IP addresses. Each address is represented as four groups of decimal numbers that range between 0 and 255, where each number represents eight bits (i.e., one byte) of information. An address is normally written in a format called *dotted decimal notation*, for example 192.123.12.1, and each decimal number represents eight bits of the total 32-bit IP address. The designers of IP divided available address space into three main groups: Class A, Class B, and Class C, with each class offering a different amount of address space. Address ranges from Class A each offered over 16 million addresses; Class B ranges offered 65 thousand addresses, and Class C ranges offered 256 addresses. This structure is often referred to as "classful" addressing. Organizations could request allocation of an address range within one of these groups free of charge by applying to an appropriate Internet registry. Note that the terms Class A, B, and C are now no longer used, having been replaced with a more flexible system known as a "classless" addressing system that uses network suffix notation to depict address ranges: for example "/8," "/16," and "/24" are the equivalents of Classes A, B, and C respectively. An example of this notation is an IP address range 195.128.4.16/30 that contains just four addresses.

Growth Problems

The architects of the Internet did not envisage the extent to which it would grow in later years. It soon became clear that their original addressing model could not be sustained from a network scaling perspective because it could not satisfy the insatiable demand for IP addresses. The original model would have exhausted available address space in a relatively short time had not various measures been introduced to conserve and deploy IP addresses more efficiently.

The introduction of "private" addressing in the mid-1990s was a key measure in the conservation of IP address space. Certain address ranges were set aside for use in private networks. Comprised of the following: 10.0.0.0 (equivalent to one former Class A), 172.16.0.0 to 172.31.0.0 (equivalent to sixteen contiguous former Class Bs), and 192.168.0.0 (equivalent to 256 contiguous former Class Cs), these are often referred to as "unregistered" addresses. They do not have to be requested from an Internet registry and may be used by anyone for any purpose, *as long as they aren't used on the Internet*. The great benefit of private addressing is that the same address space can be used on many different networks at the same time, hence alleviating the demand on public address space.

Network Address Translation (NAT)

Along with the introduction of private addressing came a technique called *network address translation* (NAT). NAT enabled privately addressed networks to access the Internet despite the prohibition against using such addresses on the Internet. NAT maps the addresses in a private network to a small number of public addresses. It is typically implemented at the point where a private network interfaces to the Internet, usually combined with firewall security functions in most commercial implementations. When a device in the private network wants to send packets of data to, for example, a Web server on the Internet, the data packets traveling to the Web server will be routed via the NAT function. NAT replaces the source IP address of the originating device with a public IP address—usually that of the NAT device itself—and then forwards the packet on to the Internet for routing to the Web server. The NAT device receives the data packets in the return direction from the Internet because its IP address is contained in the packet as the destination address. Because many devices on the private network share the NAT function, there might potentially be a problem deciding how to route those packets which come back from the Internet since they all have the same destination IP address—that of the NAT function itself. The NAT function therefore keeps track of the devices that have sent packets to the Internet by building tables in memory containing the device's IP address along with the port number information from the data packet. With return packet in hand, the NAT device can now replace the destination public address in the data packet with the private address of the true destination device, and forward the packets appropriately.

Figure 12.1
NAT in a Web
request-and-response
situation.

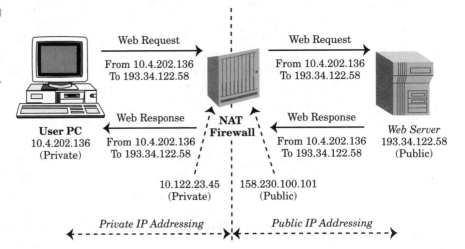

The Story Today...

Techniques introduced to extend the life of IPv4 have been very successful, but the belief that public IPv4 addresses are in short supply prevails. In reality large quantities of public addresses are still unused today. However, the Internet registries have tight control over the release of these addresses, and will make them available only where requests are justified in accordance with the criteria defined by the registries. (We'll look at those criteria in more detail later.) Today, more than half of the total available sixty-four "/8" address blocks (formerly known as Class A) remain unallocated, representing over 530 million available IP addresses.

The Light at the End of the Tunnel: IPv6

Although many techniques have been introduced to extend the life of IPv4, a longer-term solution is clearly required. The next version, IPv6, is widely seen as the answer. Recall that IPv4 uses a 32-bit number format for its address structure and can theoretically provide up to nearly 4.3 billion (2^{32}) IP addresses. In comparison, IPv6 uses a 128-bit IP address structure that could theoretically provide up to 3.4×10^{38} (2^{128}) IP addresses. This is an absolutely huge number, i.e., 340 trillion, trillion, trillion IP addresses. IPv6 will eventually replace IPv4 and be deployed throughout IP-based networks. The transition to IPv6 is covered in more detail later in this chapter.

Internet Registries and Requesting IP Addresses

Any organization may need public IP addresses for their IP-based network(s), and every such address issued has to be unique in the world. The Internet registries are tasked with managing and administering the IP address space on a global basis, with policies and procedures in place to receive requests and process them appropriately. The Internet registries form part of a much a larger body, organized in a hierarchical structure, which is responsible for many operational aspects of the Internet. The overall co-ordinating body for this organization is known as the Internet Corporation for Assigned Names and Numbers (ICANN).

Today there are three main registries, known as regional Internet registries (RIRs), that control the global public IP address space, with each RIR being responsible for specific geographic areas as follows:

1. **Réseaux IP Européens Network Coordination Centre (RIPE NCC)**—Serves Europe, Middle East, Central Asia, and African countries located north of the Equator.
2. **Asia Pacific Network Information Centre (APNIC)**—Serves the entire Asia Pacific region, including 62 economies/countries/ regions in South and Central Asia, Southeast Asia, Indochina, and Oceania.
3. **American Registry for Internet Numbers (ARIN)**—Serves North America, South America, Caribbean, and African countries located south of the Equator.

One objective of the Internet registries is to control allocations so as to ensure that the public IP address space is used in a conservative and efficient manner. To assist with this task, they have laid down various criteria[1] for approving requests for public addresses. As a general rule, the applicant organization will need to justify and demonstrate at a minimum:

1. A genuine need for public addresses. Supporting information to show that it is not practical or feasible to use private addressing will be beneficial to the application.
2. A planned high utilization of the requested range over a specific timeframe. For example, showing how 1000 IP address will be deployed and used over, say, a two-year period.

As a part of the initial GPRS network design activities, the GSM Association worked with all the RIRs to generate IP addressing guidelines[2] for use by mobile network operators. These guidelines assist network operators by recommending where public and private IP addresses are used in their GPRS network infrastructure, and providing advice on the addressing options available for GPRS terminal devices. Where public IP addresses are needed, the guidelines emphasize that all mobile

---

[1]Public IP address request details for each respective RIR can be obtained from: www.ripe.net, www.apnic.net, and www.arin.net.

[2]GSMA IREG PRD IR.40 "Guidelines for IPv4 addressing and AS Numbering for GPRS Network Infrastructure and Mobile Terminals," August 2001, J Malra & K Fullbrook.

operators have a shared responsibility to conserve and make efficient usage of this address space. These guidelines are available to anyone from the GSM Association's public Web site.[3]

IP Addressing for the GPRS Network

The GPRS network can be roughly divided into two main parts: the GPRS network infrastructure and the mobile terminals. IP addresses will be required for both. Figure 12.2 illustrates the general IP addressing requirements for the network.

Figure 12.2
IP addressing for a GPRS network.

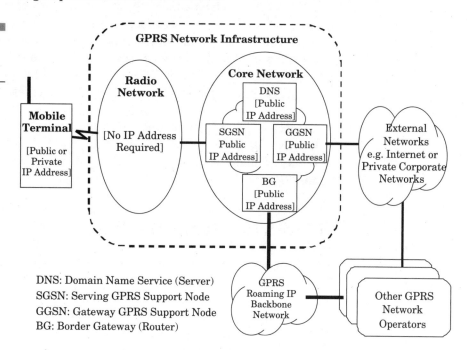

DNS: Domain Name Service (Server)
SGSN: Serving GPRS Support Node
GGSN: Gateway GPRS Support Node
BG: Border Gateway (Router)

GPRS Network Infrastructure

The GPRS network infrastructure can itself be divided into two further parts: the radio network and the core network.

---

[3]GSM Association's Web site: www.gsmworld.com.

Radio Network

The GPRS radio network is comprised of the network elements responsible for connecting the GPRS terminal device and the operator's GPRS network over the radio/air interface. These network elements include radio base station controllers and radio transmitters/receivers. The techniques used for network communication over the radio network are not based on the IP model, and so there is no need for IP addresses in this part of the GPRS network.

Core Network

The GPRS core network is comprised of network elements responsible for the management and control of the end-to-end connection between a GPRS terminal device and external networks such as the Internet. Communication and connectivity between these elements is based upon the IP model, and thus require IP addressing. The following core network elements of the GPRS network will need IP addresses:

- **SGSN**—Serving GPRS support node
- **GGSN**—Gateway GPRS support node
- **DNS**—Domain name server
- **BG**—Border gateway

IP Addressing for the Core

When a GPRS user is roaming, he will typically require connection between the core networks of two network operators in different countries or regions. The connection must be established between the SGSN in the visited country and the GGSN in the user's home country. Connectivity will be provided via the border gateway and routed across the GPRS roaming backbone network (otherwise known as the inter-PLMN backbone or GPRS Roaming eXchange), which links GPRS network operators.

For roaming to work, the IP addresses assigned to the network elements involved must be unique, not just within each operator, but also *among all operators*.

The International Roaming Experts Group (IREG) within the GSM Association represents the interests of all GSM operators worldwide. IREG created a task force of representatives from the operator and Internet communities to investigate the problem of infrastructure addressing. They concluded that guaranteeing this level of addressing is only possible with public addresses. This recommendation was then ratified globally between the Internet registries and the GSM Association, authorizing the registries to allocate public IP addresses to each core network element involved in GPRS roaming, upon application by the GPRS operator.

Reasons for not using private IP addressing.   Private IP addressing for the core network had also been considered during the GPRS roaming design phase. It was soon ruled out for a number of reasons, including:

- Network operators at some time in the future may want to use the Internet instead of the roaming backbone network for connecting the SGSN in one country and the GGSN in the other. In this case, public IP addresses will be required.
- Although one large private address range could have been used for the purpose (for example 10.0.0.0/8 can provide up to 16.8 million IP addresses), it is not intrinsically scalable. Such a solution would require a centralized body to establish new administration and control procedures and to co-ordinate operators on a worldwide basis. The procedures for governance would have taken time to be designed, agreed upon, and tested before being implemented. Therefore the option to designate a private address range was deemed impractical.
- Network address translation (NAT) devices could not be used across the roaming backbone network because the connection between the SGSN and GGSN uses a GPRS tunneling protocol (GTP) not supported by NAT.

Mobile Terminal Addressing

Mobile terminal addressing is a crucial subject because of the need to have clear addressing policies to support the hundreds of millions of GPRS terminals likely to be deployed worldwide.

Terminals and PDP Contexts

Every GPRS terminal device requires an IP address. Some terminals potentially need more than one. How does this work? There are at least three basic types of terminal:

1. A typical GSM phone (or similar device) with the addition of a built-in WAP browser that operates over GPRS
2. A PDA device
3. A PCMCIA data card plugged into a laptop PC

Some terminals, however, are multimode: for example, the Motorola Timeport is basically a category 1 terminal that also allows a PC to be connected via infrared or serial cable. Both WAP and PC connections can take place simultaneously and probably link to different access point names (APNs).

To be more precise, it is not the terminal itself which needs an IP address but the packet data protocol (PDP) context of the connection (you can review how the PDP context works by turning to Chapter 11). Note that IPv4 and IPv6 have *different* types of PDP context so that the SGSN and GGSN can correctly forward the relevant type of data to and from the mobile. The Motorola Timeport just described allows two simultaneous PDP contexts—one for WAP and the other for a connected PC. You could imagine a future PDA device allowing two simultaneous sets of Web sessions, one to a corporate network, being paid for by the user's company, and the other to the Internet, being paid for by the PDA's user. In this case there would be two PDP contexts active, each with its own IP address. Although GSM standards allow a single device to have up to 14 simultaneous PDP contexts, in practice there is little use for this many. For simplicity, the descriptions in this section assume that a mobile terminal device requires only one IP address.

IP Address Allocation Mechanisms

The GPRS standards specify IP addresses as being allocated to mobiles in one of two different ways—fixed and dynamic. These methods are described in the following sections.

Fixed Addressing

The fixed addressing method always allocates the same IP address to the mobile. The chosen address is set up as part of the mobile's subscription in the home location register (HLR). Whenever the mobile opens a PDP context to a particular APN, the address specified for the subscription is sent from the HLR to the SGSN, and thence to the mobile. In this method the IP address assigned to the GPRS terminal device is enforced by the mobile network itself.

Figure 12.3
Fixed IP addressing.

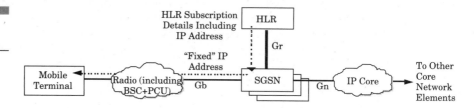

The situation where fixed addressing makes most sense for a GPRS terminal device is in the telemetry market where, for example, a soft drink vending machine might communicate with its head office via a GPRS link and a private IP network. Fixing the address to the soft drink machine eliminates any further need to verify the machine's identity, since the same address is always used. This scenario is a good example of how a private network and private IP addressing would be executed if that option had been viable. Accessing the Internet via a permanently fixed IP address is an inefficient use of a public IP address; although this is not forbidden, it is discouraged by ICANN as a matter of course.

Dynamic Addressing

Dynamic addressing avoids the problem of permanent address allocation by allocating addresses on the fly. The SGSN assigns an address to the PDP context when it is opened, and retrieves the address when it is closed. Although the SGSN allocates the address to the mobile, it is the GGSN that obtains the information using one of the following mechanisms:

■ Via RADIUS from an external RADIUS server
■ Via DHCP
■ From a local address pool on the GGSN
■ From the customer network via an L2TP tunnel from the GGSN

The DHCP/RADIUS server(s) may either reside in the external customer network to which the GGSN connects, or be run by the network operator and located internally. Examples of this are shown in Figure 12.4.

Figure 12.4
IP address allocation
methods.

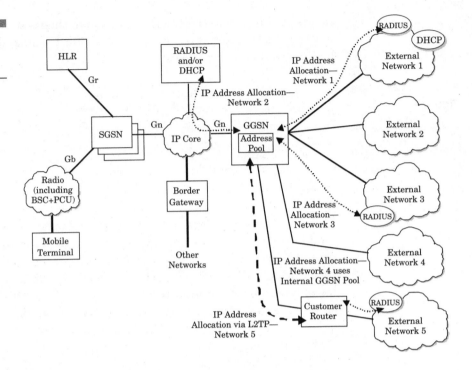

Note that equipment manufacturers do not necessarily support all the dynamic options in their products.

IP address allocation from local address pools on the GGSN. This option operates much like address pools on a standard public switched telephone network (PSTN) dialup server. Pools are defined on the GGSN—for example 1.2.3.1 to 1.2.3.255—and the GGSN allocates an address from the pool when required. One complication is that, to be fully effective, the address pools need to operate on a per-APN basis. In other words, each APN has its own address pool(s). This allows the GGSN to simultaneously support different APNs that use the same IP address space, as will be the case with corporate networks since today most use the same private IP ranges. When the customer network chooses to have IP address allocation performed by the network operator, this option (where available) will be preferred over other methods as it is the simplest to set up and operate.

IP address allocation via RADIUS. A RADIUS server allocates the IP address in response to a request from the GGSN. Although this server can either be in the operator's network or customer's network, in practice it is only likely to be used in a customer network because it is more complex for an operator than allocating IP addresses from pools on the GGSN. When RADIUS is used by a customer network, it is likely to also be used to perform the function of user authentication as described in Chapter 11. Note that the CLI of the user's terminal (e.g., +447802220004) is normally passed through as part of the RADIUS request to ensure unique identification. RADIUS can be configured to provide a "fixed" IP address so that it always provides the same address to a particular terminal every time it makes an address request. In this case the "fixing" is entirely under the control of the external network.

IP address allocation via DHCP. A DHCP server—in either the customer network or operator's network—allocates IP addresses in response to a request from the GGSN. DHCP can be configured to provide a fixed IP address, so that it always provides the same one to a particular terminal when it makes an address request. In this case, fixing is entirely under the control of the external network.

L2TP tunnel from the GGSN. L2TP stands for layer 2 tunneling protocol. A tunneling protocol is used to carry IP user traffic on a shared IP network in a way that hides the IP addresses in the user traffic, allowing lots of different and potentially conflicting flows to share the same underlying IP network. L2TP is commonly used in the PSTN for dial up to Internet service providers or corporate networks. It was designed to carry IP traffic from devices such as a PC via PSTN dial-up to a remote access server. An L2TP tunnel then carries the user traffic across an "insecure" network such as the Internet to a remote network such as a corporate network. In GPRS, the L2TP tunnel runs between GGSN and a router in the customer network. There are two significant benefits to using L2TP. First, network operators can build an IP network to carry GPRS traffic to their customers in a way that ensures the user's mobile data is completely independent of the distribution network. Second, IP address allocation and user authentication to the mobile terminals is carried out by a local router under full control of the customer (the router which terminates the L2TP tunnel) rather than by the GGSN controlled by the network operator. This arrangement may be preferred by external networks that are particularly security conscious, as they do not then need to trust a third party with their authentication. The operation of L2TP is described in RFC 2888.

Transparent and Nontransparent Access

The GPRS standards use the terms "transparent access" and "nontransparent access" to describe particular combinations of IP addressing mechanisms and authentication.

In transparent access, the network operator provides direct access to the Internet. There is no user authentication, and IP addresses are allocated to the mobile terminal by the network operator, using any of the methods already described. Public IP addresses will be used since access to the Internet is involved. Because transparent access as described does not practice conservation of public IP address space, it is not recommended. The same functionality may be provided using alternative configurations.

In nontransparent access, the network operator provides access either to the Internet or to a private network. User authentication (username+password) is carried out on a server run by the customer network or by the network operator on behalf of the customer network. IP address allocation is done using any of the methods already described. RADIUS or DHCP servers can be operated either by the customer network or by the network operator.

Customer Network IP Address Ranges

With the widespread use of private IP addressing, it is possible that multiple customers could all be using the same IP address range, for example 10.0.0.0. This is not a problem because GPRS has been designed so that the various external networks *can* use the same address ranges without conflict. In practical implementations, there can be minor addressing problems to deal with, like ensuring that the DHCP servers in two external networks connected to the same GGSN do not have the same IP address. These types of detail are completely dependent on the equipment manufacturer's implementation, and each GPRS network operator may solve its addressing issues in different ways, all based upon the same concepts.

The Terminal Addressing Problem

We've established that all GPRS mobile terminals need at least one IP address. Allocating them in advance, however, is not an option. Doing so would rapidly exhaust the available supply, and there are simply not enough addresses in the existing IPv4 stock to meet the needs over the next few years. Some commentators have seen this as a significant problem facing the mobile network industry. To understand why, let's look at the dialup Internet service provider (ISP) industry, and then at some history of the Internet and its methods of addressing.

Dialup Internet Access

Historically, Internet access has required the accessing device—generally a PC with a PSTN phone connection—to have a temporary IP address for the duration of its connection, typically for a period ranging from a few minutes to a few hours. Once the user disconnects, this IP address is released for reallocation to someone else. ISPs, rather than reserve an address for each subscriber, rely on customer behavior statistics. These suggest that no more than 1 in 20 dialup subscribers will want connection at any given time. Therefore, even a large ISP with a million customers may require only 50,000 IP addresses.

Usage Pattern of IP Addresses by GPRS

With GPRS, the connection pattern is expected to change dramatically. Once users have switched their terminals on and are connected to the network, they are likely to stay connected for much longer than a dialup customer would. This behavior is encouraged by the typical GPRS billing package, which charges users according to the amount of data they send rather than the duration of connection. Hence any temporarily allocated IP address will stay allocated to an individual user for a much longer period than with the dialup ISP, and operators will require more addresses to meet subscriber needs.

Number of IP Addresses for Terminals

How many IP addresses will GPRS terminals actually require? Let's break the problem down by first looking at the scale of the problem in a typical GSM/GPRS network, and then extrapolating the findings globally. Taking the example of the O2 network in the United Kingdom, historical behavior shows more than half of all subscribers have their GSM voice phones switched on during the busiest period. Today O2 has over 10 million customers and the number is still increasing. Terminal device (handset) manufacturers have indicated that within two years virtually all their products will have WAP browsers and be capable of GPRS service. Even if only half of the O2 customers (i.e., 5 million) sign up for GPRS Internet services, this means a potential demand of at least:

> 5 million customers $\times$ $\frac{1}{2}$ of them switched on and connected = 2.5 million customers and hence 2.5 million IP addresses required in the busiest period!

Looking globally, the figures from the GSM Association show 455 million GSM subscribers worldwide in December 2000 and 646 million in December 2001. This is a dramatic growth rate (42 percent in 1 year) and the number of subscribers is expected to grow further over the next few years, with an estimate of 846 million customers by December 2002. Applying the O2 experience to the world figures leads to a figure of:

- $\frac{1}{2} \times \frac{1}{2} \times 646$ million = 161 million addresses based on December 2001 subscriber figures
- $\frac{1}{2} \times \frac{1}{2} \times 846$ million = 211 million addresses based on December 2002 subscriber figures

These figures assume one address per terminal, allocated only to terminals that are actually switched on. Consult the figures again:

- Addresses required based on December 2002 subscriber figures—211 million
- Available public IP address—530 million

You could look at these numbers and say "What's the problem? 211 is less than 530!" However, there are many other worthy applications competing for public IP addresses, such as fixed-line Internet customers using PSTN dialup, ADSL, and cable modems.

This point is extremely important. Our intention is not to say whether the number of unallocated public IP addresses is 400, 500, or 600 million or some other figure. Nor is it to say that the number of IP addresses required by GPRS customers will be 200, 300, or 400 million. Rather it is to show that comparing the approximate number of GPRS users with the number of available addresses, and taking into account the fact that there are many other Internet engineering applications which require public IP addresses, suggests that we'll generate a major problem within 2 to 3 years if we proceed on the basis that every on-line GPRS phone requires a public IP address. Historically, forecasts for cellular phone and Internet use have been too low, so we might hit an addressing problem sooner. Neither customers nor network operators want to be in a situation where they are restricted due to a lack of public IP addresses. Therefore some alternative method must be devised and applied by all GPRS operators in order to conserve the available supply.

Tackling the Terminals Addressing Problem

How do we solve the addressing problem? The solution requires that we look beyond the immediate situation and examine the services that will be offered to users. This involves identifying sectors within the user groups and forecasting those services that will be in most demand based on what we know today. As difficult as this may be, it's viable because mobile terminals are much more limited than the PCs presently used by the majority of ISP fixed-line customers.

Summary of Grouped GPRS Users and Applications

GPRS users can be divided effectively into the following groups:

1. Business users accessing their company network via a direct link
2. Users accessing the Internet. This group needs to be further divided:
   a. Those using WAP only. It is expected that the overwhelming demand will be from users in this group.
   b. Those using WAP plus a popular service such as Web or POP3 email, which is compatible with NAT.
   c. "Open" access users, or those whose applications need an open pipe to the Internet. Probably involves using a PC-type device. One example would be someone using VPN encryption services to connect to a corporate firewall via the Internet.
   d. "Internet" service APN users. This is a special service intended only for roaming customers who are visiting another network

away from home and wish to gain local access to the Internet. It is defined in the GPRS standards and provides open access. Technically it is identical to b above.

The addressing scheme for each of these categories is shown in Table 12.1. Segregation between services is achieved using different access points on the GGSN.

TABLE 12.1

Types of GPRS Service and Recommended IP Addressing Scheme

Service	Address Type	Notes
Direct Corporate LAN access	Company	Uses company address scheme (normally private)
Internet access—WAP only	Private	No justification for public
Internet access—WAP with other popular NAT-compatible applications	Private	Private addressing works well with this option
"Open" Internet access	Public	Service is defined as open access without NAT
"Internet" service APN	Public	Used only with international roaming

A typical GPRS terminal can do WAP and only WAP. Therefore, the network operator can't encounter any uncomfortable surprises from these customers. With PDAs and especially PCs, the majority of users will use popular NAT-compatible services and never load an unusual application requiring open Internet access. Users who buy a NAT-based service and then find they need open Internet access must be able to obtain a straightforward service upgrade from the category "Internet access—popular NAT-compatible applications" to "Open Internet access" on demand. The number of users expected to require upgrades is small, particularly since the industry is trending towards applications where mobiles always interface to land-based servers and not directly to each other. The land-based server will be part of the same private addressing scheme as the mobile terminal. (The trend has emerged from the fact that land-based servers can gracefully handle the occasions when a mobile goes out of normal coverage range and cannot be reached. Compare this with a mobile on its own. If the mobile goes out of coverage it is unable to signal this to other terminals or applications and so any communication between them breaks down in an unstructured way).

The segregation technique described in this section substantially reduces the number of public addresses required by GPRS operators with the consequence that the scale of the perceived addressing problem is reduced to something manageable.

Direct Corporate LAN Access

This option provides user access to a corporate LAN for such everyday applications as document retrieval and corporate email access. The network operator will typically provide direct Gi-connectivity (i.e., from the GGSN) to the user's corporate LAN via a leased line. All addresses used within the corporate LAN are provided and maintained internally by the corporation, and this address space is simply extended out to the mobiles. Most corporations use private addressing for their internal network.

Note that some companies provide access for customers via the Internet, using some sort of encrypted tunneling software such as IPsec. Although IPsec software has historically been incompatible with NAT, hence requiring "open Internet access" in GPRS, new NAT-friendly IPsec implementations are now available and so would be compatible with a GPRS Internet connectivity option that featured private addressing with NAT.

Internet Access—WAP Only

In the early years of GPRS service the majority of users are expected to fall into this group. Gi-connectivity is provided to a WAP gateway, with onward connection to local and Internet-based WAP servers. This service can be supported using private addresses, with onward access to the Internet provided either by use of a proxy on the WAP gateway, or with NAT. Use of public address space cannot be justified for this group.

Internet Access—WAP and Popular NAT-compatible Applications

This service can be supported using private addresses, again via use of proxy on the WAP gateway or NAT. All the widely used Internet applications—Web, FTP, POP3, email, and newsgroups—are compatible with typical NAT implementations. In addition, versions of IPsec are now available that are compatible with NAT.

Internet Access—Open Internet Access

This service is defined as providing connectivity to the Internet via an open Internet port without any proxy servers or NAT in the path. Public addresses *must* be used. Typical users are corporate employees accessing a corporate gateway via the Internet, using older (non–NAT-compatible) IPsec VPN software. The number of customers expected to take up this service is currently estimated to be less than 20 percent of the total GPRS users.

Internet Access—Internet Service for Roamers

Technically this service is the same as for the open Internet access group above, but it is used differently. Intended for customers roaming into a visited network who wish local Internet access, this service *must* be served using dynamically assigned public addresses.

Note that this service only applies to the case of "international" roaming—that is, when customers traveling abroad roam onto another network in a different country. It does not apply to national roaming where customers roam beyond their home network coverage onto the network of a "partner" competitor in the *same* country.

The number of customers estimated to want this service is expected to be a small percentage of the total GPRS roamers (currently estimated to be less than 10 percent). The majority of customers are expected to use the standard roaming service to connect back seamlessly via the interoperator backbone to the services offered by their home network, in which case the addressing scheme of the home network is used and the addressing considerations are the same as if the customer was actually in their home network.

Addressing Policy

The GSM Association worked closely with the Internet registries to create addressing policies for GPRS infrastructure and mobile terminals. You can find these policies elucidated in document IR.40, publicly available from the GSM Association Web site (www.gsmworld.com).

Infrastructure Addressing

For roaming to work, as we've argued earlier, GPRS operators must use a common addressing policy. In summary, it says the following:

- Public addressing is used where necessary in the key components of the GPRS network, principally the SGSN, GGSN, and DNS. Elsewhere, private addressing must be used
- Standard Internet registry guidelines, request policies, and procedures are used when requesting public addresses.

Most operators will need to apply for public addresses to use in their core network and, in the majority of cases, will need to run a combination of public and private addresses in their core network. Although this may sound complex in practice it only adds minimal extra overhead.

Mobile Terminal Addressing

The policy for mobile terminal addressing was created in response to two immediate needs:

- GPRS operators needed common guidelines on the use of public addresses and how to requests and obtain the addresses.
- Registries needed common guidelines on how to respond to address request from operators.

The policy that resulted is based on existing Internet registry guidelines, request policies, and procedures, with GSM Association extras. In summary it says the following:

- Only use public addresses where mandatory for the service, or where operators can demonstrate that private addressing is not feasible or practical.
- When requesting public addresses, operators should demonstrate conservative and efficient usage of the requested address space, e.g., by comparing the number of public addresses requested to the quantity of private addresses being used for existing/planned services.
- Share a pool of public addresses among the user community that needs them by means of dynamic addressing.

The guidelines in IR.40 are based around the principles described earlier in this section.

Autonomous System Number (ASN)

Another element associated with IP networking is the autonomous system number (ASN). It enables networks to identify one another when exchanging routing information between neighboring networks.

Use of ASNs is essential on the roaming backbone network that interconnects GPRS network operators. There are potentially two types of ASN that could be used for this purpose: public or private. The *public* ASN range is between 0 and 64511. This range is administered and maintained by the Internet registries. The *private* ASN range is between 64512 and 65535, which provides up to 1,024 values.

The ASN policy on the roaming backbone network is that either public or private ASNs can be used. Operators can use their existing public ASN on the backbone network even if already in use on the Internet, or if they don't already have one they can use a private ASN. Although the private ASN range is similar to private IP address ranges in that an organization could use any value that they wished, given that the roaming backbone will interconnect hundreds of GPRS networks, just choosing an ASN at random would cause complete chaos! Therefore, for roaming purposes the range of private ASNs is administered and maintained by the GSM Association for its members.

In summary, the network operator has the option to use either public or private ASN for their network, the only proviso being that a private ASN must not be advertised on the Internet.

Transition to 3G

The first GPRS services were launched in mid-2000 and started to take off in late 2001, when a wider range of terminals became available. Because GPRS is an overlay on the existing GSM system, most operators could very quickly achieve the same level of GPRS coverage as for their circuit-switched services. In contrast, when 3G is first launched in 2002/3 it will inevitably have restricted radio coverage because it will take time to build the thousands of radio sites needed for full coverage. The first "useful" 3G terminals will be need to be dual mode, operating in 3G mode when in 3G coverage areas and switching over to GPRS where 3G is not available. The nature of the 3G radio system is that while it offers much higher bandwidth than GPRS, a typical site serves a smaller area than a similar GPRS radio site. For network operators it will therefore be more costly to expand 3G coverage than GPRS, and so it is inevitable that the high levels of population coverage that customers want will be several years in coming. GPRS will have to co-exist with 3G for most of its life. Fortunately, 3G was designed with this in mind. It uses exactly the same SGSN and GGSN technology as GPRS but with a different radio system, so the IP-based Gn and Gp interfaces described in Chapters 10 and 11 are compatible. Because of this compatibility it is fairly straightforward to engineer smooth handovers between the two systems.

A prerequisite for 3G-to-GPRS handovers is that the same GGSN must be used by both systems, and must be a 3G GGSN capable of working with both GPRS and 3G SGSNs. The only complication here comes with the level of functionality on the Gn and Gp interfaces. Today's GPRS systems use Release 97 functionality, whereas the first 3G systems use Release 99. Therefore, the 3G GGSN needs to support both Release 97 and Release 99. Even if a network operator upgrades its GPRS system to Release 99, it is likely that some of its roaming partners will still be running on Release 97; therefore it is essential to retain dual-release capability.

As a maturing technology, GPRS can provide data service at lower cost than 3G and will remain attractive to those customers for whom cost savings are more important than high throughput. Thus, GPRS will have a useful role to play alongside 3G for many years.

Transition to IPv6

One of the biggest questions in the networking industry is if and when it will migrate from IP version 4 to IP version 6. The first implementations of GPRS will all be based around IP version 4, so migration is an issue that must be carefully considered by nearly everyone who expects to be doing business a few years from now.

The design of GPRS cleanly separates the IP addressing of the terminals from the infrastructure. Therefore, with one minor exception, the migration issues affecting terminal devices can be separated from those affecting the infrastructure. It is important to remember that GPRS networks will be interconnected to the GRXs and roaming backbone, so this need to maintain connectivity to roaming partners must also be considered as a key element of the infrastructure transition strategy.

When?

When will migration to IP version 6 start? Nobody knows for sure. On the one hand, we still don't have a full range of IPv6 products available, so operators don't have any choice except to go with IPv4. On the other hand, there aren't any overriding reasons to migrate to IPv6, so operators are not putting pressure on manufacturers to deliver IPv6 products. Without the demand there won't be an early supply of IPv6 products. Similarly, the absence of IPv6 services to purchase means that no great demand for IPv6 products has developed. And so the cycle continues....

One likely scenario lies with future releases of 3G. Release 5 of the 3G standards—finalized in late 2002—mandates IPv6 for the IM (interactive multimedia) domain. Therefore when products supporting these standards are eventually deployed, perhaps from 2004 onwards, the vicious circle will be broken. Since 3G "falls back" to GPRS in areas where there is no 3G coverage, GPRS must have acquired v6 capability by that point—otherwise the users' services will completely stop instead of handing over. Although some IM products may not work well on GPRS because of its lower bandwidth and higher latency, users will still want to continue using less performance-critical services while on GPRS. Thus, it is likely that 3G IM will lead to a critical mass of IPv6 services on GPRS and this will in turn create the demand for IPv6 in other areas, causing a snowballing effect towards universal IPv6 deployment. Key barriers to this are the extra complexity of IPv6 and the need to build a base of skills, experience, and IPv6-capable equipment.

IPv6 Features and Benefits

In the GPRS mobile environment, many of the built-in features of IPv6 such as autoconfiguration, QoS, and security have already been adopted in their IPv4 flavors, so IPv6 offers small advantage. IPv6 does offer vastly increased address space, but nobody knows for sure when this will be needed. The techniques described earlier show how proper allocation and use of IPv4 public addresses can alleviate the pressure mobile terminals place on the system. In turn, we can expect to see the pressure on the available IPv4 address space reduced, and the drive to migrate to IPv6 will be less urgent.

With the increased address space of IPv6 comes a problem we need to tackle. Suppose there are sufficient addresses available to allow any IPv6 device to be placed on the Internet. With Internet connection comes a much higher susceptibility to malicious attacks or probing. This is already a problem in the "always-on" world of ADSL and cable modems, and even crops up with long-duration PSTN access. Note that users of always-on GPRS services often pay for service by volume of data sent, and such users may end up paying for attacks. This problem is avoided with those IPv4 designs implemented using NAT in conjunction with basic firewalling techniques.

Impact on Terminals

When the IPv6 migration starts, it must build up from a base of zero and exist side by side with IPv4 for a fairly long period. IPv6 terminals will need to access, for example, Web servers running on both IPv6 and IPv4. Consider that a Web page consists of multiple individual objects that are fetched separately by the terminal and only built into a single page once received and assessed by the terminal. Potentially, the objects will reside on a mix of IPv4 and IPv6 servers and will therefore need to be fetched with a mix of methods. For example, a customer with an IPv6 Web browser may access a Web page located on an IPv4 server, but where the page displays an advertisement from a company operating an IPv6 environment. There are several alternative ways of doing this:

Dual stacking. This means having both IPv4 and IPv6 capability in the terminal. In the GPRS mobile world this is achieved by having a terminal capable of both IPv4 and IPv6 contexts, either separately or simultaneously. To match this requires parallel pairs of APNs—one each for IPv4 and IPv6. If a user wanted to access both IPv6 and IPv4 Web sites in the same session he would need to access both IPv4 and IPv6 APNs simultaneously by activating both types of context at the same time.

Tunneling and interworking. The terminal activates just an IPv6 or IPv4 context, depending on the capability of the network. If it has an IPv6 context and needs to access an IPv4 system, it tunnels the information over IPv6 and it is then "untangled" by an interworking device in the IP network. Similarly, if it has an IPv4 context and needs to access an IPv6 system, it tunnels the information over IPv4 and it is then untangled by a different interworking device in the IP network.

Application-specific interworking.   The terminal would activate only an IPv6 context and communicate with, say, a Web proxy server for Web access. If the user wanted to access an IPv4 Web site, the problem would be sent to the proxy server to sort out over IPv6, and only the proxy server would have to actually communicate using IPv4. The results would be sent back to the terminal over IPv6. A similar principle is used if the terminal can only activate an IPv4 context, in which case if it needs to access an IPv6 Web site, it sends the problem to the proxy server to sort out via an IPv4 connection.

Figure 12.5
IPv6 terminal accessing information on both IPv4 and IPv6 servers.

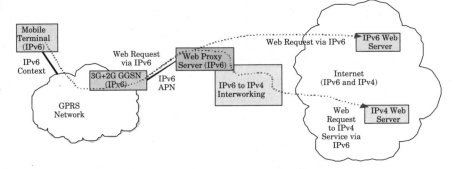

GPRS roaming adds to the complexities. If a user roams on to a network that doesn't support IPv6 PDP contexts, what should operators do? Customers won't accept loss of service, so the only alternative is to provide a service over an IPv4 PDP context.

It is presently early days in the industry for considering how to migrate to IPv6, and there is no consensus over which approach to adopt. Meanwhile, GPRS network operators are remaining with IPv4 and watching developments within the telecommunications and Internet communities.

Dual Stacking

Dual stacking (i.e., dual context types in a terminal) has some advantages:

- Straightforward to understand and implement at the network level
- Able to support any application
- Able to roam on to a network that doesn't support IPv6 context types

And likewise a number of disadvantages:

- Both contexts need an IP address. If the IPv4 context needs to be activated the terminal is still contributing to the use of IPv4 addresses (which could be a public address), nullifying one of the main advantages of migrating to IPv6!
- The applications on the terminal become more complex due to the need to decide whether to send their traffic to the IPv4 or IPv6 context.
- Increased use of network resources due to the need for two PDP contexts per terminal. For the network operator, this increases implementation costs because more SGSN and GGSN capacity will be required.

Tunneling and Interworking

Tunneling and interworking mean carrying IPv4 traffic over an IPv6 network by the process of tunneling as far as an interworking server to break out the IPv4 traffic, or the converse—carrying IPv6 traffic over an IPv4 network. This approach has the following advantages:

- Able to support any application
- Only requires one PDP context—IPv6 or IPv4, depending on the network capability
- Minimizes the number of IPv4 addresses required when an IPv6 context is used

Its disadvantages are:

- The applications on the terminal become more complex due to the need to decide whether to send traffic in IPv6 native form or tunnel the IPv4 over IPv6
- The land-based IP network needs to provide interworking equipment

Application-specific Interworking

Application specific interworking has the following advantages:

- Simple to implement on the terminals
- Only requires one PDP context—IPv6 or IPv4 depending on the network capability

Its disadvantages are:

- The land-based IP network needs to provide the application specific interworking equipment
- Minimum number of IPv4 addresses required
- Only feasible to support major applications such as Web, email, and some types of messaging/chat

Roaming Considerations

Many of today's GSM customers make use of roaming when they are away from home. GPRS customers will similarly take it for granted that they can access their usual services via roaming. The introduction of IPv6 adds some interesting problems to be solved. Once the first GPRS operators have implemented an IPv6 PDP context capability, it will inevitably take several years for all operators to upgrade their systems to include a similar capability, despite competitive pressures. Consider the plight of a customer with an IPv6-only terminal who finds himself in a country where only IPv4 services are available. For these circumstances, terminal devices must provide a dual-context capability simply to ensure access to GPRS services. In this situation—an IPv6-capable terminal in an IPv4-only network—it is likely that only a subset of the customer's normal services could be available. To make IPv6 services available, terminals would need to tunnel IPv6 traffic over the IPv4 network to an interworking server. What we see here is effectively the reverse of the tunneling and interworking scenario already described. But there is an option: application-specific interworking could also be used, and it makes things overall much simpler.

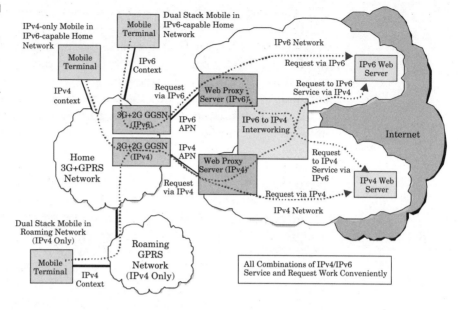

Figure 12.6
How IPv4 and IPv6 in GPRS can operate together via interworking.

Core Infrastructure

Although IPv6 terminals may be introduced and SGSNs and GGSNs upgraded to support IPv6 PDP contexts, there is still no compelling reason to introduce IPv6 in the core of the GPRS network. The core consists of the SGSNs, GGSNs, DNS, and the IP routers that interconnect them. Each operator's core is relatively small and few would benefit from any of the IPv6-specific features. The situation is different on "Gi" networks, i.e., those networks to which the GGSNs provide access. We've suggested that IPv6 is most likely to appear first in the 3G IM domain. It can also appear inside other Gi networks without affecting either core or terminals, so long as interworking is provided.

IPv6 will probably start to appear in the core when general migration has begun in the business world and we've accumulated both the experience and the products to make it viable. At that point new capabilities will be introduced. From a standards perspective, that point is likely to occur from 3G release 5 onwards. With IPv6 in the core, the interoperator backbone, GRX, and APN DNS must all be upgraded to support IPv6 or roaming connections between networks will not work properly. Backward compatibility is mandatory so that operators who haven't yet deployed IPv6 can continue with their IPv4 network unaffected.

The exception in the infrastructure where IPv6 *must* be deployed is the Gi interface on a GGSN interfaced to an IPv6 network. IPv6 may also be deployed in the GRX network. At this time, work is underway by the GPRS operator community to determine the merits of such a move away from IPv4 to IPv6 for the GPRS roaming backbone network. The advantages are marginal, and there are complexities in implementation, but the opportunity does exist to bring IPv6 to this segment of the GPRS network configuration.

Summary

The introduction of IPv6 creates some practical difficulties. These are not just due to the need to gain experience in the new technology of IPv6, but to the complexities of interworking with IPv4 and the need to ensure that services will be accessible irrespective of what type of terminal a user has. The various migration and implementation scenarios need to be considered carefully by operators and suppliers in order to adopt the best approach.

The authors consider that for GPRS terminal devices, the optimum approach is a combination of application-specific interworking for popular applications like Web, POP3 and IMAP4 email, and instant messaging, and tunneling and interworking for other applications. To ensure full compatibility, operators will need to implement both IPv4 and IPv6 versions of APNs for access to popular Internet services. Inside the operator's core network there is no justification for upgrading to IPv6 until such time as new equipment requiring IPv6 is installed.

References

RFC 1918 "Address Allocation for Private Internets," February 1996, Y. Rekhter, B. Moskowitz, D. Karrenberg, G. J. de Groot, E. Lear.

GSMA IREG PRD IR.40 "Guidelines for IPv4 Addressing and AS Numbering for GPRS Network Infrastructure and Mobile Terminals," August 2001, J Malra & K Fullbrook.

"Tackling the Mobile Addressing Problem, A White Paper on IP addressing for GPRS Mobile Terminals and the Implications for Network Operators," August 2000, K Fullbrook.

CHAPTER 13

Access Point Names

Carsten Otto

T-Mobile

If you've read this far in Part 3, you already know that numbering and addressing in the GPRS world is not easy. In order to save mobile operators from reinventing the wheel, the access point name (APN) was based on the naming conventions of the Internet from the first.

Services

An APN is the logical way to name a GPRS service that a subscriber wishes to use because it uniquely identifies the subscriber's location in the application space. For roaming, the service structure, and therefore the associated name, can be divided into two groups:

- Services that are offered through local access
- Services that are offered through home access

An example of a *service offered through local access* is basic Internet access. Just as a roaming GSM user currently uses the local operator's telephone network for making voice calls, a GPRS user would use the local operator's Internet access for surfing the Web. This arrangement is comparable to accommodations made for an Internet user visiting an Internet cafe in another country. The quality and availability of the Internet connection at the cafe determine the user's satisfaction. In the case of a GPRS user, the local operator is responsible for the satisfaction of any GPRS user roaming in its network.

An example of a *service offered through home access* is corporate intranet access. On the public Internet, the required level of quality and security cannot be satisfied via current service levels and encryption solutions. They can only be assured through virtual private networks (VPNs). The GPRS Roaming eXchange (GRX) offers the perfect solution for intranets or any personalized information. Any request from a roaming GRPS user will be forwarded to the user's home network, and the user will have the same experience as if she were actually in her home network. In our Internet cafe example, this solution is just like letting the user take his entire computer abroad and logging on to the Internet just as he does at his desk. Without, of course, actually moving anything in the way of computer equipment.

Because of the obvious complexity and obvious advantages of the GPRS Roaming eXchange, most operators focus on the second solution. They offer their full range of services through home access. This is the first step to a *virtual home environment*. The concept allows users access to their personalized services anywhere, at any time, in the same way they are accustomed to using them. At the same time, it enables operators to serve their users with customized solutions and therefore to concentrate on their core competencies.

Functionality

The access point name has the same functionality as an Internet domain name (e.g., www.3gpp.org). As an Internet domain name refers to a page on a specific server, an access point name refers to a specific connection point, the gateway GPRS support node (GGSN), of a specific GPRS network. In the Internet, the domain name server (DNS) is used to translate a domain name into an IP address, and in a GPRS network the domain name server is used to translate the access point name into the IP address of the GGSN. Therefore we have the option of using easily understandable alphanumeric access point names instead of hard-to-remember numeric IP addresses. The detailed structure of an access point name is described in the following section.

Both the home location register (HLR) and user's mobile handset store the APN. The HLR contains all GPRS subscription data, including all access point names and roaming destinations allowed. Corporate services, for example, will obviously be accessible only for a special user group, and some roaming destinations might not be available for certain services or to the users of prepaid cards.

Here are the technical steps of accessing a GPRS service with an APN:

- The mobile user activates a service on his handset.
- The handset immediately sends a request message to the GPRS network's SGSN.
- The SGSN compares the requested service against subscriber data in the home location register to check whether the service has been authorized for this user.
- Then the SGSN tries to acquire the IP address of the GGSN supporting the requested service from the local GPRS domain name server.
- The request will be rejected if it cannot be found in the local GPRS domain name server and the user is within his home network. If the user is roaming in another network, however, and the SGSN can't

find the corresponding IP address in the local GPRS domain name server, it will query the user's home network domain name server using the APN operator identifier.

- As soon as the IP address of the GGSN is located, the SGSN will connect the user to the service by creating a GTP tunnel to the GGSN.

Subscriber data in the HLR may also contain the so-called wild card APN (*). This functionality either provides a default access point name or allows access to any access point name (and therefore any service) that is administrated and known in the network. The GGSN, on the other hand, can restrict access by blocking any user who is using wild card functionality.

Structure

Just like an Internet domain name, the access point name consists of a combination of labels separated by dots. These labels can be divided into two main parts:

- APN network identifier
- APN operator identifier

As the names suggest, the APN operator identifier specifies the operator's GPRS network while the APN network identifier specifies the connection point (GGSN) in this GPRS network. The APN network identifier is also called the *service indicator*, because it is used to access the service associated with the GGSN in question. The APN operator identifier is an *optional* part of the access point name. Only the APN network identifier is mandatory. But a unique APN operator identifier must be created to handle the scenario in which different operators choose the same name for completely different services. (That would still work for our example of worldwide services, but these hardly exist anyway.) An APN operator identifier is also used during roaming to link the service to a certain home network. The GSM Association has established an internal naming convention for assigning access point names, and these definitions coordinate the complex network of hundreds of potential future roaming partners.

The network identifier consists of at least one label. Labels are not case sensitive and should contain only numbers, alphabetic characters, and the em-dash. To ensure uniqueness, GSMA recommends using existing Internet domain names registered by the operator (e.g., t-mobile.de). In the following example, T-Mobile Deutschland (Germany) and VoiceStream Wireless (United States) use a combination of service label and Internet domain name to achieve unique network identifiers:

Figure 13.1
Structure of the APN network identifier.

The first label in each case, *internet* and *wap*, stands for the service offered to the user. The second label, *t-d1* and *voicestream*, is an abbreviation of the operator name used in the national registered Internet domain name. It is common to use a range of names to reduce typing mistakes and accommodate a variety of brands (e.g., t-mobile, tmobile). The third label, *de* and *com* respectively, specifies the country of the mobile operator, as it is used in its Internet domain names. Please note that in some countries the common Internet domain names consist of more than two labels: xxx.co.uk in the United Kingdom, for example, or xxx.com.sg in Singapore.

An APN operator identifier is also made up of three labels: the mobile network code (MNC), the mobile country code (MCC), and "gprs." The mobile network code and the mobile country code are internationally allocated numbers found in the international mobile station identity (IMSI). Using the previous examples, the operator identifier for T-Mobile Deutschland (MNC=001; MCC=262) and VoiceStream (MNC=026; MCC=310) would be as follows:

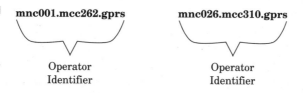

Figure 13.2
Structure of the APN operator identifier.

Using our two examples, the full access point name of T-Mobile Deutschland and VoiceStream Wireless are as follows:

Figure 13.3
Full APN for T-Mobile
Deutschland.

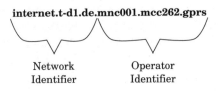

internet.t-d1.de.mnc001.mcc262.gprs

Network   Operator
Identifier   Identifier

Figure 13.4
Full APN for
VoiceStream Wireless.

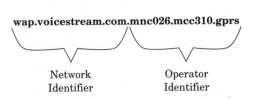

wap.voicestream.com.mnc026.mcc310.gprs

Network   Operator
Identifier   Identifier

Remember that the APN operator identifier is not mandatory. It will only be requested if the user is roaming in other networks. Typically, users will only have to know the APN network identifier, which should be easy to remember, since the user is already familiar with the brand name and Internet address of his operator.

Value

Apart from the obvious technical necessity of an access point name, it offers several benefits and hidden values that should not be underestimated. In the minus column, GPRS' often cited "always-on" capability creates new challenges for billing which have to be acknowledged.

The main value of the access point name is the possibility to differentiate services. Operators can use APN network identifiers for different services: hence WAP services use wap.voicestream.com and Internet services internet.t-d1.de. They could base their billing on access point names by taking a monthly charge for each different service the user requests. Additionally, the subscription check of the access point name adds a further layer of security for corporate clients with their dedicated company access point name. This method of billing is also very familiar for the user, because it is comparable to the pricing models of cable TV. GPRS users would pay an additional amount on top of their current monthly fee for their WAP package, their sport channel, or their corporate intranet access. On the other hand, this method cannot be applied to every single service, since the number of access point names is limited and should be limited to avoid confusion.

Billing based on access point names also yields new branding possibilities. Since most GPRS users will be comfortable with the Internet already, it will be easy for them to understand the structure of access point names. Service and operator branding can be underlined with access point names.

Last but not least, it is important to mention that if the access point name is used for differentiation, billing, and branding, all the usual business and marketing rules apply. Access point names should be integrated into all marketing and branding activities. As important as it is to use a number of APNs to achieve differentiation, it might even be more important to stop short of overbranding, and creating unnecessary confusion through too many services or very complicated names.

CHAPTER 14

GPRS Security

Charles Brookson

CEng FIEE AFRIN

"Quis custodiet ipsos custodes."—"Who will guard the guards?"

Juvenal's *Satires*, circa 120 AD

GPRS offers a number of security enhancements to operator and customer. The standards themselves furnish technical security options for network operators, such as authentication, anonymity, and encryption of the data and some of the signaling. Beyond these features, there are reasonable third-party and industry-developed security solutions. Much more important than the underlying technical features, however, is the corporate will and workflow to ensure that they are used correctly (or even used at all!), and that all the other aspects of good security are also put in place.

Naturally, since the only secure system is one that is never turned on, and since our goal is to have a working system, both the operator and users of GPRS must decide what security measures they will use against attacks and frauds. These measures will have to be based on a risk analysis of the security threats, and the operator should attempt to identify cost-effective solutions to the various security issues.

Security Features of GPRS

The technical security set offered by GPRS is very similar to that offered by GSM:

- **Identity confidentiality**—Customer identity is protected from eavesdroppers on the radio and path to the serving node.
- **Identity authentication**—The customer proves his identity to the network by the use of a challenge and response.
- **Confidentiality** of user data and part of the signaling (i.e., signaling between the mobile and the GPRS serving node), where the information is protected by the use of encryption to make it unreadable by an eavesdropper.

In addition to the sorts of security defined by the GSM standard, GPRS also secures the backbone.

The inner workings and details of these features are contained in the 3GPP standards,[1] but an overview is useful for those who are planning

---

[1] 3GPP TS43.020 3rd Generation Partnership Project; Technical Specification Group Services and System Aspects; Security Related Network Functions. *Note:* This was called GSM 03.20.

or embarking on implementations. Here's a summary of GPRS security services.

Identity Confidentiality

Identity confidentiality provides privacy to the subscriber by making it difficult to identify him from his signal over the radio or his connections to the SGSN. Good security practice avoids using the customer's unique international mobile subscriber identity (IMSI) wherever possible, and instead assigns a temporary identity known as a temporary logical link identifier (TLLI). The TLLI is accompanied by a routing area identity (RAI) to rule out ambiguities caused by a duplicate TLLI. The actual relationship between the TLLI and IMSI is held within a database in each SGSN.

When possible, signaling encryption is also used to protect dialed digits and addresses, as these can also betray a customer's identity and compromise the call's anonymity.

Identity Authentication

Identity authentication is specified in the GSM standard.[2] Authentication is performed in the SGSN, where pairs of random numbers and signed responses obtained from the home location register are stored. As we'll see shortly, the SGSN uses a challenge-and-response procedure to determine whether the smart card to be identified has the correct authentication algorithm for A3 and the correct key Ki.

User and Signaling Data Confidentiality

The user data and signaling key GPRS-Kc are derived by using function A3/8, as in GSM. Derivation yields a 64-bit key called GPRS-Kc. Synchronization depends on a ciphering key sequence number, GPRS-CKSN, which is also described in the standard.[3]

GPRS ensures synchronization by requiring the encryption algorithms to be driven by INPUT and DIRECTION bits.

---

[2]42.009 Security Aspects. *Note:* This was called GSM 02.09.

[3]44.008 Mobile radio interface layer 3 specification. *Note:* This was called GSM 04.08.

The Algorithms

Seven GPRS encryption algorithms (GEAs) are allowed for in the GPRS specifications, of which two (GEA1 and GEA2) have already been defined by ETSI SAGE. GEA2 was defined about a year later than GEA1 and was able to improve upon it in part because of the easing of export control legislation. Another algorithm (GEA3) based on the 3GPP algorithm, Kasumi, was developed in July 2002.

In use, the mobile and the network negotiate the GEA at the start of the call. New authentication and encryption algorithms were anticipated in the design of GEA, and can be introduced within the lifetime of the standard via considerations similar to those in GSM for the A5 algorithm.

The Authentication Process

All of the above is easier to grasp if we trace what actually happens in the security system from the time a user initiates a message to the time he is admitted to the network.

Figure 14.1
Authentication and key derivation.

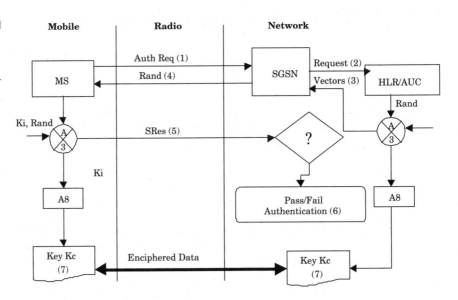

By reference to Figure 14.1 above:

1. The mobile station (MS) sends an authentication request to the network, which arrives at the SGSN and is sent on to the...

2. Home location register (HLR) and authentication center (AuC). The AuC then generates *triplets* consisting of a random number (RAND), signed response (SRES), and encryption key GPRS-Kc. The first two are used as a challenge and response to authenticate the smart card in the mobile station (or more precisely, to ensure that the card has the right key Ki associated with the IMSI).

The key Kc is used to encrypt all the data between the MS and SGSN. GPRS-Kc and SRES are calculated from RAND using the authentication algorithm A3/8. This algorithm may be unique for each operator, because only the derived triplets are transmitted during roaming and other operators need not know the authentication algorithm itself.

3. The triplets of vectors (RAND, SRES, and GPRS-Kc) are sent to the SGSN, which next sends RAND on to...

4. the MS. When it arrives, the MS (and smart card or SIM) uses the same authentication algorithm, A3/8, to calculate SRES and GPRS-Kc. SRES is returned to...

5. the SGSN, which compares the SRES it has just received to the SRES in the authentication triplets.

6. If comparison shows them to be identical, then logically the MS must have the correct authentication algorithm A3/8 and Ki, and therefore is judged to be genuine.

7. Both the MS and the SGSN also have GPRS-Kc, and use this key to encipher the session between the MS and SGSN.

It is interesting to note that if the MS does not have Ki or the authentication algorithm, it cannot calculate Kc, and the encryption will fail. This should prevent someone hijacking a call in progress, as there is "implicit authentication" throughout the session, assuming that no other party can derive Kc. Not all the data on this link will be enciphered, since some of it, such as the routing area update request information for the call, will need to be exchanged between the MS and SGSN.

End-to-End Security Connections for Services

GPRS does not itself offer end-to-end security but many other standards, such as those from the WAP Forum,[4,5] do specify methods of end-to-end encryption compatible with GPRS. Other methods, such as the

Internet security protocols covered by IPSec, may also be used.[6] Organizations have the additional option to set up virtual private networks to be able to access company information over encrypted links.[7]

Some of these techniques do not protect information from the operator or user, so if you are a third party offering valuable services, you will want to devote some critical study to the various possibilities for a scheme that suits your application.

If end-to-end encryption methods are used, a user may have different identities for different services—like shopping, banking, or booking a golf tee time. This is so that if one identity is compromised, it does not compromise others: the identity that accesses your bank account must clearly be separate from the identity that sets up your next golf game. Note that these several identities will probably be stored in a terminal, and therefore will need some form of protection themselves.

Network Security

A GPRS network is based on an Internet packet system. As such, it displays all the same security vulnerabilities as any IP network, including the Internet.

In designing the network, you will need to take care to ensure that sensitive parts of the infrastructure (such as billing data, encryption keys, and identities) cannot be compromised. There may also be legal considerations, such as data privacy or lawful interception, that may require you to include encryption, access control, or interception within the infrastructure.

Pay particular attention to key management issues such as transportation and storage of keys. An attacker who manages to uncover, say, the individual user key Ki, may be able to clone a terminal or service, and bill to it.

---

[4]WAP Transport Layer End-to-end Security WAP-187-TLE2E-20010628-a from www.wapforum.org.

[5]WAP Transport Layer Security WAP-261-WTLS-20010406-a.

[6]See the Internet Engineering Task Force, www.ietf.org.

[7]See Certicom on www.certicom.com.

The Internet

GPRS is specifically designed to connect users to the Internet. That means that gateways and serving nodes—not just customers—will need Internet addresses, and other networks will need to know what they are. Easily available network mapping software can be used to look at network configuration, and it's a good idea. By way of example, remember that terminals in an operator's network are inside the firewall; unless the network is properly configured, they may be able to access infrastructure and let customers access sensitive information on the network.

Figure 14.2
Internet security issues.

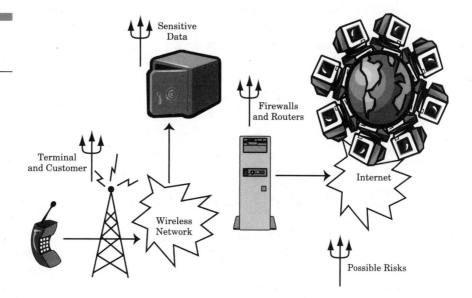

Figure 14.2 illustrates some of these concerns. Customers' terminals are routinely threatened by rogue code in the form of viruses or Trojans. Firewalls and routers must be kept updated to stop intrusion, and sensitive information such as encryption keys and customer information, must be protected. Finally, network information flowing over the Internet must be secured.

All the attacks to which the Internet is susceptible are also possible in GPRS networks. These include denial of service attacks against network nodes and customers, and hence the threat of compromise to both customer and network data stored on the terminals and infrastructure. Standard preventive measures for GPRS will include firewalls, router control, address translation, and hiding and encrypting sensitive data.

In effect, the operator will now have to become an Internet security specialist to deflect the attacks and threats that can be expected. Operators will also be tasked with updating the software and firewalls continuously to protect the integrity of the network and the information it carries. Software version control will become more important, as will the ability to test new hardware and software before bringing it into service on a live system.

Customers have the right to be fully informed about the specific new risks they face when downloading untrusted software and updates to their terminals, browsing untrusted sites on the Internet, or opening email attachments. As the nature of the operator's business changes, we can expect to see some changes in the relationship between operator and customer. Effectively, wireless operators will become Internet information service providers and will have to partner with the customer to identify the risks and reasonable countermeasures acceptable to both. Customers will have to be coached on how to secure their terminals against possible viruses and attacks, and operators must help them by making updates to software and services as seamless as possible to support any extra security functionality required.

Viruses and Trojans

Since GPRS terminal equipment will be probably "always on," and will also be equipped with software like Internet browsing and email services, it will be continuously open to attacks from viruses, Trojans, worms, and so on. Smart applications and software that can execute computer code are increasingly attractive to the virus writers of the world, and they allow for automatic download of code to terminals. Such software can perform tricks analogous to common Internet attacks—monitoring usage, downloading files, making calls unknown to the user, and more. Expect to see a range of anti-virus software developed to deflect the hacker (some initial offerings for handheld PDAs show the direction we can expect the technology to take). Sadly these hacks are particularly attractive because perpetrators now have the capability to generate revenue from them; we'll look at how they're doing this in the upcoming section on potential frauds.

If terminals have the ability to update their operating systems (and we envision that some will), they can be tampered with to perform functions unwanted by the authorized user. However, this is more likely a problem for the future, with software radios and other devices using software managed by computer.

Cable modems and other always-on devices have fostered an increase in certain kinds of attack, one example being the infamous Back Orifice, which lets someone run your computer remotely without your knowledge. GPRS terminals, as previously indicated, are in this category. You will need to implement firewalls or similar technology to protect terminals from this type of incursion, and even to provide protection from attacks from other terminals on their own network!

General Security Considerations

One of the most important parts of corporate security is ensuring that all of an organization's security mandates are defined and handled correctly. This generally entails setting up an internal unit responsible for overseeing security, and investing it with sufficient authority from the management board to be effective. The areas to be covered by the security overseers are multilayered. They include:

- Policies
- Information
- Electronic security
- Business security
- People security
- Physical security
- Investigations
- Fraud

Discussing each of these areas in any depth will take us well beyond the scope of this chapter. Readers who want to consult a useful guide can obtain the International Standards Organization (ISO) standard called "Information technology. Code of practice for information security management."[8] This document addresses:

- **Compliance**—To ensure the systems and processes comply with legal requirements, such as data privacy regulations.

---

[8]BS ISO/IEC 17799:2000, BS 7799-1:2000 Information technology. Code of practice for information security management.

- **Personnel security**—To ensure personnel are adequately informed and educated in the use and management of security.
- **Security organization**—To ensure that adequate resources and policies exist for the education, ownership, and outsourcing of security.
- **Asset classification and control**—To ensure that information and systems are appropriately marked and protected.
- **Business continuity planning**—In case of interruption, to ensure that plans and procedures are in place that enable business processes to resume and continue.
- **Security policies**—To have supporting policies in place for information protection.
- **Computer and network management**—To provide adequate protection of network assets and computers within the business.
- **System access control**—To manage access to information and electronic resources throughout the business for authorized users.
- **System development and maintenance**—To ensure secure development of software, and the use of authentication, privacy, and integrity techniques.
- **Physical and environmental security**—To protect technology assets and information from physical harm.

Fraud Issues

Fraud exists where there is monetary gain to be made. It is difficult to predict what frauds will prove practical in GPRS, as they will depend on variables such as:

- Charging mechanisms
- Billing mechanisms
- Services offered
- Technical weaknesses in the processes
- Business weaknesses in the processes

There are quite a few ways of gaining revenue from frauds, and this bounty may make GPRS attractive to a fraudster—telecommunications crime is a very big business. For example, if the service provider charges customers for content when they view a page or use a service, then the provider may be tempted to load rogue software into customer terminals that redirects them toward unrequested content. In another example, a rogue service could prompt customer terminals to connect to premium rate or infoline services, where the provider typically gets a cut of the revenue generated. Software could also be introduced to seize the identity of the terminal user for malfeasance in banking or e-Commerce services.

It is possible to develop fraud detection systems (FDSs) that will notify you when a fraud is taking place. But to do this you must have first found the fraud and identified its "fingerprint" as a series of rules (for example, short calls to certain addresses). So all FDSs share a common weakness; namely, they can't recognize a particular kind of fraud until it has already happened, been detected, and patched. Far better to build protection into the process itself, for obvious reasons! So far, it's none too clear how an FDS will behave in a packet-based system (development work is just starting in this area). Some FDSs can try to work out the rules on their own, by identifying patterns that may predict a fraud. Probably a combination of rules-based detection and learning is the ideal solution.

Interoperator billing in GPRS networks will be based on either the duration of the connection or the amount of data sent. This will make it harder to ascertain the types of fraud. For example, GPRS billing does not require the destination of addresses used to be transferred between operators.

Another likely concern is the use of dynamic addressing, where an Internet address is allocated to a user from a pool when it is required, and returned when it is no longer in use. This will make it hard to identify a user, especially when the link between the user and address is brief.

A Final Word

When developing a product or service, make sure to study the whole business process for the potential of misuse. GSM systems in the past have suffered from problems such as prepaid subscriptions based on insecure mechanisms. The same will be true of GPRS. It is predictable that GPRS will experience more security risks than GSM, if only because it connects to the Internet with its millions of other users. So GPRS can be only be secure if you have thought through the issues, taken the right precautions, and put management and technical systems in place to activate them.

It's the Application, Stupid

As these pages have already confirmed, GPRS does nothing. It provides no service, does not open doors, or leap tall buildings in a single bound. However, it does facilitate such activities—well, perhaps not the leaping of tall buildings—and is therefore known as a bearer service. A better definition still is to think of GPRS as a facilitator, something that enables other things to happen. Because GPRS is a bearer, when the technology is implemented it basically just sits there, waiting to enable something else—like a motorway waiting to transport vehicles.

This is a vastly different matter from the GSM technology that underlies GPRS. GSM was developed for a specific purpose—to carry voice calls. It is largely a single-application bearer and therefore, from the moment GSM is first activated, it is ready to do something—namely process and deliver voice calls. Everything required to make a call exists within the basic technology. That's not the case for GPRS, which has the ability to carry a vast array of applications.

Here is where applications and services come in. A GPRS service provider must decide what applications and services to add to the GPRS bearer in order to bring the technology to life and give value and meaning to customers. It is in the choice and provision of services that mobile operators will really begin to compete, and where GPRS becomes an exciting technology. While it is generally accepted that most, if not all, GSM operators will install GPRS networks over their GSM service platforms, the services they choose to develop and offer will be a true test of their marketing abilities.

Historically, factors such as distribution, coverage quality, and even retail pricing have differentiated wireless operators. With the introduction of GPRS, that relatively simple marketing matrix is blown wide open and complicated beyond recognition by the sheer range of choices. Buried within this challenge is the quest for the Holy Grail, a "killer application." Some people simply don't believe such a thing exists. Some believe it has yet to be developed, while others are putting their faith in basic Internet access. Then there is a further group that does not think the Holy Grail is in the packet data field at all but is in fact in the basic GSM technology, where it is disguised as wireless voice telephony. And already exists.

It is easy for pundits to rationalize that GPRS' success will be limited without a killer application. To identify a killer application is to deliver certainty into a world of uncertainty. Therefore, their belief may well derive more from the fact that the future is too difficult to predict than from an iron-clad belief that GPRS can succeed or fail on the back of a single application. But that takes us back to trying to forecast the future market for technology, something that is notoriously difficult and is epitomized by the oft-quoted Thomas Watson of IBM who predicted that the worldwide demand for computers would be "five." Today, we all know that five is considered the absolute minimum number of computers required by a family of four to survive, so long, that is, as they are networked together and connected to the enormity of the Internet via a high-speed link.

Clearly, Mr. Watson erred on the conservative side, but the mere existence of the PC did not guarantee its incredible success. Indeed, it's possible that the explosive growth of the personal computer never would have happened without the advent of inexpensive mass-marketed devices.

GPRS may be facing a similar scenario right now. Without such devices running a plethora of applications tailored to specific consumer needs, GPRS may be limited to the corporate sector. This notion is all the more important because the early forays of mobile telephony into data hardly set the imagination of the masses alight.

The development and launch of wireless application protocol (WAP) was originally heralded as the beginning of the mobile Internet. Widely hyped, the service is generally viewed as falling well short of expectations. Where were WAP's shortcomings? First, it was rolled out with great fanfare by device manufacturers and operators alike. But the buying public quickly determined that WAP was too complicated and, in some cases, too expensive to use on a regular basis. Second, the devices were incapable of delivering Internet content that was in any way comparable to that which users were accustomed to seeing on laptop or desktop units. Add to that the fact that the set-up time was slow and the connection circuit switched. Every delay was highlighted by per-minute billing, and customers generally felt that WAP wasn't worth the effort. The value proposition didn't meet the needs of the mass market.

Lessons learned from WAP were hard but very helpful to the launch of GPRS. The technology is not being launched by operators until it has been fully tested and the end-to-end service functionality is better understood and communicated. Less hype and more meat is the order of the day; no black-tie launches. GPRS service providers are also careful not to let customer expectations exceed technology capabilities. While this was not always the case (particularly in terms of speed of packets), reality is slowly creeping into the GPRS marketing campaigns of today. Like elephants, wireless customers have long memories and WAP has provided what some consider to be a benchmark of how not to launch a new service. GPRS operators today have the opportunity to leverage these lessons in order to introduce services that the buying public will embrace. The new mantra is *underpromise and overdeliver*. Thank you WAP.

Although WAP has been knocked down more times than the average opponent of Mike Tyson, it has not been counted out. Indeed, WAP Release 2 promises to rectify many of the issues that have plagued the protocol to date. Customers will benefit from the migration to xHTML, which better mirrors the Internet mark-up language used by application developers. No longer will Web sites have to be rewritten to allow access from WAP devices. Many also believe that WAP benefits from the move from circuit- to packet-switched transmission that comes with GPRS. So, should the industry wait until WAP Release 2 is widely available before beginning the move to GPRS? Most of us believe that this would be a huge mistake, because to wait is to waste valuable time. We need drive now and that is why the GSM Association's mobile services initiative was born.

The mobile services initiative, or M-Services for short, arose from the recognition that end-to-end service provisioning of the mobile Internet could not wait for a receptive market conditioned by WAP Release 2. The initiative, initiated by the vision of Mauro Sentinelli, CEO of Telecom Italia Mobile, defines the operators' requirements for terminal device manufacturers and applications developers to meet market demands. Device manufacturers no longer have to guess what operators want to bring into the marketplace. At a time when uncertainty is the norm, the M-Services guidelines provide specific services and release dates. These set a timetable for device manufacturers to meet in the reasonable knowledge that their time and effort is being exerted in the right areas and that there will be a demand for their products. In a similar vein, application developers will no longer have to develop applications for one browser and then have to start from scratch for the next.

I've heard it said that M-Services is too little too late. Maybe so, and maybe it would have helped drive the data market if developed earlier. But we have to work with the real situation as it exists today, and the mobile Internet running on the GPRS bearer will still benefit from this initiative. We don't know yet if the operator requirements will be transformed from words on a page into commercial GPRS devices, but we do see that a common set of requirements has emerged. These will form the basis for competition and may be the only way to exploit the GPRS technology.

The mobile commerce sector is monitoring the fate of the M-Services requirement guidelines very closely. M-Commerce is viewed by some as the electronic purse of the future or, in other words, the ability to conduct financial transactions through wireless devices. The theory is to exploit the existing rating, retail and wholesale billing and payment functionality of service providers and extend it to a micro-transaction level. In this way, wireless devices can become the preferred medium for future transaction payments. Today, financial institutions are occupying the macrotransaction space through the use of checking, credit cards etc., but microtransactions have remained the sole domain of cash.

Can M-Commerce enter this field successfully and provide a better service to the customer? Many obstacles remain before this dream becomes a reality, among them global agreements on security, device capability, transaction processing, and settlement. However, if obstacles can be addressed, M-Commerce may become the de facto global standard for microtransactions in a few years' time.

This chapter has meandered away from the topic of why applications are the key to GPRS success, but with good reason. Unless we steer developers towards the applications people are waiting to use, we are putting the cart before the horse. As with all applications, a "one size fits all" approach is unlikely to yield the desired result: the variety of data is its greatest asset. Therefore, customization at the user level may be key to the success of GPRS applications. Toss in the user demand for voice, global roaming, and continued security protection and it is clear that developers wanting to enter the GPRS space have many issues to address. It is to provide direction through the maze of possibility that M-Services was created, and the operator community believes that the guidelines will facilitate faster and easier time to market for new applications.

But wait—why do we need others to develop the killer apps for GPRS? Can't the service providers develop them in house? Some will and indeed are trying. Others believe that the true measure of success in the mobile applications sector—just like the Internet—will be a raft of inspired outsiders working in garages and basements on code that can unleash the potential of GPRS. Since the vast majority of these code writers have never seen the inside of a wireless operation, they will require certain access points and agreed specifications to ease their entry into this new market sector. GPRS provides the backbone structure, M-Services helps define user guidelines and M-Commerce holds out the hope for microtransactions to unlock a new, mobile-specific market. The developers will then have to invent and provide what the customer cannot do without.

In the longer term, the wireless sector may find ways to bring location awareness into the applications picture through what is generally known as location-based services (LBS). One of the great things about

LBS is that almost everyone involved has their own views as to what LBS is and what it should be. One thing all have in common, however, is an ability to pinpoint the customer's whereabouts, to some agreed level of granularity, in order to enable a whole suite of applications the customer has never seen before. They could be travel related, commerce related, or information related—but they will be built on the premise that as a customer travels around, their service needs change.

How LBS will be realized remains mostly conjecture at this time. Not only are there technical challenges ahead, but also prickly privacy and security concerns to be addressed to ensure that the negative attributes of the Internet (such as spam) do not find their way into the GPRS world. Integration of LBS into GPRS is therefore thought to be somewhere down the road, probably in the next 18 to 36 months. Until then, trials continue.

Before you delve into the insights of the subject matter experts, I leave you with this parting thought: Consider *quality* as the killer app for GPRS. Consider that no matter what the application, the quality of the end-to-end customer experience will be what drives or stalls the adoption of packet-based data services. We begin from the premise that the customer definition of quality in the Internet sector compares unfavorably to that in the fixed and (in most cases) wireless telephony worlds. Ultimately, it is how quality is measured when these two worlds collide that could make the difference between a tepid acceptance of GPRS by the marketplace and its ability to transform the way we live our lives. Remember, you read it here first.

CHAPTER 15

WAP
and Lessons
Learned

Philippe Lucas

Orange France

WAP and GPRS: are they made for each other? Many people will say yes. Browsing the Web on a handset is great, but paying our beloved operators for a circuit-switched connection, even if it's only a matter of minutes, is less agreeable. Knowing you'll need up to 45 seconds just to connect (when all you wanted to do was check the evening's TV schedule while waiting for the bus): not fun. Customers want to pay for what they consume. For data this means paying for volume, not connection time. Here is where GPRS is particularly interesting for WAP. It enables customers to pay only for what they download and not for the time you're simply reading your screen. It is therefore very likely that WAP will take off with GPRS. But this means that some optimization of the two protocols will have to happen, and that's what this chapter is about.

Finally, many people say that GPRS will provide speed. Others predict bandwidth, and everyone agrees it will be very useful for browsing. Well, it is unlikely that GPRS will automatically deliver significant improvements in bandwidth. What will probably push people to browse WAP content with GPRS is a much more suitable pricing structure.

WAP Forum

The WAP Forum is an organization of corporate members in the wireless industry that developed the WAP specifications with the goal of providing Internet-like services over low-end mobile devices. It was created in 1996 by a U.S. company called Unwired Planet, renamed Phone.com in 1999 and, after merging with software.com in 2000, recently rebranded as Openwave. Unwired Planet was assisted in the development of these specifications by the three main handsets manufacturers: Nokia, Motorola, and Ericsson.

The WAP Forum itself is incorporated as a U.K. company. Currently the Forum has hundred of members including representatives from all 2G and 2.5G technologies, operators, manufacturers, application providers, etc.

Over the years the WAP Forum has succeeded in becoming a *de jure* standardization body, with the industry's general agreement to implement its specifications.

Objectives

The Wireless Application Protocol Forum has as a governing objective the development of a global protocol specification (1) to work across differing wireless network technologies and (2) to adapt Internet protocols to the wireless environment to match the limited characteristics of wireless technologies. These characteristics can be grouped as follows:

- Limited air interface bandwidth with high latency
- Limited display capabilities
- Limited processing power
- Limited keyboard size and usability
- Limited battery life

These characteristics, common to devices in the mobile environment, precipitated the design of new protocols for Internet-like services on handsets. Actually, the main differences between the WAP and the "classic" Internet protocols is the addition of a completely new element, the WAP gateway, and a new language to define pages displayed on handset screens.

General Architecture

Let's look deeper into the differences between WAP and IP. Figure 15.1 compares the pure WWW client/server model to WAP architecture.

In the classic Internet, a device makes requests directly to a server in order to retrieve the information it needs. It must do so using the protocols defined by the standardization body of Internet community, the IETF (www.ietf.org). In a WAP configuration, communications between a device and a server require mediation. A gateway is provided to convert information from a Web server on the Internet into a form that the handset can accommodate, despite its limitations. This new architecture allows wireless access systems to use the Internet protocols, once adapted. The adaptation consists of:

- Defining a new programming language able to be displayed on handset browsers
- Converting the protocols understood by the Web server into protocols better adapted to the wireless technology at the gateway level

Figure 15.1
Typical architectures for the Web and WAP.

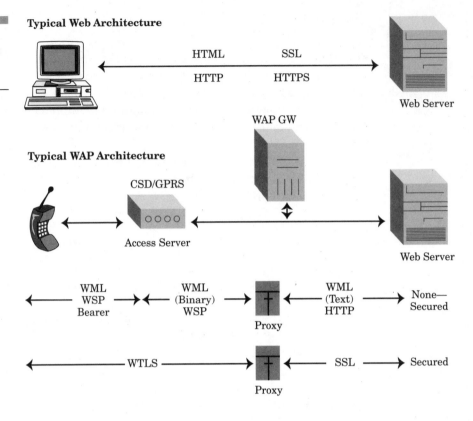

Protocols

The WAP implementation is defined in layers, each of them accessible to the one above, as illustrated in Figure 15.2.

Recall that the WAP specification mission is to be global, to adapt IP protocols to the wireless environment, and to operate across all wireless technologies. This last objective is often expressed as "wireless technology agnosticism," but in practice may devolve on becoming as independent as possible of radio access methods and the bearers used to convey information. Getting "agnostic" means that the higher protocol layers must be as generic as possible.

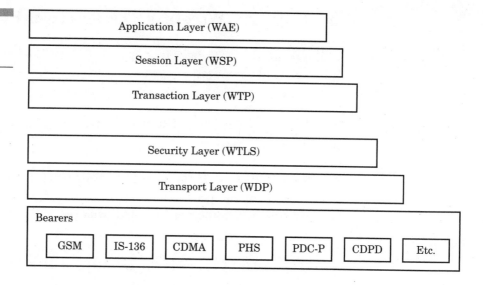

Figure 15.2
Layers of the WAP protocol.

WAE Layer (Wireless Application Environment)

This layer defines the application environment of the WAP specification. As such, WAE interacts with two elements of the handset:

- The WAP pages to be displayed on the handsets (data presentation). The handset must have a WAP-enabled microbrowser and be able to display the new page format defined by the WAP mark-up language (WML), the language that expresses Web pages in small-screen, small-footprint form.
- The telephony capabilities of the handset. In a WAP session, it is possible for the user to switch from a WAP session to a telephone call simply by clicking one button.

WSP Layer (Wireless Session Protocol)

WSP controls the active session under connection-oriented or connectionless service. This layer is based on the HTTP/1.1 functionality and is optimized to match the long latency and low bandwidth of wireless technologies.

WTP Layer (Wireless Transaction Protocol)

WTP provides a lightweight transaction protocol adapted from HTTP for thin clients like handsets.

WTLS Layer (Wireless Transport Security Layer)

WTLS translates the IP security layer SSL to TLS protocols adapted to the wireless environment. This layer is also used for implementing data integrity, privacy, authentication, and other security-related functionality.

WDP Layer (Wireless Datagram Protocol)

WDP adapts the lower layer (bearers) to the transport layer. This is where interoperability across wireless access methods is ensured for the whole protocol stack.

Interfaces

Using WAP over GPRS requires us to decouple the two levels of service access: bearer usage and application level. GPRS provides the bearer layer under the application that is using the WAP specification to browse the Web.

Interfaces between layers are essential to finding and differentiating services, applications, and bearers. GSM/GPRS bearers are determined by ETSI and the 3GPP, whereas the WAP Forum develops equivalent WAP specifications. Unfortunately, these two bodies never worked in close relationship, leaving some adaptation open to proprietary implementations and making optimization all but impossible.

How to Access WAP Services with a Handset

In WAP as in IP, connectivity, transport, and translation are three separate processes. That's why they each involve different standard bodies and protocols. If WAP is running over GSM, users can access WAP content with a circuit-switched data connection (i.e., a permanent CSD connection), a packet connection, an SMS, or a USSD. All of these will transfer information from a handset device to any web server using WAP pages. These methods are illustrated in Figure 15.3.

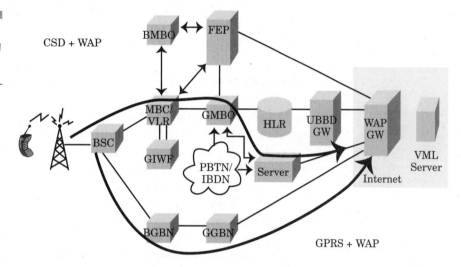

Figure 15.3
Methods of accessing WAP service with a handset.

CSD connections use the PSTN lines and dialup. The modem must be connected to a WAP gateway to guarantee protocol transmission. This process can also be decoupled, making dialup independent of the WAP gateway. To that end, any customer can use a classical ISP and program the IP address of the WAP gateway into his handset.

Using a CSD connection, the WAP session is initiated by calling a modem number (ISP number) that provides access to the Internet. Then the IP address configured in the WAP parameters of the handset will be used to access the WAP gateway. Once connected to the gateway, the handset can receive any WAP service using a WML server.

In a GPRS connection, the WAP session is initiated by "dialing" an access point name, which represents an exit door from the mobile operator's network to WAP services and in particular the WAP gateway. Prior to starting WAP, of course, the handset must open a GPRS session to request the bandwidth needed for browsing. Without an open GPRS session and the appropriate allocation, the WAP session cannot be launched effectively.

Success Factors

You have probably seen many articles about WAP—its weaknesses, its potential, and finally its notorious failure to penetrate the mass market. Since we don't need to rehash all of that here, I would like to take the opportunity to consider some factors that might have made a difference in the outcome.

Customer Demand–Market Demand

We must never forget that customers buy a product because it will yield an application they want. It seems that technology blinded us in the past few years, leading to a wave of new products and technologies, but in the end not many more services than we had before.

WAP is an example where the mobile industry was not in line with market demand. We promised the Internet on the mobile, when only a few lines of text were accessible. We told customers that better wireless bandwidth would make anything possible over the fixed Internet achievable on a handset. These promises were wrong, not because of the technology per se, but because we were looking at a potential El Dorado instead of what the customer said he wanted. Wonderful technology can't create a market if the offer doesn't match the demand. "Mobile Internet" is not the Internet on the mobile: services have to be thoroughly adapted first. Not all Internet services will migrate over mobile to begin with. It is highly probable that the most successful ones will exhibit the following characteristics:

- Immediacy
- Real-time delivery
- Easily personalized
- Able to access information remotely

Operators, content providers, service providers, and manufacturers are now working on developing the right applications with which to launch the mobile internet soon.

Ease of Use

If the technology is well adapted to handset limitations, developers have not spent enough time improving the man–machine interface to encourage widespread use of even simple services. The size of the keyboard is a key to the difficulty of using WAP services. Ease of use must be the number one priority if we want to see GPRS services used by most customers instead of a select few.

One problem for developers today is that browsers do not all work in the same way. Services must therefore be developed, tested, and specifically designed for the particular handset. This rapidly becomes unrealistic, of course, as more and more handsets arrive on the market. Browsers perhaps constitute the biggest challenge to WAP adoption in the developer community.

Customers, too, have paid a price: the need to configure WAP handsets to access mobile Internet services over GSM. Because these configurations are painful for the customer, it is now possible for operators to predefine configuration parameters with the handset manufacturers, or even better, to update parameters as needed simply by sending an SMS over the air. If remote configurations are not fully standardized and supported by most manufacturers quite soon, it is likely that the second option will prevail.

Connection Time and Bandwidth

Much customer disappointment with WAP services can be traced to waiting times for access and the duration of connections. With CSD connection, most services (including authentication) are achieved in 30 to 45 seconds, which is comparable to classic voice connections. Once she has accessed a service, however, the customer is paying for the time connected and not the amount of data transferred. The customer experience can be very negative when, after some minutes of waiting for service, she is presented with an error message saying that the page cannot be displayed!

In France, developers have already been down this road with the Minitel. In retrospect, the joke is that a Minitel customer always looks at two things: the screen, to get information, and the off button, to pay as little as possible for that information.

GPRS should provide customers with a much better experience. First, the access time is much faster, with under 10 seconds expected to be generally achievable. Second, pricing structures reflect the information transferred rather than the duration of a connection.

You'll note that I do not promise the customer a third advantage: speed of connection. Actually, even if handsets can provide more than 1:1, up to 3:1 (that is, three times faster from the network to the handset than the other way round), it is very unlikely that the networks will allow it when GPRS traffic increases. But the main two advantages previously described should prevail over this one. It is likely that we'll have to wait for UMTS services to see significant increases in bandwidth.

Improved Standardization

One problem WAP has consistently faced is inconsistent implementations—on the part of manufacturers, on the handset side, and on the network side (gateway). This has sometimes made it impossible for users to read WAP content on handsets that use gateways from other manufacturers. The principal fallout from this situation is that software now prevails in the development of such systems, making it very difficult to ensure that all implementations conform to the same specification. Yet the industry has always been aware that standardization is absolutely necessary for defining test procedures and getting the third-party agreement to guarantee interoperability between manufacturers as well as technologies.

Will GPRS Renew WAP?

GPRS is undeniably WAP's last chance to become a part of everyday life. As many chapters have explained, GPRS is the natural GSM bearer for data services. Since the emergence of WAP, GSM has had to use CSD connections, despite long setup times and its tendency to raise the price of usage.

Finally, beyond these technological considerations lie the eternal verities that content and services will have to be available to grow and to meet market demand.

CHAPTER 16

M-Services

Jörg Kramer and Yves Martin

Vodafone and Orange France

M-Services (short for mobile services) is an initiative launched by the GSM Association in early 2001. The goal was to expedite the development of mobile handsets by documenting operator requirements and encouraging device manufacturers to meet them. M-Services is not a new technology, but rather a joint definition of the handset features and behaviors that operators are assuming in new service offerings. Think of M-Services as a framework based on standards and protocols that exist today, outlining an end-to-end solution for interactive services. To make its run at the goal, the GSM Association set up an M-Services Special Interest Group consisting of experts from the Services Group (SerG) and Terminal Working Group (TWG).

The original target in Phase I was to provide M-Services–compliant terminals by the end of 2001. Within a very short period of time, as measured against typical standards creation cycles, this group was able to define requirements for mobile browsing, provisioning, messaging, and downloading in the mobile world. The GSM Association delivered a Phase II requirement document in February 2002.

The M-Services Recommendations document was approved by all GSMA members in June 2001 and presented to the public. GSM terminal vendors unanimously supported the concept behind the initiative and unanimously committed to developing products in accordance with the recommendations. The first M-Services Phase I-compliant terminals appeared on the market in October 2001.

Why M-Services?

Mobile terminals were enabled to access Internet content with the introduction of early WAP-based services. Compared to our PC-based experience of Internet services and content, text-based mobile WAP screens were unimpressive. No color displays, no icons, no animations. Lots of hassle with complex WAP and GPRS provisioning settings, and interoperability problems caused by imprecise WML definitions and ambiguous browser rendering. In sum, not a very sexy user experience.

On the other hand, the world had since taken note of the enormous success of Japan's I-Mode service, which attracts a huge number of customers and reliably generates additional data services revenue. The reasons for the great success of I-Mode and the failure of WAP are manifold, and we can't rehash them here, but they did not dissuade the GSMA membership from pursuing the development of a strong mobile service offering around WAP and other interactive services. The focus of the M-Services work was put squarely on the *user experience* and the *enhancement of usability* for a range of services.

As mobile telecommunications converge with the IT world, the standardization environment for 2.5G and 3G is getting more and more heterogeneous and complex. Today's very successful short message service (SMS) was specified bit by bit by ETSI for GSM. SMS's successor, multimedia messaging service (MMS), was created in a mixed environment by 3GPP and the WAP Forum with input from standards bodies like IETF, W3C, ISO, and ITU. In most current standards, received wisdom is to make as many key features as possible optional, which leads to freedom of implementation. The consequences of this approach are imperfect interoperability and uncertain compliance of terminals and network systems. The aim of M-Services is to overcome these predictable difficulties by providing an end-to-end solution for mobile interactive services.

If you look at the typical development path for a handset introducing a new technology, you're likely to find something like Figure 16.1:

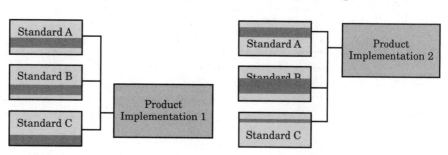

Figure 16.1
Standards are too open and lead to different product implementations depending on manufacturer's implementation.

First, the industry collects or invents relevant standards from interested entities (3GPP, WAP Forum, etc.) Let's call them Standards A, B, and C. Then the handset manufacturers try to map out a product implementation strategy by selecting among technology and standards options. Because the wireless data market is the source of a lot of innovation, there are a lot of standards and a lot of choices to be made. On top of that, standards are young and relatively immature; developers often must provide a measure of interpretation when getting into implementation details. There is no mystery in the fact that product implementations tend to differ from one manufacturer to another, even though they are all targeting the same features based on the same standards.

For example, WAP is a combination of optional features that are all related to different services. Therefore, when defining WAP 1.2.1

requirements, protocol designers must go deeper into options and sub-features to get agreement on what the user will experience. As the development of a browser (or any client) in the handset takes time and resources, manufacturers need help prioritizing in order to build products that address the user's needs from the outset. Unfortunately, without a strong signal to the market the natural desire among manufacturers to differentiate themselves in the competition for "most innovative product" will manifest itself as different implementations.

In unambiguously declaring what the operators see as their needs, M-Services intends to be a strong signal to the industry. Although these guidelines don't technically have the status of standards, M-Services are best understood as imperatives from the community of GSM and 3G operators who sat together and tried to put aside competition in order to build a wireless data market.

Figure 16.2
M-Services reduces the gap between the standards and operator's needs, thus leading to one single product implementation.

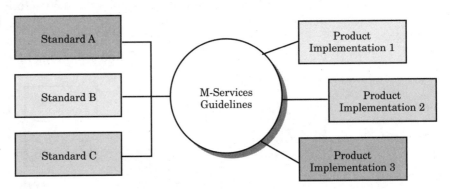

A second but no less important reason for M-Services is the industry's need to define consistent user interface components on the handset for conducting the business end of a mobile data session. The handset is a key component in the value chain and also in service delivery to the user. It thus impacts the operator's ability to offer, deliver, and bill services properly. For example, when a user downloads an object, the handset must offer him or her:

- A menu of downloadable objects
- Information on charges for the download
- Acknowledgment of receipt by the user
- Security controls to avoid fraud

As a consequence, M-Services is also the occasion to analyze the business interactions for future services and how they are translated in the handset.

What is M-Services Phase I?

M-Services Phase I guidelines were publicly released in June 2001, targeting handset roadmaps for September 2001 and June 2002. Phase 1 provides an end-to-end solution for interactive services and includes a framework based on existing standards and protocols. In the following sections, three key areas are described. Guidelines on usability include requirements for *Graphical Browsing* and *User Interface*, *Over The Air Provisioning*, *M-Services Access Key*, *Display and Memory*. The Download discussion describes methods for bringing content onto the terminal: *WAP GET* and *Download Fun* (optional) are proposed, and a list of recommended *Formats* and *Codings* is furnished. Additional requested features are *SMIL* (presentation format), *Java* (J2ME), and *vcard*. Messaging includes requirements in the area of *MMS*, *EMS*, *Long SMS*, and *email*.

For all services *GPRS* is identified as the preferred bearer.

Usability

As the complexity of services and applications mounts, usability becomes a key element for success. The M-Services recommendation placed great emphasis on it, specifying the following features:

WML 2/XHTML Browser

M-Services recommends a WML 2/XHTML basic browser that also supports WML 1.2.1 (in other words, it's smart to implement a dual-mode browser right from the start). Although this is an interim solution, it offers a smooth migration path to full WAP 2.0 and standardized content generation.

Figure 16.3
Browser interface
evolution.

WML Extensions

If you decide to use WML 1.2.1 as an alternative, the required functionality should be provided via a series of WML extensions. These GUI extensions could bridge the gap between WAP 1.2 and WAP 2.0, allowing users to gain the benefits of graphical user interfaces while WAP 2.0 is being finalized and implemented.

Human Machine Interface (HMI)

The guidelines document detailed *HMI Requirements* for the usability of a WAP browser on the phone. They say, for example, that hyperlinks shall appear as underlined text and not bracketed, and that selection lists shall be numbered with their appropriate access key. Additionally they put forward requirements on usability in the areas of browsing, messaging, messaging client capabilities (SMS and MMS), the personal information management (PIM), games, and the personalization of HMI.

Configuration and Provisioning (OTA)

A solid concept for configuration and provisioning over the air (OTA) has to minimize hassles with complex configuration settings and has to ensure the best user experience available under all circumstances. In the early days of WAP, configuration and provisioning made problems. Some terminals did not support the configuration of the setting parameters over the air and some supported only one set of parameters. When a customer accidentally changed the settings, it was a nightmare to restore the parameters.

The M-Services recommendation requires the use of the WAP 2.0 OTA provisioning standards delivered by the WAP Forum. In addition, the SIM card may be configured and updated via OTA, and it shall be possible to lock one set of parameters against overwriting. At least five profiles shall be available.

Readers should note that configuration and provisioning is relevant not only for the WAP parameters but also for all GPRS, MMS, and email client settings.

M-Services Key

To give users *one-click* access to WAP Services, an *M-Services key* is defined. This access key shall be realized either as a dedicated hardware key or as a softkey—for example, a long press on the "8" key. This is a much-wanted feature among customers and leads to higher service penetration.

Display

M-Services defines minimum requirements for the number of lines, characters per line, and number of pixels in a display. Graphical black-and-white displays are basic, but gray scale and color displays are highly desired. Icons are recommended at least on the top level of the user interface.

   A color display and an icon-driven user interface enhance navigation options, give better service visibility, and make life easier for the customer (who is much more likely to press one button than she would be to press several).

Memory

Requirements on persistent storage memory include at least 20 ringing tones and 20 wallpapers. Fifty kbyte of cache memory is recommended.

Additional WAP Requirements

The following WAP elements are recommended:

- WAP push
- Wireless transport layer security (WTLS2 and WTLS3)
- User agent profile (UA Prof)
- Wireless identification module (WIM)
- Wireless telephony application interface (WTAI Public)

Download

M-Services defines solutions for downloading multimedia objects on the mobile terminal. The Download option provides a new range of innovative products with content types like screensaver, ring tones, applications, wallpapers, and animations.

WSP/HTTP GET in WAP1.2

This is the preferred M-Services solution because it is based on WAP standards. During navigation, the browser shall offer an option for storing media elements of the wml deck (e.g., WBMPs). The terminal must implement a segmentation and reassembly (SAR) feature, to ensure that the size of the object is not unduly limited. This method provides a smooth migration to HTTP GET in WAP 2.0.

Download Fun

An alternative, optional solution, Download Fun, requires a special download agent on the terminal. It incorporates an end-to-end solution on the network side to include authentication, personalization, billing, blacklisting, and whitelisting.

Download Object Definitions

M-Services defines a set of objects and the appropriate MIME types and codecs. Definition is necessary to ensure interoperability.

Presentation

The presentation format provides users with a scene description of downloaded objects like pictures, audio clips, and text. A *scene description* might be a spatial and chronological synchronized presentation of two related objects. Synchronized multimedia integration language (SMIL) shall be supported for presentation.

TABLE 16.1

Some Definitions for Download Object

Object	MIME TYPE	Codec Specification
Ringing tone	Audio/MP4; Application/MIDI; Audio/WAV; Audio/iMelody	MIDI, 8-Level Polyphonic MIDI
Wallpaper	Image/(W)BMP; Image/gif (optional); Image/jpg; Image/png	GIF 87a, 89 (with animation), PNG, WBMP, JPEG
Screen Saver	Image/jpg; Image/gif (optional)	JPEG, GIF 87a, 89 (with animation)
Picture	Image/(W)BMP; Image/gif (optional); Image/jpg; Image/png	GIF 87a, 89 (with animation), PNG, WBMP, JPEG
Game	Open	—
Java application	Application/jar; Application/jad	—
Audio	Audio/MPEG4; Audio/mp3; Audio/WAV	MPEG-4 AAC, PCM and optionally MP3
Skin	Application/skin	—
Video	Video/mp4 (MPEG4)	Video codec: H.263 baseline (Profile 0, Level 10), MPEG-4 Video, Simple Visual Profile, Level 0

Java

Java offers an intelligent, terminal-based services platform. For mobile devices, Java compliant with MIDP 1.0/CLDC (J2ME) standards is recommended. For downloading Java applications, OTA provisioning is selected. Operations available for the Java applet are download, store, play, and delete. To furnish the appropriate security environment, M-Services calls for the 3GPP standard MexE.

Messaging

Messaging entails a client in the handset to receive and present the message. The standards and technologies are numerous, as this is a convergence point between mobile services (SMS) and Internet services (e-mail, presence).

Multimedia Messaging (MMS)

MMS is an evolution of SMS and enhanced message service (EMS). It provides functions similar to the ones users are used to, like *message push*, and it also adds support for attaching images or objects to the message. MMS gives the customer clear benefits and a migration path from SMS to MMS that includes picture, text, audio, and video messaging. It makes use of WAP features and packet bearers like GPRS and 3G.

M-Services defines MMS as the preferred messaging solution. The MMS implementation shall be built using 3GPP (TS 23.140) and WAP Forum (WAP 209) standards, and the presentation format language SMIL (W3C). For maximum interoperability, M-Services V2 will specify the terminal and network compliance requirements in detail.

The M-Services *Messaging Customer Experience* recommendations describe requirements for the application launch, the messaging client, and for the composition, sending, forwarding, writing, reading, and deleting of messages.

Enhanced Messaging Service (EMS)

EMS is a 3GPP standard (TS 23.040) based on SMS. As a service, EMS is a halfway between the SMS and MMS, providing ring tones, operator logos, business cards, picture messages, and animated screensavers. The *User Data Header* makes it possible to include binary data such as:

- **Text formatting**—Text alignment, font size, and style
- **Animation**—User- and pre-defined animations
- **Pictures**—16 × 16, 32 × 32, or variable pixel size; black and white
- **Sound**—User and predefined melodies

Long Message

M-Services recommends support for this feature. Users must be able to write a long SMS (up to 640 characters) in a single text-entry dialog.

The receiver must be able to recreate the long SMS by assembling the short messages in the right sequence, and must be able to present the message as one SMS. Before a long message is sent, the terminal must notify the sender, tell him how many SMSs are actually scheduled, and give him a chance to cancel the sending operation.

Email

M-Services defines email as an *optional* messaging method and recommends implementation guidelines for the communication protocol and specific functionality. If email is supported, the email client must implement the POP3 protocol. Support of IMAP4 is optional. Additional requested email features are offline read, offline queue, authentication, download of (just) message headers, and selective download.

M-Services: What Impact on the Market?

At the time of the first press release, M-Services was publicly endorsed by most of the manufacturers, but we had to wait at least 6 to 9 months to see real benefits from it.

Phase I results vary with respect to the criteria used to evaluate success. First of all, Phase I has been criticized for its lack of clarity and traceability. In some parts, the signal sent out was blurred because we failed to prioritize among possibilities or because bundling a lot of different requirements makes it difficult to determine if a handset is in full compliance to M-Services guidelines.

Second, the document was released very late with respect to the target dates, and very few manufacturers were able to meet the timeframe.

Despite those problems, we were gratified to see the arrival of more than 30 products compliant to M-Services—at least for browsing and downloading—in first half of 2002. One of our most important aims was to make the operator community's voice heard by the industry, establishing the possibility of further collaboration, and we succeeded in doing that.

M-Services Phase II Requirements

Between October 2001 and February 2002, the GSM Association has pursued the work of evolving M-Services requirements for the handset. Phase II was released on the February 20, 2002, targeting products for the September 2002 and April 2003 timeframes. The major enhancements are the following:

- **Segmentation of the market**—M-Services Phase II requirements address only voice and smart phones.
- **Quality of the document and readability of the requirements**—All the requirements are numbered and divided into comprehensive parts.
- **Extension of user interface**—M-Services Phase II continues to emphasize the importance of a unified set of ergonomics for specific services.
- **Avoidance of proprietary features**—Phase II relies only on existing standards, and points out where standards are incomplete or missing altogether.
- **Focus on browsing and messaging transition**—As the handset acquires more capabilities, browsing, and messaging features will rely on Internet protocols for Internet-based portions of the service. This is also true in the case of WAP 2.0 and perhaps MMS release 5. Because IP represents a major technical transition, M-Services phase 2 tries to define the steps forward as precisely as possible.

Future

M-Services are an ongoing mission with enhancements, extensions, and revisions to come. Therefore, it is difficult to detail what their future will look like. Nevertheless the GSM Association has found in M-Services a great way for the operator's voice to be heard. It is generally accepted that the first phase of M-Services was far from complete and left a lot of important topics for future iterations. Phase 2 has improved upon these omissions, but there is still a lot to be done. Future work includes:

- Digital right management (as soon as a large-scale standard is approved)

- Detailed Java requirements
- Security
- User interface
- Standardized download solution
- Presence and immediate messaging
- Media players

More and more technologies are pending, and each will require the industry to make determinations between options, parameters, settings, etc. Above all, a market-driven approach must emerge to define a sustainable business model.

Last, it is important to consider the testing aspect of M-Services. As features get more complex, a clear testing scheme will help all players to have a measurable way of qualifying M-Services compliance. The GSM Association has not yet reached a conclusion on that topic, but it has plainly put handset or end-to-end testing on the critical path to new data services.

The GSM Association will take on the responsibility for maintaining the momentum developed in the first version and extending it to new features. A new phase of the M-Services guidelines is slated for production, targeting the areas mentioned above, with the same mandate as the first one: to close the gap between the standards and the operators' needs.

CHAPTER 17

M-Commerce

Stella Penso

Turkcell

First, let's demystify mobile commerce. M-Commerce is just good old commerce—the difference being that this time around it's transacted with a mobile gadget. It's an opportunity to enjoy more convenience in daily life. The opportunity lies in our ability to make purchases on the move and to make them without carrying any money. M-Commerce will encourage users to spend money in situations where they cannot now— just as mobile phones have made it possible for people to place calls when they're away from home or office. Mobile communication has even absorbed a proportion of fixed communication's business, since the mobile phone has become a substitute for the fixed phone. Therefore, there is an undeniable potential for M-Commerce to garner a considerable proportion of current commercial activity, and an even stronger potential to expand virtual commercial activity.

Just like unwired commerce, M-Commerce can accommodate many different forms of payment unit, payment tools, payment schedules, and payment environment. What sets mobile commerce apart is the way transaction information is transmitted (i.e., over the radio networks). The mobility dimension is revolutionary not only for end-users but also for merchants, financial institutions, service providers, and mobile operators. While the mobile technology used in transmitting transactions is obviously a key component from the perspective of all these transaction participants (and is the central theme of this book), mass market acceptance depends not on the merits of transmission technology but on the adoption of procedural, logistical, and legal mechanisms that satisfy the requirements of consumers, regulators, and the M-Commerce parties listed above.

So M-Commerce is really a world of its own—and a potentially dangerous world at that. At the end of the day, commerce is about spending money—essential or frivolous. Leaving aside the social phenomena, this chapter will take the premise that M-Commerce is an advantage we all wish to pursue.

One note about technology: this book is about GPRS, its flexibility, and merits. GPRS facilitates M-Commerce by allowing transactions to be completed faster and without being interrupted by disconnections. The circuits that carry our voice calls or the signaling links that carry our messages in data packets (SMS) have already opened the door to some M-Commerce uses and applications, but the flexibility inherent in GPRS transmission will likely expand the scope of M-Commerce and instill more trust in the transaction process.

In the next several pages, I will stay away from "transmission" and close to "transaction." First, we'll consider M-Commerce as a subset of commerce. Under the heading of enablers, I clarify the "musts" of any transaction and the ramifications of these "musts" on mobile operators. Then, I move on to the payment experience in M-Commerce. Finally, since I am from the GSM operator world and de facto occupied with roaming, I also present a picture of M-Commerce from the roaming perspective. I conclude my chapter with some comments on open issues and some thoughts about the future.

Expanding Commerce through M-Commerce

The mobile phone[1] has become a tool with which users access a remote world from wherever they are—for communication purposes. In order for this wherever/whenever access to draw commerce, paying-by-mobile must bring both more convenience and a new dimension to commercial activity. In other words, mobile commerce should fill the gaps and expand the boundaries of today's commerce. The ideal form of payment from the customer's point of view, and the undeniable hook from the marketer's point of view, is a single transaction process, available wherever you go, and simple, flexible, intuitive, and secure.[2]

It All Started with "Micro"

In terms of filling a gap in commerce, mobile payment methods are often seen as the first promising alternative to small cash transactions or, in other words, micropayments[3] (since macropayments are already handled by credit cards). Other contenders—such as debit cards or tokens—

---

[1]Out of convenience, I refer to mobile devices as mobile phones from this point on.

[2]Joanne Taaffe, "Operators move ahead on m-commerce," *Communications Week International*, April 1, 2002.

[3]"DoCoMo backs mobile wallet for m-commerce future," *Total Telecom*, April 10, 2001. "Sonera launches pilot for mobile payment," *Sonera Press Release*, March 4, 2002. (A research by Tower Group reveals the intention of 118 million Europeans, 145 million Asians and 22 million Americans to use their mobile phone for micro-payments.)

[4]"Strategy: Micropayments: Small is the new big," *ci-online*, May 1, 2001.

[5]The survey is mentioned in: Anne Young, "M-commerce set to fly—when services arrive," *Total Telecom*, June 12, 2001. The results of the research, conducted in the last quarter of 2000 in six markets (United Kingdom, South Korea, Italy, United States, Brazil, and Finland) and involving 11,295 end-users, can be found in the report *3G Market Research mCommerce: An end-user perspective*.

have failed to catch on.[4] In a survey[5] conducted at the end of 2000, respondents considered it key to be able to make micropayments with their mobile phone. Some familiar micropayment purchases identified by the study are movie tickets, sodas, parking meters, gas pumps, sundry shops, jukeboxes, bill payment machines, restaurants, vending machines, and video rental stores. These items are in the category of "local" payment—payment for a physical item within reach. A survey conducted by Nokia in March-April 2000[6] reports that respondents find the ability to make local purchases particularly practical while on the road (tolls), in post offices, in health-related offices (hospitals, dentists, pharmacies), and in places where it is inconvenient to carry a wallet (the gym or swimming pool). One of the most recent surveys[7] on the topic reveals a global user base relatively receptive (44 percent of respondents) to utilizing their mobile phone for micropayments.

Here, There, and Everywhere

The potential convenience of mobile payments for micropurchases is not limited to the local realm however. We can also find items with micro-prices while shopping online—this constitutes remote payment. Similarly, local payment can be made for an item that has a macroprice (airline tickets, an expensive meal, or tickets for Carnegie Hall). Whether it is macro- or micropayment for a local or remote item, M-Commerce is really about putting the user's "wallet" in a trusted place, enabling him to make secure virtual payments to merchants in real time or ex post facto and receiving a service in return.

In this capacity, M-Commerce can partly replace cash and other types of payment units, reducing endless trips to the ATM. It can transfer money or pension payments from accounts to the readily available m-wallet, and make purchases on credit cards without using the physical card for many types of commercial activity.[8] All of that depends upon establishing a trusted process through necessary security schemes and the SIM. The SIM may be your trusted connection, or it may be your payment platform, offering a choice of payment methods. The convenience of choosing

---

[6]*Demand for M-Commerce—Local payment & Online Shopping*. Market Study of VAS. Nokia. June 2000.

[7]Mobinet Index #4, An A.T. Kearney/Judge Institute of Management Collaboration, February 2002.

[8]While it may be feasible, the market for insurance, mortgages, loans, and asset management are beyond the scope of our discussion.

from a menu of payment methods will push M-Commerce forward, as customers begin to feel the same ease that they feel pulling out a credit card or the cash from their pockets. The same ease, but less hassle.

It May Be Even More Fun if You Can't Touch It

The shopping opportunities of M-Commerce are not limited to physical items. The mobile medium adds two new factors to the area of commerce, which can also hasten its adoption. The first factor is the mobile phone itself. The mobile phone is a new geographic location of daily life. Yes, it is small, but it is a geography in which there are new reasons to spend money: downloading music, jokes, ring tones, logos, streaming video, and games to the mobile phone itself. Consumers can enjoy these entertainment items anytime they want through their mobile phone, and *only* through their mobile phone. The second factor is just as new: because of the SIM card, which in effect is the operator's link to the customer, the operator always knows where you are. Customer location information is valuable to the customer and to third parties who would like to serve the customer (as long as the service is provided with the customer's consent). Location-based services, which are the topic of Chapter 21, represent novel opportunities for establishing links and relationships between the customer and the merchant. Most obviously, when the customer is in the merchant's vicinity, the merchant can reach out to him through the aid of the mobile operator. The same kind of relationship can be established to communicate traffic reports or emergency news. Financial transactions, trading stocks, topping up prepaid accounts, bidding in mobile auctions, and even person-to-person payments are all possible applications of mobile commerce.

The "Musts" of M-Commerce

Local versus remote payment? Real time versus credit? Micro versus macro pricing? Physical versus nonphysical goods? All of the above? Whatever the purveyor's choice is, M-Commerce is enabled by technical, logistical, and trust infrastructures. Commerce is based on trust, or more precisely on mechanisms that allow the parties to trust each other and exchange goods and money. A trusting environment amounts to a secure environment, particularly with respect to how the payment authority allows the right customer to withdraw funds from his m-wallet for transfer to a merchant. An adequate infrastructure will always be one that fulfills the peculiar requirements of all parties (customer, operator, content provider, visited network in the case of roaming). However, there are a few shared concerns which can be considered the "musts" of commerce whether electronic, mobile, or physical.

- Transaction content must not be altered during transmission; this is called *data integrity*.
- The transaction must be safe from eavesdropping; that is, no unintended parties should be able to view content or data from the transaction. This is the confidentiality "must." Storing credit card data in the payment platform in incomplete form, or never revealing your customer's credit card number or the name to the merchant, are ways to alleviate confidentiality and privacy concerns.
- The third must is *authentication*—ensuring the customer is who he says he is, and the merchant is the intended and authorized merchant. Authentication of user can be automatic (WAP gateways use the MSISDN number) or interactive (username, password, PIN codes), providing flexibility in determining the select method appropriate for the desired security level.
- The *nonrepudiation* "must" protects merchants and customers from post-transaction regrets: if a customer decides to deny that he gave his consent, the transaction record logged in the payment system will attest to the truth. A merchant might send transaction IDs to track service delivery and confirm that it has been accomplished. The digital signature mechanism is a tool to satisfy the nonrepudiation "must;" SIM-based payment applications support nonrepudiation, but the current generation of handsets does not.
- The final must is *authorization*. The payment authority authorizes payments after ensuring that the customer is in good standing and has funds that cover the transaction amount. Customers authorize payments to be deducted from their m-wallets. And merchants are authorized to deliver the goods or services by payment authority (bank) after customers authorize payment. The customer may choose to confirm the value of the transaction or have it be confirmed automatically when he sets up his M-Commerce account with an operator. Payment authorization can also be tiered, some payments requiring no confirmation, others a basic level of confirmation, and the remaining extra security and password.[9]

---

[9]Mbroker™ Security, MoreMagic, Security White Paper, page 1.

The Role of the Mobile Operator

The only permanently present party in M-Commerce, and therefore the party most exposed to satisfying these "musts," is the mobile operator. Even in instances where it is oblivious to the fact that a transaction is taking place, the operator is transmitting the traffic and the transaction information. That is the operator's de facto presence, but it can hold other roles as well:

- It can create the mobile marketplace (e.g., the portal)
- It can authenticate the end user and/or the merchant
- It can host a mobile wallet
- It can go as far as authorizing payment from the mobile wallet
- And of course, it can also collect the transaction amount from the customer, along with transaction costs, via its monthly billing system or through prepaid cards, and then distribute the money to other parties. The role the operator assumes vis-à-vis other parties will determine the location of interfaces and how/when each party "talks" to each other through these interfaces. As the authenticator, the operator uses the SIM and an additional verification method (PIN number or password) to confirm the identity of the user. If the operator manages the service, it has to confirm through the aid of a merchant management system that the merchant belongs to its partner portfolio. Authorizing payment is the trickiest of tasks in the commercial world because the responsibility to authorize brings with it the responsibility to accept liability. Operator authorization of the transaction requires a link between operator and merchant, either directly or through the payment service provider.

How Does the M-Commerce Loop Stay Live?

If the "musts" are the skeleton of the M-Commerce loop (starting with the customer intention to purchase and ending with the remittance of payment and the delivery of the requested good to the customer), viable logistics is the flesh: a "back office" that can track, record, and store each phase of the transaction, clear the transaction, and facilitate the settlement process[10] is vital in the resolution of post-purchase disputes, the clarification of inaccu-

---

[10]"Telcos need cut of transactions industry to survive, says Intec," *Total Telecom*, July 6, 2001.

rate transaction execution, and setting straight any errors in the maintenance of customers' mobile wallets and virtual accounts. The use of a single system to authenticate all the necessary parties may increase transaction speed as well as detect more consistently any fraud activity.

By the way, if you think that all these key requirements seem like too much diligence or unwarranted care, remember that we go through these steps unconsciously in everyday life, at the grocery store, the movie theater, or the ticket counter. The difference is that all the "musts" are already taken care of directly by consumer and merchant in real time. And even then there is room for fraud.

Once the skeleton and the flesh are in place, the loop will remain alive in the presence of win-win outcomes. In M-Commerce, the customer wins the opportunity to make his usual purchases at times when he could not before, to purchase items that did not exist before, and payment options that free him from carrying his wallet around. He may end up paying more for it (which he already does in the case of a mobile phone call compared to a fixed phone call), but moving the transaction to the mobile medium may also decrease the cost of providing goods. Operators always make money from network usage, regardless of whether they are involved in the transaction, and if they are involved, they may get a cut of the transaction. Banks, too, may experience lower costs when their customers handle financial transactions from mobile phones, and they also get a cut from the M-Commerce partnership. The win for content providers is access to a larger customer base (larger because of the penetration of the mobile phone). The gain for merchants is similar. Business partnerships will determine who gets what, but at the end of the day, everyone can gain because the commerce pie gets bigger.

For everyone to gain, however, the customer first has to pay.

So How Does the Customer Pay?

Consumers can be offered a selection of payment methods during any transaction to evoke the ease of cash or charge. Depending on the total transaction value, the purchase price may be reflected on a phone bill; deducted from a cash/virtual account[11] (stored value account) in the

---

[11]Virtual accounts, often residing on a server, get fed from traditional bank accounts. Virtual accounts may be offered by non-banks, which allows the operator to be in the position of hosting and/or creating the account and of handling authorization. (Gérard Carat, *E-payment Systems Database—Trends & Analysis*, EPSO, March 2002.)

m-wallet, which is located somewhere on the payment platform;[12] paid upon delivery at the door, added to the credit card statement, or deducted from a checking account by instant debit. Given a specific transaction value, the transaction leader has to decide whether to charge the customer on credit (phone bill, credit card, private cards) or real time (prepaid account, instant debit, stored value or virtual account, or cash upon delivery). Three considerations may influence decision:

- The threshold values for each payment method
- The creditworthiness of the customer
- The personal preference of the customer

The payment infrastructure should be flexible enough to support making interactive payment decisions and be compatible with installed security mechanisms.[13] A flexible payment platform:

- Is modular and scalable
- Collects, manages, and clears payments
- Allows the interactive pricing of digital content
- Has an interface with the customer so that she is able to define some parameters of her desired payment method
- Accommodates different pricing types, such as flat and usage based

What Is Out There for Customers?

A recent announcement (Spring 2002) by Vodafone and T-Mobil of the imminent launch of their joint m-payment solution may signal the recognition, on the provider side, of the need for flexible solutions: the details are not yet out, but their solution will facilitate payments for macro purchases by national credit/debit cards and for micropurchases via phone bills and prepaid cards. Phone bill payments, while quite convenient, can pose some problems: in some jurisdictions, mobile operators are not allowed to bill for non-telecom goods and services. While Sonera opted for the phone bill option in initial trials, it now has abandoned that effort (and moved on to the "Sonera Shopper") since employers, who

---

[12]"DoCoMo backs mobile wallet for M-Commerce future," *Total Telecom*, April 10, 2001.

[13]Examples include payment platform solutions by MoreMagic and Nokia. Payment Transaction Platform—Technology White Paper by MoreMagic. Nokia Payment Solution materials.

cover a quarter of the mobile phone bills in Finland, were not thrilled to finance the micropurchases of their employees.[14]

Another dimension of phone bill purchases is the exposure of the operator to payment fraud: these purchases put the SIM in the position of a credit card[15] and expose operators to the risks of late payment. Customers' incentive to honor their phone bill payment obligations (so that they can continue making phone calls) does mitigate late payment risk; therefore, for postpaid customers, phone bills are still an appealing venue of payment collection. Given the predominance of prepaid customers, however, a more relevant venue is prepaid accounts. The use of prepaid accounts for the use of third-party goods and services may put these accounts in the category of e-money,[16] the issuance of which is a regulated activity in Europe. How mobile operators may be regulated for allowing their customers to use their prepaid accounts for m-payments is still an open issue.

There are quite a few M-Commerce trials by operators in progress. In terms of market acceptance and visibility, however, they are still in initial phases. What seems to be most widespread now in the dawn of M-Commerce is Paybox, with its 750,000 subscribers in Europe (as of March 2002).[17] In the absence of standardized payment platforms and mobile payment procedures, the Paybox initiative has been able start filling the vacuum with an operator-independent system. Half owned by Deutsche Bank, Paybox obtains customers' confirmation via their mobile phone to make direct debits from their bank accounts for payments to merchants. The mobile phone number and the PIN number are used to link the user with the registered bank account; in other words, mobile phone and PIN numbers are used as authentication tools. What does the mobile operator do in the meantime? It carries traffic during every transaction.[18] The next few years will yield the results of the play-off between mobile operator payment systems and Paybox-like solutions.

---

[14]Joanne Taaffe, "Operators move ahead on m-commerce," *Communications Week International*, April 1, 2002. "Sonera launches pilot for mobile payment," *Sonera Press Release*, March 4, 2002. Sonera Shopper

[15]"Strategy: Micropayments: Small is the new big," *ci-online*, May 1, 2001.

[16]Under the EMI (Electronic Money Institution) Directive, e-money is a monetary value "stored on an electronic device...issued on receipt of funds of an amount not less in value than the monetary value issued...accepted as means of payment by undertakings other than the issuer" (Gérard Carat, *E-payment Systems Database—Trends & Analysis*, EPSO, March 2002).

[17]Boris Groendahl and Braden Reddall, "Vodafone, T-Mobile to launch 'mobile wallet'," *Total Telecom*, March 14, 2002.

[18]http://www.paybox.co.uk

Can I Use My Mobile Phone to Shop while Abroad?[19]

Good question. Well, mobile shopping while roaming internationally can be tricky. If the desired purchase is on the home operator's portal and the transaction is to be made remotely, then the location of the customer is no barrier at all. It will require no special arrangement, except the presence of GPRS roaming capability on the visited network. The merchant may be located outside the national boundaries of the operator, but transaction-clearing agreements would be adjusted accordingly. So the location of neither customer nor merchant affects their transactional relationship.

In the case of local purchase, as long as the home operator has a direct relationship with the local merchant in the country where the subscriber is roaming, transaction details can be sent back to the operator via a broker or the payment service provider. In the GSM jargon, this is called "home M-Commerce roaming" since the consumer is taking advantage of the *home* operator's arrangement with a local merchant. An alternative is "local M-Commerce roaming," where the roamer has access to the *visited* network's remote (i.e., portal) or local M-Commerce services. In this case, who authenticates the user? Who authorizes the payment? How can the user make a payment in the first place? Can the home operator act via the phone bill as a collector of the transaction amount? How can the visited network monitor the parameters of a transaction by a user who is not its subscriber? These questions aren't peculiar to international commerce, but they arise in this context because all M-Commerce–related data resides with the home operator and would have to be transferred to the visited network for decisions on the questions above. There is no method at this point that enables the home operator to monitor, online and in real time, the costs incurred by its subscriber while she is using a visited network's M-Commerce service. User authentication, customer approvals, and communication of the final price to the end user are still beyond the ready capability of a visited network. So, if the roamer is going to be able to buy flowers or pay for the parking lot, home operators must provide visited operators the necessary information for authentication and authorization for fraud-proof m-payments.

---

[19]Discussions with Eduard Ebbink of KPN Mobile.

It is possible that service variety will be greater when a user has access to both home and local M-Commerce services while roaming, if only because there will be many more relationships between local providers and merchants. Assuming the visited operator is able to trigger the payment process and deduct payment from the customer's virtual account, the customer would pay the local VAT and be done. However, if the transaction amount is transmitted to the home network for the home operator to bill, double taxation may be unavoidable—the first tax is the local tax on the purchased good and the second tax is what is on the home operator's bill. In the short term, operators should ignore these currently insurmountable problems for "very micro" payments and allow roamers to place premium-rate mobile calls to vending machines. Yes, there is double taxation, but for a can of soda, it is probably worth it.

One of the findings of the aforementioned 2000 Nokia study on local payment was that the mobile payment alternatives would make life easier for the consumer while traveling. The convenience of paying for small transactions with the mobile phone is undeniable. However, at least for the next couple of years while M-Commerce is spreading, it will be more feasible to take advantage of the home operator's remote M-Commerce services. Service selection and availability while roaming can be enriched, however, if the visited network acts as a conduit between local merchants or content providers and the home operator.

A Few Last Words on M-Commerce: Variations on the Theme of Roaming

M-Commerce is actually a type of roaming (I like to call it intersector roaming). Therefore, its success depends on interoperability.

Those of you who have read the preceding chapters are well aware of the concepts of roaming and interoperability. For mobile operators, these are the two magic words. The beauty of GSM is the ability to use your mobile phone wherever there is a GSM network; the user is not confined by the national boundaries of her operator's network. While traveling abroad, she can take advantage of local networks without having to establish a new relationship. This is an outcome of interoperability between networks of the same technology. Its success depends entirely on the implementation of standard technologies and standard agreements between operators. Therefore, by owning one network, an operator can actually make money from its access to all GSM networks with which it has a roaming agreement. The mobile industry has observed that consumers appreciate and value the convenience of using the same services in the same way wherever they go.

Roaming is not limited to geographic freedom; we can also talk about intergenerational roaming (2G–3G); interstandard roaming (GSM–CDMA) and intra-standard roaming (GSM 1900–GSM 900). M-Commerce itself is actually a type of roaming: intersector roaming. For the successful delivery of a transaction service, it requires *interoperability*, *functioning interfaces*, and *agreements* among parties of different sectors (including operators, payment service providers, content providers, application service providers, hardware suppliers, software developers, system integrators, banks, and credit card issuers).[20] The really interesting dimension of M-Commerce (at least for those inside the industry) is therefore not really what can be bought, but determining which sector or party will lead the transaction, how the transaction will be designed and carried out bug-free, how each party will "talk" to each other during the transaction, and, of course, how each party will split the spoils. The interesting dimension from the customer point of view is the consistency of the experience and how much she has to learn to use the service. With these questions ahead of them, industry participants are convinced that the market will favor a standardized solution or customer experience,[21] perhaps under the umbrella of a global brand and ideally in the framework of a global infrastructure for authentication and authorization.[22]

The Other Mysteries of M-Commerce

In this brief chapter, I attempted to present M-Commerce as an extension of commerce and mention some of its mechanisms and capabilities

---

[20]Sanjay Kalluvilayil, "Owning the Wireless Customer Experience," Edgecom, White Paper 2001.

[21]Boris Groendahl and Braden Reddall, "Vodafone, T-Mobile to launch 'mobile wallet'," *Total Telecom*, March 14, 2002. Joanne Taaffe, "Operators move ahead on M-Commerce," *Communications Week International*, April 1, 2002. Timo Poropudas, "Common interest smoothes mobile payment waters," *Mobile Commerce Net*, April 18, 2002.

[22]George Malim, "Mobile & Satellite: M-Commerce—Lack of global standards spurs regional groups..." *CWI Online*, June 18, 2001.

but have left out some crucial dimensions since only a dedicated chapter would do justice to each of these topics. One dimension is the discussion on M-Commerce transaction security—security is one of the obvious cornerstones of any commercial activity, as security breeds trust. Another dimension is the impact of mobile devices on M-Commerce and, alternatively, the potential impact of the M-Commerce vision on mobile devices. An additional missing dimension (one that I am particularly interested in) is the related regulatory environment: in which cases can operators authorize payments? Do operators need a banking license to move funds around? Can there be a global e-money directive specific to mobile payments (probably not and if so, what do we do)? In Europe, how will local legislations coincide with EU legislation, and will the emergence of country-specific implementations inhibit the creation of a standardized infrastructure or logistical process for M-Commerce? Can double taxation be avoided? And so the list goes on.

And Finally...

Predictions are meant to be wrong, so I will refrain from making any. However, I will share with you my perception of future M-Commerce.

My SIM card should contain all my M-Commerce ID information in secure form. Most shops will support "m-pay." In these shops, I want to be able to pay simply by scanning my phone at the cash register. When I scan, all transaction data will be transmitted for either my operator or another party to authenticate (based on my M-Commerce ID) and then to authorize. Finally, it will send a signal to the merchant and to me that it is all clear. And when I travel, my M-Commerce ID will travel with me; I won't need to do anything different just because I am abroad. I will be willing to enter a secret code to activate the transaction process before I scan the barcode with my mobile phone—however, that will be the extent of it, no more key punching.

CHAPTER 18

Applications in a GPRS Environment

Richard L. Schwartz

SoloMio

It doesn't even sound right. A chapter about "applications" in a book about phones? What do applications have to do with phones anyway?

Let's start slow. Applications run on computers (like Microsoft Word), right? If you stretch the term a bit, the World Wide Web is filled with applications. When you want to buy a book, Amazon.com provides an application to search for a book, order it, and have it delivered. Running on a server somewhere at Amazon (probably many of them) is a big honking application that is connected to a monstrous database of information. As a user, you are connecting to that application using your browser (which is, by the way, another general-purpose application). Without thinking about it, you already access many applications from your phone, even today. Every time you leave a voice mail message for someone, you are talking to a server software application, processing and storing your voice as data and responding to your button presses on the phone to navigate you through an application menu.

What about the phone itself? The phone is a communication device. You pick it up, dial a number, and hope you hear "hello" at the other end. If not, you probably sigh, then decide to leave a voice mail message asking for the other party to call you back, in hopes that you in turn will be available when they do. When *your* phone rings, you probably stare at the caller ID (somewhat recently) and decide whether to answer it or not. If you are in a meeting, you might actually turn off the phone to avoid being disturbed. Unless you are waiting for the birth of your first child—in which case you likely monitor the phone for urgent calls. This kind of phone behavior is what the mobile phone is all about. Everybody understands it without really analyzing it. It is obviously a communication device, and obviously for voice purposes, with maybe a few bells and whistles like voice mail or caller ID to make communication easier.

Although you don't see it this way, the phone itself is actually a small special purpose networked computer—with a display, buttons, a bit of "firmware" or software running inside, and a very sophisticated communication protocol to transfer instructions and "control flow" between the handset and phone network switching. Fancier phones today have additional capabilities such as automatic lookup of incoming phone numbers in your local phone address book (if they're there, the screen will show your familiar name on the screen: "Mom's mobile," "Marc's office"). Also available is a menu that pops up for an incoming call to let you decide how to handle it: answer it on the built-in speaker phone? Add that caller to your local phone book?

Illustrations by holly@large communication design (www.hollyatlarge.com).

What about these WAP applications the industry was talking up, then talking down? WAP got people comfortable with the idea of a new set of phone applications to do everything from ordering taxis to sending flowers. Is there a "killer application" for the phone or, as some suggest, a "killer cocktail" of many smaller applications? The concept of applications for the phone wasn't new with WAP—what is new and important is the higher degree of standardization and openness, which provides a rich opportunity for third-party application software companies.

Applications in Phones— What's Already Here?

At the back end, buried deep inside the mobile network operator's switching network, are very large-scale network applications. They can be well established like the voice mail system, or evolving like the pre-paid systems that handle subscribers whose accounts might be running out of money in the middle of a call. Those applications hook into the actual switching systems for "call flow." Modern phone networks, whether mobile or fixed-line, have intelligent network interfaces that allow new applications to be added to the call ringing and call answering. With these interfaces, network operators can purchase network software applications from outside telecom software vendors. Number portability is probably a good example of such applications. It lets the customer own the same phone number even as he switches between carriers. Another is VPN, a popular corporate telecom service application that lets colleagues dial your normal office phone extension and reach you on your mobile phone.

So far, we've talked about voice interfaces and voice applications for the phone. Not long ago a tidal wave swept through most of the world in the form of text interfaces for "calling" people with short message service (SMS). Simple text messaging has been a hugely successful phone application. Rather than ringing someone up with the intent of speaking to her, you can tap a message into your phone and have it delivered with a beep to the screen on her phone. Why do it? It's quick, it's pretty easy, it's cheap, it's hip, and this way you can probably get through to her even when she's busy.

Of course, when you call someone, particularly on their mobile phone, you don't really know whether they are busy or not. With the deep penetration of mobile phones, people are pretty *reachable* no matter where they are. That doesn't mean they are always *available*. Most people find voice particularly intrusive when they're otherwise occupied—it's a personal contact after all. SMS applications address this. If you suspect the person you're calling won't want to talk, send a quick message ("Running 10 minutes late" or "Meet you in the pub?"). The person addressed with a text message is pretty predictably going to glance at it, and can respond discretely or tap a quick reply when the meeting is over.

SMS turned out to be a very profitable application for mobile network operators. It doesn't take a lot of network bandwidth; it can be scheduled or delayed depending on network load; customers are willing to pay a handsome fee for it; and it is very popular, particularly with 18- to 24-year-olds.

Naturally operators love it—and are looking for other applications that have the same feel and value proposition. In this category are the Blackberry devices from Research in Motion, which allow email to flow back and forth via a small pager-type device. The device vibrates to let you know that a message has arrived, or perhaps there's an instant messaging alert. Instant messaging as an application has started to appear in the mobile phone as well. Why? For a lot of the same reasons as SMS. You can get to people at times when they wouldn't "come to the phone." We now know that there are other motivations for people choosing to interact with text rather than voice as well. Delivered text is light and less emotionally charged than voice. It's a great way to tell someone something (e.g., "Sorry, I'll be late") without being challenged, and it's a safe way to ask for a date. (There's another youth application!)

Figure 18.1
Built-in phone
applications today.

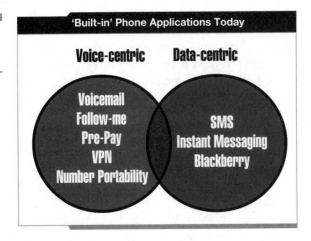

Instant messaging can reach out beyond the cell phone itself. It ties in (subject to many industrial tugs-of-war for market share dominance) to large existing PC and Web communities. You can talk to people from your phone IM who are themselves sitting at their computer IM (just like with voice calls, you might be thinking...hold that thought!)

So far, we are talking about applications literally built into the phone by the phone vendor. Or built into the network, often by the network hardware manufacturer who supplies the switch or the infrastructure for voice mail phone ports. In short, we're talking about software applications developed by hardware vendors, and defined almost exclusively by the telecom service providers. Their interfaces were proprietary, well guarded, and not available for third-party software application vendors.

But times, they are a-changin'.

Open Standards Hit the Accelerator

Open standards are systematically being introduced into the phone network. They're in intelligent network switching and routing systems, they handle traffic between the network and the handset, and they control the user experience within the handset itself. End to end, applications can now be defined by third-party software developers and hooked into the network.

Applications are being developed and sold to the mobile network operator, network infrastructure vendor, terminal vendor, and most importantly, directly to the end customer. End-users could be the mass-market consumer, the business phone user, or the corporate enterprise accounts—on behalf of employees.

The standards are in place and the floodgates have opened. No actual floods yet—but definitely a steady trickle. There have also been a few stops and starts as too much money and enthusiasm produced the usual unrealistic first applications crop. There is no substitute for learning by trial and error. The trick is to let others err and try something else. Being first is a double-edged sword...either the pioneers win by grabbing market share or they lose with arrows in their backs. Today's application software companies will fall into one of these categories perforce...but which one?

Was WAP a Good Thing or Not?

Let's talk about WAP and WAP applications. By themselves at first, and then in conjunction with GPRS. Starting a couple of years ago, the phone got a browser and the phone network got data. The browser was a bit crude, the screen was kind of small (probably always will be), and the data connection was circuit-switched (in other words, a physical phone call must be made each time a data connection is needed). The browser used a Web-like language (WML—wireless markup language), which meant that the browser was tantalizingly close to the desktop computer's Web browser. The WAP browser was like a blank slate in the phone ready to be filled by the user or even by the network.

At the same time hype reawakened. Glory to the phone! The Internet goes mobile! Every Web developer and online service saw stars and $$$. One billion additional users for their service. All they had to do was make a few changes to let the phone Web browser surf their site, and riches beyond their dreams would flow. Then a funny thing happened. Users didn't seem to care about the overwhelming majority of the new services and applications for a couple different reasons.

Figure 18.2
Mobile phone users
like to talk.

The most fundamental problem is that users think of a phone as a phone. They don't want to do the same things with it that they do with a PC in their office or their desk at home in the evening. That is better done at a large screen, when they have the *time* and *focus* to surf the Web, research books to buy, or read up on dinosaur bones excavations in Africa. The thing about the mobile phone is, people are by definition doing something else when they use it. People are out and about, buying something, driving, in a meeting, or whatever. Data is not top of mind.

This is not to say that people wouldn't want to order a taxi to pick them up when they leave the restaurant or be alerted on their mobile phone when their plane flight is late. Those are great applications and great services when you need them. The whole concept, though, of data-related services on the phone has to be conceived as a new experience meeting new needs. Nothing to do with the computer and the Web—at least until proven otherwise for a particular application need.

PC Internet applications are feature-rich and versatile. That's not a model we can transfer to the phone. I think of the mobile phone user as demanding *10-second applications*. Whatever they have to do, the application to do it better not demand more than 10 seconds of attention. A glance at the screen, pressing a button or two, and that's it. No navigation, no surfing...nothing extraneous. In order to be that kind of fast and simple, requiring no thought on my part, the whole thing had better be deeply personalized, crisply tailored, and anticipate my needs. The phrase "right to the point" comes to mind. Otherwise the pain overcomes the gain. Forget it!

Figure 18.3
10-second
applications.

10-Second Applications

10-Second Application Criteria:

- Meet the needs of the **mobile** user

- Minimum distraction, minimum attention

- Get in, get out, get success – in 10 seconds or less

- Learn, and don't keep asking

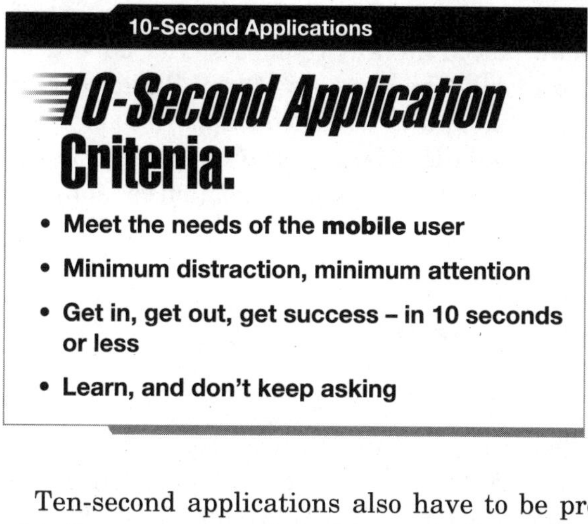

Ten-second applications also have to be pretty smart—smart about who the users are, what they typically want, and their preferences for getting it. Ten seconds isn't very long. And, by the way, a person's tolerance for telling the phone application the same thing twice, even a week later, is pretty low. Phone applications have to have "memory" and learn how to discern and adapt to patterns—to fit like a glove.

Can an Application Learn?

In the Web world right now, some of this learning goes on. "Personalization" is a kind of analytic engine technology sometimes used by the best Web sites, like Amazon.com. You might have noticed that after buying a few books there and coming back a few times to search for more, the site seems to recognize you. The home page for Amazon.com literally changes to talk to you about book or video offers that fit *your* taste. You may not even notice, but you certainly buy more books. It never actually asked you to fill out a 12-page questionnaire about your taste, it just kind of learns about you slowly by watching what you buy and what you look for when you visit. This same technology, borrowed from Web business applications, will be a fundamental building block in delivering intelligent phone applications. In a mobile 10-second application, this facility will be a necessity, not a nice-to-have. Mobile users will not stand around and wait (no pun intended)!

WAP applications have yet to make this transition. Many of the first WAP applications were adaptations of Web services and applications. In fact, many of the first WAP technology elements focused on "transcoding"—how do you literally re-purpose the same Web content and applications for mobile phone access? Ouch! While there are some applications that might "cross over" from Web applications to the mobile world, developers must do a lot of *fresh thinking* about the mobile user, his distinct needs, and what kind of new application capabilities will appeal to him.

Phone Meets Data. Is It True Love?

Let's bring the two threads of our application discussion together. The phone is a communication device, perceived mostly as a voice communication device. People have started to understand, somewhat independently, that forms of nonverbal communication such as SMS and instant messaging can help communicate in some contexts. The mainstream question, then, is how can the new data network and WAP-in-the-phone change how the mainstream communicates using their mobile phone? It must start with an evolution in voice communication, just like with voice mail and caller ID—and the cell phone itself, relative to landlines. Much of this change will result from unifying and combining voice and data applications in a way never done before.

Think about the insights with which SMS, RIM pagers, and instant messaging surprised us. Think about the intelligent network standards that can enable applications to literally get control over incoming voice calls or place outgoing calls. Think about what the emerging true 3G phone network might mean when voice is just another data type like text or video or Microsoft PowerPoint slides. Imagine how phone calls might combine transmitting voice together with pictures, documents, or even instructions on whom to add to the call as they become available. How "phone calls" might be initiated by clicking on a company Web site, or how incoming calls from a customer might automatically pop up their account and service history.

Figure 18.4
Rich call combines voice and data.

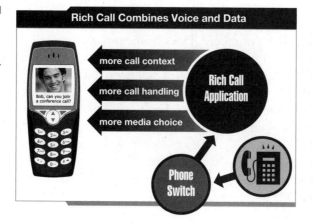

This general category of application is sometimes called "rich call" or "rich voice"—in that it adds nontraditional features to the traditional voice-calling paradigm. It is often also about extending this same paradigm to nonvoice media. Being able to initiate a call via an SMS text message but respond via voice, for example, or being able to accept a call on your cell phone that was received at the speaker on your desktop PC, or setting up a multiparty SMS "conference call" to chat about which restaurant to rendezvous.

GPRS Is the Foundation

The technology is now just about in place to build and deliver these types of applications. Open standards (we'll be more specific in a moment) can let independent software vendors build applications like these that are not vendor-specific—not dependent on the particular handset or particular phone network infrastructure. Some of these applications will be sold to the mobile network operators themselves, while others will be offered as enterprise software applications sold through the same channels as customer relationship management (CRM) or enterprise resource planning (ERP) systems.

Why did we say "just about in place"? First of all, GPRS is the first general "always-on" network technology. Sort of, at least, mostly. Circuit-switched data capabilities with WAP had a significant delay (latency) with every use and couldn't really be counted on to stay connected for longer periods of intermittent use. While GPRS doesn't quite deliver permanent connectivity in a technical sense, it *almost* feels to the user like his mobile phone has an Internet (IP) address that can be used like a "data phone number." The significance of this is that network data applications can decide to initiate a data dialog with the mobile phone user—to wake up the handset and display a data menu or ring the phone. They might do this, for example, to alert you that there has been a delay in your airline flight (press a key to rebook on another airline) or to tell you that Joe is on the phone and give you some options for handling his call.

GPRS doesn't yet treat voice as another data type. It is, strictly speaking, an overlay network that adds packet handling for data, but still relies on separate circuits for voice calls. This limitation doesn't go away until 3G networks like UMTS come about with newer Voice over IP transmission technology. Even first deployments of 3G networks, by the way, may distinguish voice from other types of data—just to make sure that voice calls never fail, no matter what. In the spirit of full disclosure, combining voice and data handling into a single application has one more limitation—the handset. (Class B phones won't be able to mix voice and data for the next couple of years.) An incoming voice call may stop data transmission, and vice versa. Be aware of these limitations and whether they affect the application you are inventing or evaluating. At the same time, GPRS can be thought of as a kind of "3G on training wheels"—much third-generation innovation will occur first in GPRS. Although they'll have to adapt to lower bandwidth and less-than-optimal voice and data integration, many pioneering services will be trialed and popularized first in GPRS, then extended to be faster and better in UMTS.

WAP does, by the way, perform much better when operating in a GPRS always-on network. It transfers data faster and is faster to get started at the outset. In GPRS session, unlike circuit-switched WAP sessions, applications can be designed to stay active longer. While the WAP session is active, the application has an assigned IP address and may receive packet updates. Furthermore, most mobile operators are adopting per-kilobyte or flat-rate charging models for use of the data network, in place of the pay-by-the-minute pricing predominant in circuit-switched WAP applications. This is ideal for longer-lasting interactions that demand only occasional attention from the user (e.g., turn-by-turn driving instructions or live sports scores).

"Soon-ish" you can expect WAP-Push to be available for applications. WAP-Push is actually quite important to real 10-second applications—it is the standard for "waking up" the handset as described above. In the meantime you may have to make do with SMS, USSD, or the SIM toolkit.

WAP is not the end-all and be-all for user interfaces for applications. Java is starting to come into its own within the phone to provide a basis for more sizzle in the user experience. Right now, although it's being rolled out in a number of vendors' handsets, Java is limited to stand-alone applications (in other words, applications not involving communication!). There are industry initiatives such as Sun's MIDP efforts to eliminate these restrictions and let Java supplement or replace WAP for mainstream uses. At the high end of the phone spectrum, so-called "smart phones" are actually full computer operating systems within the phone. The Symbian EPOC joint venture has resulted in a series of Java-powered phone/PDA combinations, as have competitive efforts from Microsoft and Palm. Stay tuned for more developments in the handset experience.

Parlay Opens Up the Voice Switch to Applications

On the network side, "intelligent network" gateway platforms are entering the operator market that allow operators to acquire enhanced services applications from outside software application vendors who have built on new emerging standards for voice call control. The real open standard for trapping phone calls and providing enhanced voice services is Parlay. Parlay is an open services architecture (OSA) that spans wireless, wireline, and corporate PBX networks. With Parlay, it is possible to

Figure 18.5

Phone applications of tomorrow (GPRS).

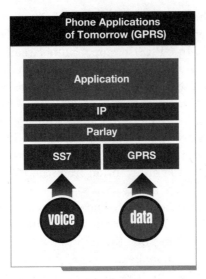

build very sophisticated applications that quite literally take control when a phone rings and provide the caller with an experience that blurs the line between being answered by a human or by an intelligent call-handling application.

Operators are behind the Parlay direction for several reasons. The first is that they absolutely *need* new applications and services for the mainstream voice market. Enhanced voice services, like the very successful pre-pay and VPN services, let the operator earn incremental revenue from its existing subscribers. By implementing open standards, operators can look to many more application software vendors to invent and supply these applications. At the same time they eliminate an unproductive dependence on the original network supplier as the source of applications. They are also looking to wean themselves from the early generation of enhanced voice service applications that were built on the interactive voice response (IVR) and voice mail systems. These first systems were quite expensive to operate and were themselves built on proprietary interfaces that locked operators to a particular vendor.

Despite the potential payoff, Parlay has quite a high barrier to getting started. It involves sophisticated call-handling concepts appropriate for the needs of the professional network service operator, but that makes it hard to build and deliver an application that integrates Parlay and voice call management. The tradeoff for Parlay's power and robustness is obscurity; you will likely need to find one of the world's few Parlay/IN gurus in order to use it in your network. You can expect many of the first Parlay applications to be "solution packaged"—i.e., sold as a complete application and including the Parlay gateway and any other necessary software infrastructure. Meanwhile, some of the same gains can be had with less elaborate protocols and interfaces coming down the pike—namely service initiation protocol (SIP) and Java object interfaces (JAIN)—that are more accessible and tailored to the enterprise software developer.

Where Do Applications Go from Here?

Where are we in the application story? We've taken you on a tour that tried to stretch your perspective on what to do with phones. Then we described some evolutionary ways in which the concept of the phone as a communications device can be pulled in new directions and forms of communication. Phones and phone networks, starting in this GPRS era, are truly malleable interfaces that can be creatively extended through open standards.

Open standard interfaces. Broad user communities who already have the capabilities on which new applications will be built. Mobile network operators who are highly motivated to promote new data applications with quick time to revenue and rapid market acceptance. These sound like the conditions for a robust new application marketplace and an independent software vendor community.

Figure 18.6
Spreading the word.

Of course, GPRS, like any other new phenomenon, will not happen overnight. Phone users will not rush out immediately to buy a new handset. Therefore, part of the success for any broad market product must come from delivering benefits one user at a time. If I buy a GPRS application, can I get the benefits even if I am the first of my circle of friends to get one? Even more instrumental is where *viral* marketing occurs—that is, where my use of the product has visible value for everyone around me, who then want to rush off and buy one for themselves. Example of a successful viral effect: RIM pagers.

The opposite of "viral marketing" is "workgroup marketing," where an entire group has to commit before anyone can receive value. Where nothing in my phone experience changes until *everyone* I call has one. These products usually take a very long time to capture the mainstream market. Example of workgroup marketing: videophones.

Conclusion

The mobile application opportunity is distinct from the world of PC and Web applications. The market is large, and the mobile network operators are heavily motivated to replace decreasing voice revenues with higher-margin data services. End users are virtually guaranteed to have in their hands, soon if not already, a powerful device that could be running *your* application.

My company, SoloMio, is an independent telecommunication software company. We develop and market a network-based mobile application for 2.5/3G mobile network operators. Our application does indeed combine voice and data in a new way to solve a mainstream communication need. We are an example of the new breed of application described in this section...we build on the full range of GPRS data technologies just described to "break the rules" of voice communication, but do it in a way that can appeal to a market of people who are, after all, phone users. We're also part of the new-era telecom software community—exploiting and advancing the opportunity of open telecom standards and the convergence of phone and data networks.

Applications in phones? Absolutely. It just means taking the phone on its own terms, and delivering the biggest change to the medium since dial tone.

CHAPTER 19

Interoperator Settlement

Axel Doerner

Vodafone

Need and Demand

Chapters 2 and 17 have variously made the point that roaming is one of the key success factors in GSM. Whilst many private users are pleased to find out that their phones (more precisely their SIM cards) are working in foreign countries, business users regard the option to use their mobile phones abroad as a necessity. Indeed most professionals working in today's global business environment deem the mobile as one of the indispensable tools for traveling, topped only by passport and credit card. The ubiquitous usability of mobile phones supports the march of globalization much the way aviation does. This, in fact, forms a solid ground for the business case of many mobile carriers.

Roaming is implemented bilaterally between the home network and those networks serving home network subscribers who are roaming. It requires physical interconnection and signaling links between the home and the serving network, so that the latter is in a position to serve the roaming user according to his subscription. Typical situations where we must invoke this sort of setup include the following:

Example 1. A mobile subscriber tries to set up a data call. The serving network needs to know whether this subscriber is actually allowed to set up data calls.

Example 2. A mobile prepaid subscriber tries to set up a voice call. The serving network must know whether he is in good standing to make the call, and when to terminate the call in case his balance falls below zero.

Technically, authentication of roaming subscribers is implemented as a natural extension of the authentication mechanisms within each GSM network. The physical signaling interconnection is used to keep the home network informed of its subscribers' actual locations. This is important so that the home network can continue to deliver incoming calls. Current practice is to route all incoming calls to a mobile subscriber's home network first, which then forwards them to the network presently serving the subscriber.

The implementation of roaming between a serving and a home network is covered by a *roaming agreement*. Wherever a roaming agreement is in place, the home network's subscribers will be (more or less) entitled to use the serving network's resources as if they were its subscribers. There is, however, no contract between the subscriber and the network he is visiting.

As a result, the serving network provides resources to the roaming subscriber without any possibility of billing him directly. Instead, it charges the home network, with whom it has a contractual relationship via the roaming agreement. This process is called *interoperator settlement*. The underlying technical process for transferring roaming charges at the call detail level between serving network and home network is called the *transferred account procedure* (TAP).

Interoperator settlement between GSM networks is governed by principles agreed on by mobile operators represented by the GSM Association. The umbrella principle is that the serving network shall apply the same tariff to all roamers irrespective of their home network. Initially, GSM operators also agreed that roamers should be treated the same as the serving network's own subscribers with respect to tariffs applied. That proved to be too much populism. With the diversification of tariffs, this principle turned absurd. Also, the serving network's business case for roaming subscribers may differ significantly from its business case for its own subscribers. These and other considerations have led to a redefinition of the interoperator pricing policy. Today the serving network's roaming tariff is equally valid for all home networks, but there is no longer a link between the serving network's subscriber tariffs and the tariff it charges roamers. The latter is now referred to as *interoperator tariff* (IOT).

This means that the following tariffs are systematically decoupled from each other:

- The retail tariff charged by the serving network towards its own subscribers
- The IOT charged by the serving network towards the home network

Figure 19.1
Retail billing of roaming usage.

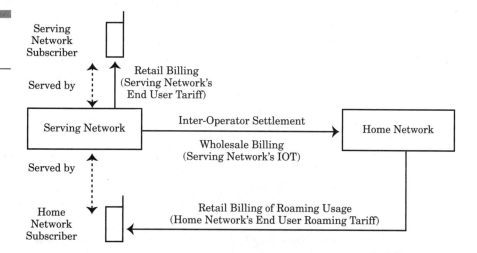

As a means of supporting special bilateral relationships between networks, operators may agree on discounts on top of the IOT.

The complete separation of end user tariffs from interoperator tariffs has an important technical consequence: *the home network must be able to reprice the call according to the tariff sold to its subscriber*. For that to happen, the technical interoperator billing interface must be ready to report all corresponding attributes which potentially drive the charge, even where they have not been used by the serving network to calculate the interoperator charge according to its IOT.

Voice versus Data Requirements

When considering mobile data services and how to bill for them, readers may wonder if there are different requirements at all. After all, many networks charge ordinary circuit-switched data calls like voice calls. The justification for this charging policy is that plain vanilla circuit-switched GSM data connections consume more or less the same resources within the GSM network as voice connections, namely a single permanent transmission channel.

This changes drastically with the introduction of 2.5G data services, i.e., high-speed circuit-switched data (HSCSD) and GPRS. Common to both services is an increased and variable quality of service (QoS). QoS is both a measure for resource consumption within the mobile network and for service delivery to the user. Indeed, service delivery to the user may change significantly with the level of QoS allocated for the data session.

For HSCSD, which works by bundling traffic channels (see Chapter 3), QoS is defined by the number of channels used and the channel coding (4.8 kbps, 9.6 kbps, or 14.4 kbps) which will be identical for all traffic channels allocated for an HSCSD connection. (For technical details, please review Chapter 10.) The overall transmission rate experienced by the user can then be calculated by the formula

Transmission rate = number of traffic channels × channel coding

This formula implies that the transmission rate for an HSCSD connection could vary between 4.8 kbps (1 traffic channel at 4.8 kbps) and 115.2 kbps (a maximum 8 traffic channels at 14.4 kbps). For GPRS, the structure of QoS is more complex. Apart from the peak and mean throughput rates, it comprises *service precedence* (transfer priority), *transfer delay*, and *reliability*. The parameterization of the QoS requested by the subscriber from the serving network depends on the user's application. The following table gives an overview of QoS requirements for some applications that may be invoked over GPRS:

TABLE 19.1

Requirements of QoS by Applications

Application	Requirements of the QoS
File transfer	High reliability
Streaming	Short delay Depending on the application (e.g., video), high throughput
Online chat	Short delay
Online games	Short dela High reliability
Email	High reliability
Instant messaging	Short dela High reliability

Two aspects of QoS are particularly important to note. Users may issue QoS for a connection towards the network at connection setup or later during the session. The network, however, may either allocate or deny the QoS requested. Also, the network itself may initiate a change in the QoS during the connection if bandwidth conditions have changed. As a consequence, each charging scheme that relies on QoS must strike a balance between the QoS requested by the subscriber and the QoS provided by the network. The tariff applied will change whenever QoS changes. This is a significant departure from standard GSM circuit switched-service usage where the tariff applied is driven only by the basic service used—which is, in almost all instances, constant during the call.

Table 19.1 also illustrates a key dilemma of GPRS: it was designed as a transparent bit type with no built-in knowledge about applications. This design in the first instance affects the user. Subscribing to GPRS services typically will not be enough to initiate service; someone still needs to provide the application logic. "Someone" may be the user himself, the network operator, or a third-party content provider. So, in the second instance, GPRS's transparent bit pipe design also affects mobile operators. It requires not only that they invent applications for their

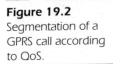

Figure 19.2
Segmentation of a
GPRS call according
to QoS.

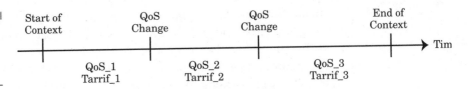

subscribers, but also that they ensure these applications will be operable given the bandwidth within the operator's network.

Last but not least, network operators have to figure out how to charge for these applications. As the GPRS network has no knowledge of the actual application, it does not support charging for applications—its built-in charging mechanisms only allow for charging the subscriber based on QoS. The first implementations of GPRS do not even support the QoS negotiation process described above; they deliver a best-effort QoS in any case to the user. Future GPRS network releases, however, will have a QoS negotiation mechanism and mobile carriers will gradually introduce QoS-based tariffs for GPRS in line with their individual deployment of GPRS network capabilities.

GPRS and HSCSD are alike in the respect that QoS can potentially drive the charge for data service usage in both. The fundamental difference between the two is, of course, that GPRS is a packet-switched bearer service. You have read in previous chapters what this means in terms of resource usage: instead of permanently allocating one or more fixed transmission channels to a session, GPRS only allocates transmission resources in the serving network when data are actually sent or received. Hence, the GPRS vision for its users of "always online, always connected, and empowered to send and receive data as the need arises."

However, in order to turn this vision into reality, mobile carriers must find a way to make end user tariffs reflect this resource allocation. In other words, GPRS tariffs must be premised on the volume of data sent and received by the subscriber. Though most observers take this for granted, tariffs based on data volume are a challenge for the mobile operator for several reasons: First, end users typically are not used to being charged on that basis, and the average end user has no feeling for the data volume involved in, say, loading Web pages. Second, the data volume measured at the application level will differ from the data volume measured at the network level, because the latter—which is the one used to calculate the GPRS charge—also includes overhead inherited from the service's layered architecture (protocol headers, etc.). Even if the user tries to trace her data volume expenditure, she will never be able to track it exactly. The volume reported by her application will not assess overheads.

To overcome these problems, operators have considered introducing GPRS tariffs that rely on session duration as primary charging parameter instead of data volume—as circuit-switched data calls do. Unfortunately there is a major drawback to duration-based GPRS tariffs: given that GPRS is designed to optimize the "always-on" connection, it does not support exact time measurements. This means that the network (depending on the actual implementation) may fail to notice immediately that a user has left GPRS coverage, and continue to record connection time for an open GPRS context, after the user's terminal indicates that the connection has been lost!

Presently the possibility of introducing time-based tariffs for GPRS seems to depend on the actual implementation of the time measurement within each GPRS network. However, operators may seek to close this gap when offering real-time services over GPRS. These services (e.g., video telephony) require high QoS throughout the session. For them, it may be argued that a virtual permanent transmission channel must be allocated for the duration of the session. With some such adjustments, charging by duration may become a reasonable way of charging for this type of service.

What we've said so far about charging for data services was primarily said in the context of end user tariffing. Getting back to the topic of this section, we have to remind the reader that tariffs may also be set quite naturally at the interoperator level. The interoperator tariffs for 2.5G data services will be a function of duration, data volume, and QoS. However, taking into account the vast diversification possibilities of tariffs for GPRS, the IOT may end up relying on a parameterization other than the one used by the home network for its GPRS end user tariffs.

One striking example is the case where the serving network's interoperator tariff for GPRS relies on connection duration and the home network's retail tariff relies on data volume (or vice versa). Of course, we have made the statement that end user tariffs on the home network are, in principle, completely decoupled from interoperator tariffs. Still, many home networks charge outgoing roaming calls to the end user by applying the "interoperator tariff plus surcharge." Where there are structural differences between how the serving network charges GPRS usage at the interoperator level and how the home network charges GPRS at the retail level, this easy scheme may lead to intransparencies for the end customer—and thus may discourage him from roaming. Accordingly, it seems that the introduction of GPRS may actually enforce the mobile operators' agreement to decouple their end user tariff from IOTs, just to keep roaming tariffs transparent to their customers.

Switching technology aside, the second major difference between GPRS and circuit-switched data services is that the user connects to a remote IP network, typically the Internet or a private IP network, rather than setting up an end-to-end connection. This means that the mobile network needs to function as a gateway between the user and the external IP network. The GPRS standard supports the roaming business in that it logically separates the functions of the access network (providing mobile access to GPRS bearer via the air interface) and the gateway network (providing access to the external network). The initial implementation of GPRS allows for the following two scenarios:

1. **Local access to GPRS services where serving network and gateway network are identical**—In this case the serving network provides interface access, the interconnection to the gateway, and the access to the remote IP network.
2. **Remote access to GPRS**—The subscriber is roaming but still uses the home network as a gateway towards the external IP network. In this case the serving network provides interface access and the interconnection to the gateway.

Figure 19.3
Local and remote access to GPRS services.

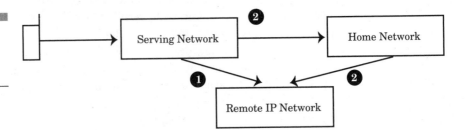

Of course, the knowledgeable reader may insist that we should employ the term "portal network" instead of "gateway network." Naturally, mobile operators hope that they will succeed in establishing the remote access scenario as the de facto standard, and thus become the customer's universal portal. The customer also benefits from accessing GPRS services via the home network as a default, because the home network may offer applications he can use seamlessly everywhere. A trivial example is WAP—still in fact the only data application built into almost every GPRS terminal. Accessing WAP via the home network even when roaming lets the user travel with a familiar, in many instances personalized, portal in a language of his choice—a portable user interface, if you will.

Depending on the scenario, it is likely that the serving network will differentiate its IOT according the resources provided. For local access to GPRS services, the serving network may want to advertise that it offers overall interconnection with the remote IP network and thus adds value to the provisioning of the pure bearer. For remote access, the serving network provides the pure bearer resource only, so a cross-network interconnection is required to reach the gateway in the home network.

GPRS provides an intrinsic mechanism to distinguish between local access and remote access scenarios, the *access point name* (APN). When configured in the terminal, the APN enables the user to choose the "logical" access point and thus the remote IP network and the gateway. Choosing the access point explicitly by means of specifying an APN may have broader benefits. It has been discussed as a potential way of distinguishing applications, the idea here being that each application would correspond to an APN. The major drawback of this otherwise attractive idea is that it would obstruct our ability to change APNs implicitly whenever the user switches applications. The user would then have to change APNs manually in the same way an SMSC address has to be changed, which is a user-inimical solution and a marketing deficit.

To summarize our conclusions: 2.5G data services allow the serving network to introduce a vast variety of tariffs, due to the increased number of parameters which may drive resource consumption and consequently charges. For GPRS, the chief factors are (1) the APN and the scenario (local or remote); (2) the data volume sent/received by the subscriber, and (3) the parameters driving QoS.

Transferred Account Procedure—TAP3

The original version of the interoperator billing interface, TAP1, provided support for GSM phase 1 services. It was launched in 1992 and officially superseded by TAP2 in 1996/7. TAP2 provided comprehensive support for GSM phase 2 features—mainly better support for standardized GSM supplementary services. A major rework of the interoperator billing interface was required when GSM phase 2+ services were introduced. The emerging data services HSCSD and GPRS required substantial changes to TAP. Those services not only added a great number of charging attributes, but added attributes that may change several times during a data call (QoS for example). This new kind of attribute implies that the format must be ready to record mid-call changes so that the home network can re-price the call from scratch, independent from the inter-operator tariff applied by the serving network. Hence, a format that's intrinsically flexible and extensible was suddenly needed, as opposed to the rigid fixed-length formats used previously. TAP3 indeed complies with these requirements via a well-established flexible format encoding mechanism, the joint ISO/ITU standard ASN.1.

Nowadays, operators are starting to use TAP3 to establish interoperator accounting for GPRS. Although TAP3 supports a number of other important services, it seems that GPRS roaming was ultimately the main reason for GSM operators to switch from previous versions of TAP to TAP3. In the meantime, the GSM Association has started work that will extend TAP3 beyond its original scope of supporting interoperator settlement for standardized network services, so that billing for non-standardized content applications will also be supported in the future. On the surface, this enhancement seems intended to support the scenario where a serving network offers non-standardized applications to roaming subscribers and bills them via the home network. Whilst this scenario is not in place so far, it is clear that billing mechanisms are needed for settlement between the home network and the content provider. The GSM Association, in principle, encourages that TAP3 be considered for this purpose.

It is hoped that a technical mechanism for interaccounting will help launch e-Business in a mobile environment. The basic idea is to apply proven mechanisms for the exchange of billing data on network service usage to content billing as well. The subscriber may then be billed via his monthly mobile statement. If interaccounting between the content provider and the mobile operator does use TAP3 as an underlying technical billing mechanism, TAP naturally evolves from a means of supporting interoperator settlement into a means of supporting interentity settlement.

Wholesale and Retail Variants

As previously mentioned, the introduction of GPRS roaming is expected to be a practical driver for separating interoperator tariffs and retail tariffs, mainly for the reason that the IOT has different charging parameters from the tariff scheme in the home network. A deviation between the two charging schemes can confuse and irritate the subscriber and put pressure on customer care centers. Operators have experienced this with respect to incoming calls when international roaming first launched. Within many home networks, the "calling party pays" principle is strictly applied, i.e., incoming calls are free. When subscribers of these networks started to roam internationally, they experienced charges for incoming calls received abroad for the first time. The home networks assessed a small fee for re-routing the incoming call to the subscriber's actual location. GPRS roaming will pose even harder challenges to operators, since more parties are involved in delivering the service to the user: The serving network provides air interface access and access to the gateway to the remote IP network. Depending on the actual scenario, access to the remote network is provided by either the serving network or the home network. Within the home network, subscribers are used to getting one price for GPRS usage. Now the home networks will have to decide whether to merge all charge components and construct a roaming pricing scheme.

Figure 19.4
Retail tariffing.

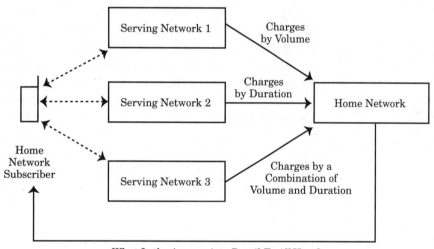

What Is the Appropriate Retail Tariff Here?

Of course, home networks are still free to apply the "interoperator tariff plus surcharge" principle to subscriber bills for GPRS roaming, and from a business perspective this may seem less risky than setting up a uniform tariff that no longer relates to the IOT. In spite of the various possibilities for setting GPRS interoperator tariffs, it is expected that most will rely on data volume and QoS as primary charging parameters. This decision should ease the introduction of the "interoperator tariff plus surcharge" formula for GPRS home retail tariffs which rely on data volume.

However, if the home network wants to offer a uniform roaming tariff to its subscribers in line with its charging policy for domestic usage, it will be forced to re-price the GPRS session. Luckily, the TAP3 billing interface provides home networks with the input from the serving network necessary to support this process.

Outlook

This chapter has given you a perspective on charging for GPRS roaming. Most of our discussion concerns pure bearer usage. For the ultimate target—seamlessly offering applications on top of GPRS and charging subscribers for them—we have barely scratched the surface. Naturally, applications will further complicate the settlement structure by adding new charge components and possibly new interaccounting relationships between operator and application provider. Finally it should be noted that the ongoing process of merging national mobile carriers under a common supra-national brand such as Vodafone, Orange, O2, or T-Mobile will help to offer uniform roaming products with almost seamless coverage. An increasingly concentrated mobile industry is likely to enable operators to resolve the issues raised.

CHAPTER 20

GPRS Roaming eXchange

Gerhard Heinzel
and Laurent Bernard

Swisscom Mobile and France Telecom

The Problem

If you're not already familiar with the term "roaming," that's probably because it's such a fundamental part of modern mobile communications that you take it entirely for granted. In simplest terms, GSM roaming allows you to make and receive calls on your mobile phone when traveling outside your home network. GPRS roaming does the same thing for the GPRS standard. However, because GPRS is a packet-switched technology designed to provide Internet access, the issues involved in carrying GPRS roaming traffic differ substantially from those faced by GSM networks.

As with GSM, GPRS roaming occurs when a subscriber to a public land mobile network (PLMN) goes to another country and connects to a different PLMN to access his usual services. While this will probably involve accessing the seemingly "ubiquitous" and access location–independent Internet, care has to be taken that other services like company intranets and home country service portals can also be accessed. Transmission security concerns must also be addressed.

GSM has been a huge success, due in large part to the "one subscription, worldwide" capability that roaming offers. As a result, we have seen GPRS roaming being established more or less synchronously with the rollout of GPRS networks around the globe. Service access while roaming can be enabled in either of two ways: Internet roaming and PLMN roaming. (Note: Since GPRS registration and authentication requirements are handled in much the same way as circuit-switched services, we don't discuss those here):

- **Internet roaming**—Traffic can be transmitted directly to the server from the visited PLMN. This is accomplished using a Gi interface that allows the PLMN to communicate with an external system via the gateway GPRS support node (GGSN). In this approach, it is the Internet that implements the roaming "leg" of the communication, which is why most people call it "Internet roaming" (the official name is "Visited GGSN Roaming").

  While Internet roaming works well for Internet servers, intranet servers that wish to use it must either connect to several different GGSNs or "tunnel" their way in via the public Internet. The first solution becomes wholly unmanageable when used on any sort of large scale, and the latter defies the very principle of an intranet— namely, independence from the Internet. Similar concerns arise with the use of designated mobile portals in the home country.

Figure 20.1
Internet roaming
model.

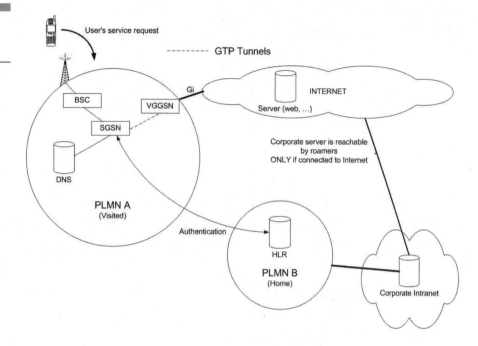

- **PLMN roaming**—Border gateways (BGs) can be used to transport the traffic to the home country first. This is accomplished via a Gp interface, which connects the GPRS support nodes (GSNs) of two different PLMNs. The Gi interface of the home GGSN is then used to connect to the Internet. In this approach the PLMNs take care of the roaming "leg" of the communication, which is why this is known as "PLMN roaming" (the official name is "Home GGSN Roaming").

   An added value of this solution is the operator's real-time control over its subscribers' roaming service usage—something operators of circuit-switched services have been seeking for almost 10 years as a hedge against rampant fraud. In looking at Figure 20.2, one might expect to see some severe tromboning effects here, with traffic carried to the home country even when it could have been more efficiently placed on the Internet from the visited PLMN. In fact, experience with current (fixed network) roaming services shows that some 80 percent of a typical roamer's traffic is exchanged with servers in the home country. The only real disadvantages of home GGSN roaming are the large number of inter-PLMN connections required to carry traffic and the need for additional components (the border gateways).

Figure 20.2
PLMN roaming
model.

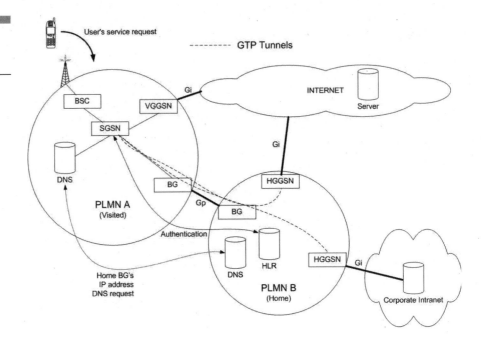

What Is "GRX"?

GRX stands for GPRS Roaming eXchange, which can be defined as: "An Inter-PLMN backbone network for GPRS inter-PLMN packet traffic."

In other words, a GRX is a centralized IP network to which PLMNs can connect to gain IP connectivity to other PLMNs. It serves as a duplicate private Internet specific to the GPRS operators, and is known as an Internet protocol virtual private network (IP-VPN). This approach is preferred primarily because of its inherent security (security being one of the major demands of operators) and cost advantages.

While several architectures have been envisioned for carrying the GPRS home GGSN roaming traffic (including international leased lines, asynchronous transfer mode, public internet/encryption, and tunneling), GRX with its circuitless, IP-based routing is the only one to be readily adopted. This is not surprising considering the advantages it has to offer.

It is important to remember that the very nature of IP is packet-based routing, which makes it impossible to provide PLMN-to-PLMN quality of service guarantees based on resource reservation (where QoS is defined as "the measure of network performance"), as with circuit-based services. Nonetheless, this doesn't stop GRX providers from using a whole set of underlying technologies—asynchronous transfer mode (ATM) or Frame Relay–based virtual circuits network, overprovisioning, layer 3 mechanisms, and others—to offer QoS.

One of the great virtues of IP is the ability to use the resources of the network to best effect in the presence of failure. Because packets are routed one by one, they can use the most efficient path at any given moment, routing around any potential problem areas. It is this ability that allows IP technology to provide enormous advantages over circuit-based services in terms of expenditures, operational costs ,and resilience.

Furthermore, GRX maximizes the bundling gain in that each physical connection carries several dedicated traffic flows and therefore can be built and maintained more efficiently and economically.

A detailed explanation of the GRX concept can be found in the GSMA document PRD IR.34—"Inter-PLMN Backbone Guidelines." (For further reference, see www.gsmworld.com).

Figure 20.3
GRX, a generic
diagram.

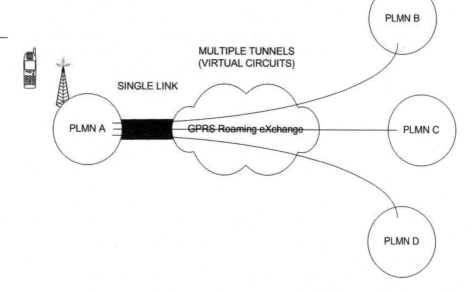

Operator's Demands

Due to the somewhat ambiguous way in which the term "GRX" is commonly used, the question arises: of the many GRX services on offer, which are basic GRX functionality and which are not? Unfortunately, the GSMA document that describes the GRX concept (PRD IR.34) also describes services that are not in fact mandatory. To blur the picture even further, different GRXs will offer different added services, and even if they are similar they will be offered under different brand names.

Still, it is possible to make sense of the muddle. Stripped of all the bells and whistles, GRX is essentially about *bandwidth*. That means we can effectively differentiate the various GRX offers by comparing them the way we would any bandwidth offer, in terms of *quality, security, connectivity, ubiquity, scalability, flexibility*, and—of course—the most important "physical" aspect: *price*.

Basic GRX Service Characteristics

Bandwidth

As indicated above, the main reason any operator would turn to a GRX is for its ability to transport what are called GTP packets (GTP is the IP tunneling protocol used in GPRS) from one PLMN to another. In order to do this, the GPRS operator must first connect to a GRX network site.

Typically, leased lines are used to connect some of the operator's sites to one or several of the GRX's points of presence (POPs). It is up to the GRX and the operator to decide between them who will order and maintain these leased lines. The party that maintains the lines will normally provide a dedicated router to the other party in order to create clearly defined boundaries for maintenance responsibilities and perimeter of service delivery.

Quality

Quality can be defined in many ways, from individual subjective perceptions (as is often the case), to quantifiable parameters like available bandwidth. The truth is, the concept of "quality of service" was introduced fairly late in the telecommunications world. This is not to say that telecommunications operators did not deliver good quality in the past; indeed, many placed service quality above all else, only to learn the hard way that for some subscribers price/performance ratio was more important than "quality" in the purest sense.

Concepts like QoS have come even later to the Internet world, where a trial-and-error approach has always been favored. The very design of IP is built around this concept, and in a network growing quickly and with only minimal controls, it is not easy to guarantee the behavior of any one set of nodes. The situation is exacerbated by the fact that the Internet was started by legally independent entities, each offering connectivity to others on an "as-is" basis, which means Internet protocol alone can not provide the kind of end-to-end resource reservation available from a CPE-to-CPE bandwidth guarantee, for example.

The problem for operators is that the public doesn't know and doesn't care about such considerations. Any operator's GPRS offer will be expected to fulfill the implicit promises made by its other services (most notably, GSM), while at the same time adapting to the rules and mechanisms of the Internet world. Subscribers will still expect to get a voice call through in no more than two attempts, and when abroad they will expect their telephone and Internet services to function "just like at home."

Quality of service for GPRS users has been defined in the following terms:

- **Priority**—The user may indicate which of his packets should be dropped first in case of network overload.
- **Reliability**—The user may ask for particular profiles with respect to packet loss, duplication, out-of-sequence delivery, and corruption probability.
- **Delay**—The user may describe the delay sensitivity of his application.
- **Throughput**—The user may negotiate the peak and mean throughput rate of his application.

In the packet routing world as it relates to GRX, however, quality typically is characterized by:

- **Delay (round trip)**—The measured time a packet needs to be transferred from one node to another and back.
- **Packet loss**—The percentage of packets lost between two nodes.

- **Jitter**—Average variation around the mean round trip delay value.
- **Throughput**—Also known as bit rate or bandwidth.
- **Bit error rate**—The percentage of bits changed during transport.

Clearly, there are differences in the way the GPRS and IP worlds approach the topic of quality. Furthermore, the GRX will only see IP traffic between GSNs; since it won't be expected to look into the GTP payload, it won't be able to prioritize user packets. That means the GPRS operator would need to either prioritize traffic internally before transmitting packets to the GRX network, or use (along with the GRX) a protocol that enables prioritization, such as Diffserv or packet tagging.

Luckily, the requirements for GPRS services are not stringent when compared with the state of the art in the IP bandwidth market and can be met quite easily. Where IP bandwidth limits intracontinental round trip delays to around 80 ms, the most stringent delay class for GPRS allows a mean value for the one-way delay of short/long packets (128/1024 bytes) of 500/2000 ms.

Typically, the QoS commitment is documented in an annex to the GRX service supplying contract—the service level agreement, or SLA, which will be discussed in detail later in this chapter. Most SLAs also contain GRX commitments for delivery lead delays for new or enhanced connections, mean time to restore service in case of failure, and overall service availability.

Security

General questions of security are dealt with in Chapter 14. Additional security concerns specific to the GRX concept will be discussed in detail later in this chapter.

Connectivity

Connectivity is defined by the IP routes provided by the GRX provider to the GPRS networks. These will be the identified PLMNs, preferably those directly connected. The existence of "peering" between GRX providers' networks (discussed later in this chapter) extends PLMN connectivity to other PLMNs not affiliated with the same GRX network. Still, for quality reasons, it is highly desirable to avoid transit through more than one GRX network when transferring data between PLMNs.

Due to different country, population, and network sizes—as well as personal travel preferences and professional travel necessities—the number of visitors and roamers any PLMN receives and sends to any other PLMN will vary from operator to operator. As a result, the priority for any GPRS operator will be to establish the highest-quality roaming possible with those operators with which it expects to exchange the most traffic... and it will choose a GRX provider accordingly.

Ubiquity

Ubiquity can be defined as a broad geographical dispersal of GRX service availability. This is important, as the prices for leased lines are distance dependent, making it generally desirable to have the shortest possible distance between the operator's site and the GRX network's POP.

The GPRS operators have an advantage here: due to the nature of their business they typically have sites in all major cities in their coverage area. As a result, operators often can decide quite freely in what city they wish to hand over traffic to the GRX, meaning ubiquity for a GRX provider will translate into coverage by country and roaming partners.

Scalability

The amount of bandwidth likely to be required between any two operators will be an important factor in determining costs. Because most GRXs try to guarantee their service quality based on a combination of traffic shaping, preallocation, and overdimensioning mechanisms, reliable load predictions are vital to reducing provisioning costs...and thereby reducing the prices charged to operators.

As usage grows, so too does the operators' need for short delivery lead times (the time needed by the GRX to provide additional "tail" and/or core network bandwidth) and the option to increase bandwidth in small increments in order to follow the usage uptake curve more closely.

Flexibility

GRXs will also need to offer the "right" type of interface, hardware configuration, and protocol—several types of FR and Ethernet interfaces, and filters on the CPE, for example—and should be prepared to adapt maintenance offers to the needs of the operators. Flexibility could also come into play in the provision of additional services: charging currency, reporting, and the like.

Additional Services

While GRXs are, of course, free to offer any type of additional service they wish, the following represent those most frequently offered to operators by the majority of GRX providers.

Domain Name System (DNS)

A GPRS DNS translates APNs (refer to Chapter 13) sent by a handset into the corresponding IP address of the GSN that leads to the requested service/server/network, so that the requesting GSN can establish the GTP tunnel. Although DNS is compulsory inside a PLMN, pan-PLMN DNS may be considered as a "comfort" service, and its characteristics are not yet fully defined by authorities (ITU-T, GSMA) as of the end of 2001.

A GRX DNS should ideally ensure that necessary information reaches every GPRS operator's particular DNS server. In one solution, the GRX DNS server is individually and manually updated whenever a connected GPRS operator informs the GRX provider of a new roaming agreement. Another solution provides for a unique "root DNS" server, which would contain records for each and every GPRS operator's DNS servers.

There are still several open issues concerning "root DNS" servers, most notably:

- Architecture
- Security
- Managing authority

Please refer to the DNS section for a more complete description.

IP Addressing and AS Numbering

Note: Please refer to Chapters 9 and 12 for a more complete description of addressing principles.

A GPRS operator is required to be registered as a local Internet registry (LIR) near a rgional Internet registry (RIR) to get blocks of IP addresses for its own or its customers' use. Most GPRS operators are GSM operators who are extending their existing GSM network to support GPRS, and are new to the IP industry. As a consequence, many have not yet registered to their RIR and instead use their ISP's blocs as customers.

GRX providers may offer assistance in the registration process as well as temporary allocation of addresses from their own address space.

Please note: The use of private autonomous system numbers (ASN), while not obligatory, has been suggested by the GSMA. These are allocated by the GSMA.

ISP/Hosting/Internet Access

The GRX provider may:

- Provide the functionality needed on the Gi interface to act as a mobile ISP (typically AAA for authentication, accounting, authorization, with the RADIUS protocol). A close relationship between GRX and operator is critical in this scenario, because the operator must transfer responsibility for parts of its customer data to the GRX provider.
- Host applications on behalf of the mobile operator.
- Transfer packets between the operator's Gi interface and the Internet.

Counting/Charging/Clearing Housing

The GRX provider may offer services related to:

- Collecting call detail records (CDRs) and (in the future) service detail records (SDRs)
- Processing CDRs and SDRs—For example, combining several CDRs related by context into a single CDR, or translating between different CDR interchange formats
- Clearing interoperator money flows

Signaling Access

As explained earlier, SS7-MAP signaling access is needed for things like the registration of subscribers in foreign networks. Experience shows that intermediate networks—though they may be perfectly suitable for carrying voice traffic or SS7-ISUP signaling—sometimes are not really suitable for carrying MAP signaling.

This is due in part to the fact that MAP is used only in mobile communication, so some operators of voice-oriented networks tend to "forget" about it when doing configuration changes. GRX providers can help considerably here by providing dedicated SS7-MAP signaling connections between the operator in question and a reliable SS7-MAP signaling hub.

Roaming Activation

Before subscribers can use a roaming service it must first be activated (requiring certain configuration changes in both the home and visited PLMNs) and then tested. GRX providers may offer to:

- Manage the activation process (useful in particular when the GRX provider already has experience with the operator organizations)
- Assist in determining the required configuration changes (the GRX provider may have experience with other brands of equipment where the operator may not)
- Carry out the initial roaming tests

Roaming Tests

Roaming arrangements must be continually tested in order to ensure that customer expectations are being met. This is a routine task that currently ties up a huge amount of manpower and investment worldwide. Moreover, under the current system, roaming tests are carried out by the visited PLMN—in other words, by the very organization that may be causing the problems the home PLMN wants to have investigated. The GRX provider will be in a position to offer a neutral service for roaming tests, and can often make more efficient use of the necessary equipment.

Security

Talking about security is like talking about quality of service: it's such a broad topic that you really have to define your terms before you can have a meaningful conversation.

So let's start by identifying the types of security we'll be concerned with:

- Physical site security (whose sites : PLMN's or GRX's ?)
- Safeguarding physical access to the network (leased lines, routers, etc.)
- Risks to GRX network function
- Securing the PLMN against attack

It's important to remember that reasonable security measures always represent a trade-off between the number of threats addressed and the costs incurred in doing so:

Since the goal of security is to ensure that nothing happens that should not have happened, the optimal security measure is to switch off all components of the network. With security measures, one must choose how close one wishes to come to this theoretical limit.

Many people promise heaven in security: they may offer real high-security guarantees, or be talking about how many guards protect their network's supervision center. You must precisely define what the concept of security covers.

Site Security

Whether we're talking about the sites of PLMNs or GRX providers, the goal of security measures is the same: to keep them free of threats like intruders, electricity outages, excessive heat, fire, and floods. All GRX providers today have the means to guarantee site security. Although usually not specified in writing, here are the most common physical security measures used to manage standard risks:

- **Physical intrusions**—Access lists, surveillance, identity checks, limited access badges, visitor monitoring, etc.
- **Electricity outage**—Autonomous backup power plants
- **Overheating**—Redundant air conditioning
- **Fire**—Automatic gaseous fire extinguishers
- **Floods**—Evacuation, self-powered pumping systems

Safeguarding Physical Access to the Network

Securing physical access to the network means managing two fundamental risks.

Network Failure

Any access based on a single router, connected through a single leased line between two sites (the standard minimum access architecture) is subject to many dangers: routers can fail, leased lines are subject to breakout (the "back-hoe syndrome"), sites can suffer outages and, in a worst-case scenario, an entire GRX network could fail. The answer to all of these potential crises lies in redundant systems.

Routers can be supported via a backup ISDN; this can be up to several Mbps, but is usually limited to 128 kbps and used with leased lines

Figure 20.4
Different types of resilient network access architectures.

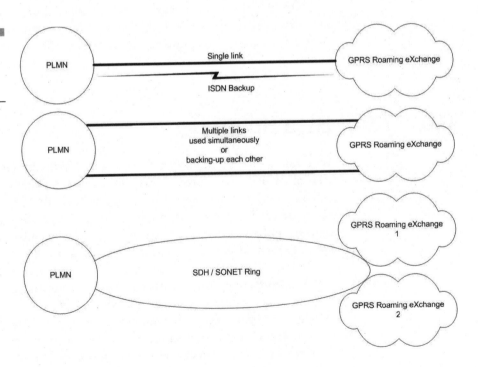

up to 2 Mbps. In the event a leased line is dug up, it can be backed up by a second leased line with a different physical path or by a second line to a different GRX site. Multiple accesses from different PLMN sites to different GRX sites can be used to overcome any site outages, the best architecture currently being an optical ring. Even catastrophic failure of an entire GRX can be compensated for by making arrangements with multiple GRXs.

Line Tapping

Protection against line tapping actually comes more from logistics than from any specific countermeasures. To begin with, it would take millions of dollars' worth of hardware and highly specialized knowledge to successfully tap GRX network lines. Even where this were possible, almost all potentially valuable information (military, financial) would be encrypted, so its worth to the tapper would be questionable. Finally, a PLMN-to-GRX link carries only roaming communications, making it less useful than tapping a target's fixed-line base of operations.

In the end, the only way to guarantee 100 percent security from tapping would be 24-hour visual surveillance of the entire length of the physical line. Even if this were possible, in the absence of any reasonable benefit to a would-be line tapper, it would make little sense.

Risks to GRX Network Function

A GRX network is a transit network, designed to carry data from a point "A" to a point "B;" as such it contains no true server, just its DNS, and that is always redundant and secured. Because the network proper is essentially a collection of routers, it is very easy to protect. No high-level protocols—such as Web-http, mail, or file transfers—ever run on the network's own equipment, so simple precautions like built-in redundancy can render it almost totally impervious to failure.

There are really only two types of attack possible in this area:

- **Denial of service attacks**—For example, flooding a node with requests so that it is completely overwhelmed and can no longer fulfill its normal function. Because the network is designed to be responsive to such requests, it can be very difficult to protect against such an attack. Fortunately, this type of attack is by its very nature limited in both scope and duration; it cannot spread beyond the first node attacked because as soon as that node is overwhelmed, it can no longer propagate the attack. In addition, this type of attack is highly visible and can be easily traced to its origin.

- **Backdoor attacks**—Taking control of a node. This kind of attack is usually intended to give the attacker either information or inheritance of the machine's access rights. Because of the lack of a true server, neither of these is of any use against a transit network like a GRX.

There are also a number of additional measures that can be used to maintain a high level of security. Since invisibility can provide protection, IP addresses of core network nodes are never displayed outside the network, rendering them unreachable. Every operation made on a node is registered; every node's configuration is saved and regular automated cross-checking between actual and saved configurations generates an alarm if any difference is detected.

There are other, even more deeply embedded measures used to protect networks. While these are kept highly secret by the operators, they tend, in fact, to be quite simple, based on fundamental capabilities of the machines. In the end, however, the best protection for any GRX network may well be the fact that there is little—if anything—to be gained from attacking it.

Securing the PLMN Against Attack

The PLMNs will want to protect themselves from—among other things—attacks via the GRX. There are several potential threats:

- The GRX might carry packets that embed protocols other than GTP, BGP, DNS and/or IPSEC. Using other protocols—for example FTP, telnet, or HTTP—the attacker might be able to access GSN maintenance functions. It is relatively easy to block this type of attack using ingress firewalls on the operator side. In addition, operators might wish to use egress packet filtering to minimize the damage an employee from—or intrusion into—their own organization could do to a partner.

- The GRX might carry an excessive number of packets that—while being perfectly legal otherwise—would lead to congestion of the GRX operator link or other resources within the operator's network. A replication of packets could create such a scenario. Here again, the operator could use ingress filtering to protect its network from the effects of such attacks. The GRX, for its part, could apply ingress packet filtering on the far end to avoid congestion of any single GRX/operator link.
- The GRX might carry packets that contain GTP content that manages to "break out" of the tunnel—in other words, to somehow become effective locally. The effects would then be similar to those described above. This, however, is by no means a GRX-specific threat; it is only possible if the SGSN does not "re-package" each and every GTP packet into SNDCP (the protocol used on the air interface), or if the GGSN allows traffic from the Gi interface to re-enter the node. The same threat exists in nonroaming configurations as well.

It should also be noted that the operator need not concern itself with the potential for eavesdropping on the GTP content, since subscribers using security-sensitive applications will tend to protect it themselves, for example by using end-to-end encryption. Those who have no such security requirements—say when browsing the Web—might prefer to avoid any additional charges for bandwidth and/or processing overhead incurred using such encryption.

In the end, the overall security of any chain of communication is equal to that of its least-secure link; there is no point in providing a maximum security/maximum cost solution for one part of the path if another is left unsecured. In a world where handset radio communications can be easily intercepted by inexpensive, readily available devices, it is highly unlikely that the weakest link will be found within the network itself.

Still, there are certain things a GRX can do. Its network links many companies, so it is up to the GRX to ensure that the traffic each operator receives comes exclusively from its roaming partners. Moreover, the GRX should make every effort to verify the trustworthiness of its own customers and those of the GRXs with which it peers.

However, a GRX can't actually screen GTP communications for possible dangers, and no miracle recipe results in perfect security. Everyone involved with open networks is to some degree at risk, and SLAs can only serve to define more precisely the limits of each participant's responsibility and liability.

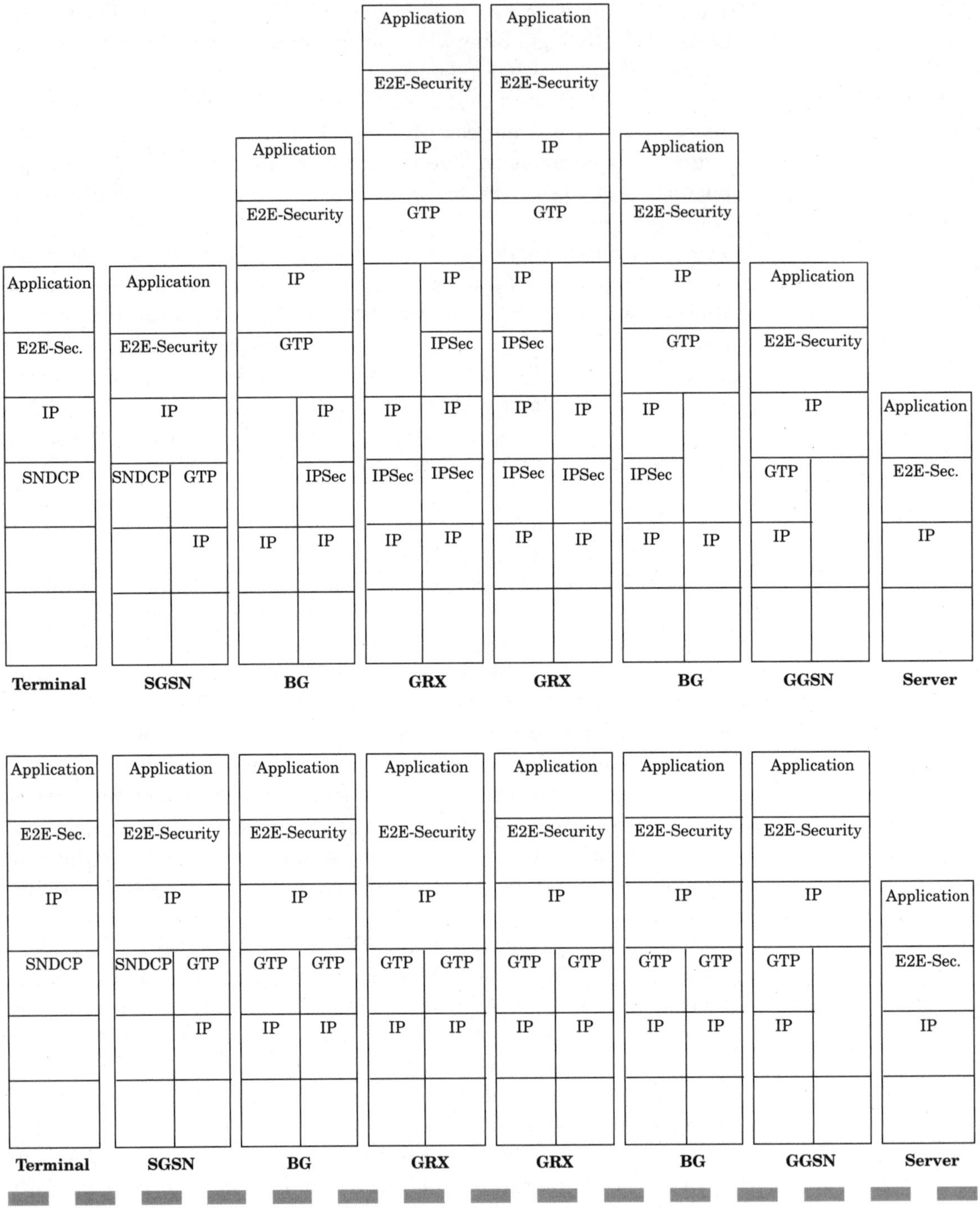

Figure 20.5

Peering: GRX-to-GRX

What happens if a customer wants to roam in a country where there is no GPRS operator connected to the same GRX as the customer's home PLMN? (Here we assume roaming agreements exist between GPRS operators.)

GPRS roaming is a matter of IP, so any sort of IP-level connectivity between home and visited networks will enable roaming. No circuit is needed; any existing logical path—such as an IP route—between the two operators will be sufficient to establish roaming…just as in the public Internet any given Web surfer anywhere can see a Web page regardless of its physical location.

There are two ways for a pair of GRXs to peer:

- **Direct private peering**—A direct connection between two routers.
- **Exchange point**—A facility in which several networks are present, connected to a common node (usually an ATM or Ethernet switch).

Figure 20.6
GRX-toGRX: direct and via an exchange point peering models.

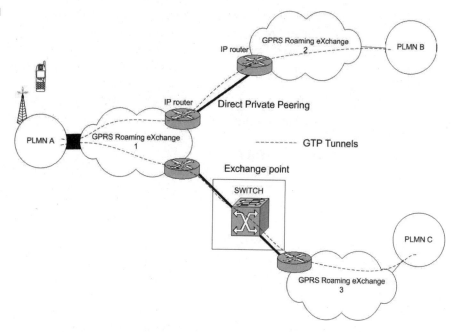

Peering arrangements raise issues of inter-GRX QoS, security SLAs and agreements, and optimal architectures.

Service Level Agreements (SLAs)

Typically, contracts assigning the provisioning of GRX services on a non-exclusive basis contain some language concerning the service level to be met. Since most of these regulations are a holdover from IP interconnect contracts, they don't always fit that well with the realities of GRX provisioning. Here are some topics that should be covered in a GRX SLA, along with some of the more popular—and more infamous—entries likely to be found in existing SLAs.

Information

- Typically the operator will want to be kept informed by the GRX provider of the actual usage of the service, and of any circumstances that could potentially threaten successful service provision.
- As long as service specifications are not changed, detailed information on how they are delivered is not generally provided.
- The parties generally agree to cooperate to prevent and eliminate any fraud or abuse that involves their respective network or carrier services. If any party suspects that such fraud or abuse exists, the parties shall co-operate and use all appropriate means to identify, eliminate, and prevent the fraud or abuse concerned.
- Notification of modifications intended by one party when those modifications could reasonably be expected to affect the other party's system.

The IP "Holdover"

Because many SLAs come from the IP world, some still contain language regarding restrictions on use of services that are not entirely appropriate, such as:

- No packetized voice
- Suspension of the service for maintenance work
- Specific peering restrictions
- Refraining from providing the service when minimal usage levels are not met
- Service to be used in accordance with reasonable operating instructions issued by the supplier
- Obligations to control the content transported (abusive, obscene, copyrighted, confidential). This can be worded thusly: "The Operator

shall not knowingly permit its customers to use the Supplier's network for any illegal or unlawful purpose."

- Some other restrictions are simply unacceptable for operators:
  - Usage of the service for internal purposes only
  - No use of service in conjunction with nonaffiliated other parties
  - Service not to be provided to third parties

Billing, Invoicing, and Remuneration

- How often shall the supplier bill the operator? Which formalities are to be observed (e.g., address, VAT number)?
- Procedures for handling of invoice disputes
- Type of pricing (usage-based, flat fee, pay as you grow, pay as you use, etc.)
- Minimum usage level (minimum price regardless of usage)
- Adjustment of prices
- Currency
- Credit notes (e.g., for periods of reduced service level)

Equipment

Which rules are to be observed with respect to the equipment the GRX provider may wish to place at the operator's premises (or vice versa)?

- Homologation issues
- Access to equipment
- Warranty/support

Default/Liability

It may be helpful to specify remedies and compensation in the event specific deadlines (e.g., for the setup of new roaming relations) are not met.

Availability

Strange as it may seem, there currently exist numerous quite different definitions of "availability": for example, it is fairly common to exclude scheduled maintenance when calculating availability. Considering the importance of service availability for users, then, it is extremely important to agree in advance on information and authorization mechanisms for maintenance work, and to make sure that the work does not affect peak traffic times.

Another (and more extreme) example is when the official unavailability period only begins after a certain number of consecutive minutes of total unavailability—say no IP packets whatsoever for a period of 15 minutes. In light of these differing definitions, care must be taken when comparing actual and contractual figures. The scope of the availability statement (router-to-router or CPE-to-CPE) can also make availability comparisons difficult.

Failure to deliver the contractually agreed availability should normally require the GRX provider to compensate, for example by giving a credit note. The compensation should correspond (at a minimum) to the amount of money the operator pays for the period of unavailability.

Quality of Service (QoS)

Target values (sampled over some meaningful interval) should be given for the:

- Roundtrip delay (both intercontinental and intracontinental)
- Packet loss (both intercontinental and intracontinental)
- Jitter (both intercontinental and intracontinental)

In addition, the method for measuring these values must be defined. Typically, a specific number of packets of a specific size are sent and evaluated for these qualities. Finally, it is important to specify compensation in the event actual values deviate from the target values by the agreed margin.

Security

As explained in the section on security, it can be difficult to say precisely what security measures should be taken. Nevertheless, the contract should attempt to specify as much as possible the responsibilities of both the GRX provider and the operator. Without such specifications, a breach of security cannot be treated as a breach of contract.

Service Status Information/Fault Management

The GRX provider should provide an interface that allows the operator to stay informed of the availability (and ideally, the load situation) of the service. Without this information, the operator may begin to speculate on the source of problems that result in the unavailability of the service.

In addition, mechanisms must be put in place for the reporting of problems by the operator and their timely handling by the GRX provider.

Domain Name Server (DNS)

A DNS is a device that translates Internet addresses into their IP equivalents. Every address or site on the Web, such as www.gsmworld.com, has a corresponding IP address that identifies its exact position on a server. This is the address the IP protocol requires in order to contact the site. In the GPRS world, the sending GSN must know the IP address of the receiving GSN in order to transfer packets from one PLMN to another (on behalf of a request from a handset).

A DNS is a hierarchical, distributed, redundant database, with delegated authority; each operator has its own DNS, with authority over its own records, which are automatically transmitted and updated to the other servers. The DNS is, in fact, the first server your Internet device (computer, PDA, WAP phone, etc.) would contact, requesting this translation.

Every GRX provider should offer some form of DNS service. The highest level is known as "root-DNS," because it handles only the addresses of the operators' devices, and not those of Web sites, companies, and the like. In the GPRS world, the "root DNS" in fact manages a private, GPRS-specific, first-level domain: the ".gprs," where in the Internet, a "root-DNS" handles the "." level, the highest layer which is above ".com," ".net," ".fr," and all the first-level domains.

The GPRS operator's DNS may obtain this information by:

- Querying the root DNS itself (if authorized when authorities will definitely define such a service for GPRS)

■ Querying the upper-level (GRX's) DNS, which is a permanently
updated copy of the root DNS

■ Being manually updated

In GSM, similar tasks (such as those related to SCCP global title
translation) have traditionally been performed by manual configuration.
However, the GSM experience shows that updating this information can
be slow and cumbersome, so it made sense for GPRS to use the DNS
mechanism rather than manual update.

Until "root DNS" is fully defined and implemented, the hierarchical
aspect of the DNS will allow GRX providers to supply GPRS operators'
DNS servers with all the information they need. The GPRS operator
need only refer to the GRX's DNS server; this will be manually updated
with information on the GPRS operators connected to the GRX, and will
retrieve information about other customers (those connected to other
GRXs) automatically from the manually updated DNS servers of those
other GRXs.

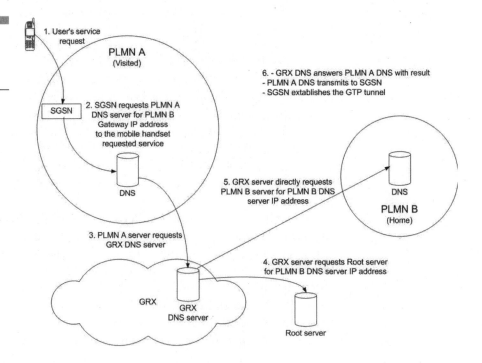

Figure 20.7
GPRS DNS: a
recursive-iterative
mixed model.

The result is an overall database capable of providing all the information customers require. This system is very powerful, quite simple, inexpensive, and has been successfully used in the Internet for more than 15 years; that's why the GSMA has chosen this approach to fulfill a number of needs completely distinct from the Internet comprehension of the DNS.

Evolution to 3G

With 2.5G, the PLMN evolved from a circuit-based to a packet-based network. The 3G UMTS Release 99 simply introduced a new radio access network and air interface into the existing networks. The introduction of the air interface has only a minimal effect on the bearer, tele and supplementary services, except of course for the added bandwidth for both circuit-switched and packet-switched data connections.

Because the interface upon which GRX providers offer their services lies deep within the core network, those services won't change when migrating to 3G—except, of course, that additional bandwidth will be needed. Since the bandwidth required for GRX services in 2.5G represents just a tiny fraction of the globally available bandwidth, extension of those services into 3G should not be a problem.

With the advent of 3G, GRX providers may find it interesting to target another market that will evolve with the introduction of the "packetizing" of circuit-switched services like voice; they will be well positioned to offer the end-to-end, QoS-guaranteed IP bandwidth required for such an interconnection of 3G operators. The relationships GRX providers will have built with the operators by then should help them better serve this market—much to the chagrin of the circuit-switched long distance carriers that handle that type of traffic today.

The ease with which the transition will be accomplished even extends to the name GRX, which will not change—its meaning will simply evolve from the current "GPRS Roaming Exchange" to "Global Roaming Exchange."

References

GPRS Doc 38/00	LS to SG "Security issues for GPRS Inter-PLMN backbone"	
GPRS Doc 46/00	Guidelines on Public Ipv4 Addressing scheme and Autonomous System Numbering scheme for the GPRS infrastructure to support GPRS roaming	1.2
GPRS Doc 49/00	GPRS DNS Usage Guidelines	0.0.3
GPRS Doc 51/00	Ipv6 in Inter-PLMN backbone	
IREG Doc 062/00	IP Address Guidelines for GPRS/UMTS Operators	2
GSM 03.02	Digital cellular telecommunications system (Phase 2+); Network architecture	7.1.0
GSM 03.03	Digital cellular telecommunications system (Phase 2+); Numbering, addressing and identification	7.4.0
GSM 09.60	Digital cellular telecommunications system (Phase 2+); GPRS; GTP across the Gn and Gp Interface	7.4.0
PRD IR.33	GPRS Roaming Guidelines	3.1.0
PRD IR.34	Inter-PLMN Backbone Guidelines	3.1.0
PRD IR.35	End-to-End Functional Capability Test Specification for Inter-PLMN GPRS Roaming	3.0.2
	GPRS/3 G Roaming Inter-PLMN Backbone Network Domain Security Considerations	1.0.0

CHAPTER 21

Location-Based Services

Jessica Roberts

Nokia

Location-based services (LBS) are being projected as the next major class of high-value services that wireless operators can bring to their customers. Wireless networks offer subscribers one distinct advantage that fixed networks cannot provide—mobility. With mobility, location becomes a critical attribute that can be used to enhance the end-user experience. Location-based services are defined as services that integrate an estimate of a mobile device's location with other information to provide added value to the customer.

In the increasingly competitive mobile communications marketplace, it is critical for operators to differentiate their service by offering services that are personalized, localized, and relevant. From the end-user's perspective, location-based services are value-added services of the first order that can provide navigation information, tracking information (for people, vehicles, and valuable property), information services, and safety and emergency services.

According to market studies, the market opportunity for LBS is approximately $500 million today and poised for rapid growth as technologies such as GPRS and 3G, both of which support higher-bandwidth applications, come to the market and evolve. Today, it is possible to locate Class A and Class B GPRS terminals (or handsets) using the circuit-switched technologies described later in this chapter. The standards for locating GPRS handsets using packet-switched technology are still under development and should be available in the second half of 2002. Figure 21.1 provides some detail on this emerging market.

Figure 21.1
Global market forecasts.

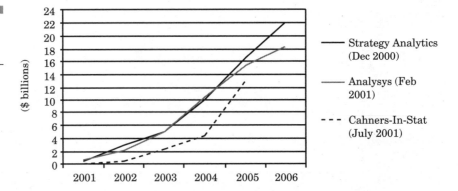

Location-based services have wide appeal across all customer segments. Strategy Analytics estimates that LBS revenues will grow to $16.6 billion by 2005 and to $55 billion by 2010. A number frequently seen in the trade press is $18.3 billion in revenues by 2006.

Location-based services are also at the top of the consumer wish list. According to Forrester Research Inc., seven out of the ten most-wanted mobile services *need location information*. LBS are important in providing a means of brand differentiation to operators who are constantly looking to increase margins, add value, differentiate services, and build customer loyalty.

Figure 21.2
Consumer wish list.

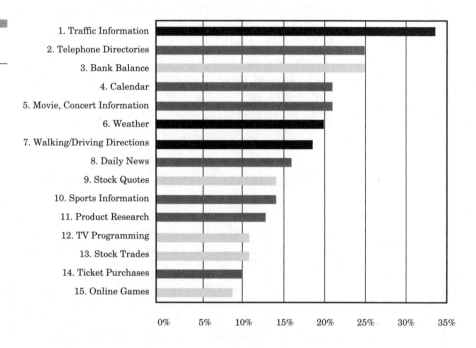

1. Traffic Information
2. Telephone Directories
3. Bank Balance
4. Calendar
5. Movie, Concert Information
6. Weather
7. Walking/Driving Directions
8. Daily News
9. Stock Quotes
10. Sports Information
11. Product Research
12. TV Programming
13. Stock Trades
14. Ticket Purchases
15. Online Games

0%   5%   10%   15%   20%   25%   30%   35%

■ Stong need for location information.

■ Possible need for location information.

□ Location information not needed.

Source: Forrester Research Inc. 2000:
Forrester's Consumer Technographics®
September 2000 Europe Study

In December 2001, in one of the first reports looking at deployed services, Strategy Analytics reported that the leading mobile location services deployed are "Find the nearest...," Yellow/White Pages, Friend Finder, "Where am I?," and location-based games. Customer anticipation reinforces the operators' goal to implement location-based capabilities adequate for MLS, and to do it soon. In Europe, operators can leverage the success of short messaging (SMS) by deploying SMS-based location services. SMS is viewed as an ideal delivery method to educate users about the potential benefits of location-aware services. Although operators are not able to provide detailed maps using SMS, the basic services listed above can be deployed and used today by most subscribers. In-car traffic and congestion information also emerges as a very marketable feature in market surveys.

In the United States, step-by-step driving instructions and family tracking are expected to become the most popular value-added services. Younger users in both the United States and Europe are interested in location-based advertising. According to analysts, location-sensitive notifications and alerts to mobile phones will be the most popular services in the next five years, achieving 27 percent penetration in Western Europe and 35 percent in the United States. Finding and guiding services, including mobile yellow pages, will attract 30 percent of United States and 21 percent of Western European subscribers. Tracking will likely remain a strong niche application.

Location-Based Applications

Location-based applications can be divided into several categories: navigation/routing, tracking, information services, safety and emergency, entertainment, and operator services. Each of these categories is described below.

Navigation applications (also called "finding applications") provide the user with information on how to find the best route from position A to position B. The service can be enhanced when combined with another application like fleet management, or location-aware databases: "Where is the nearest Chinese restaurant?" Examples include route description, turn-by-turn directions, and dynamic route guidance with maps. "Find the nearest..." is the top location-based application launched to date.

Tracking applications trace people, vehicles, or valuable property via positioning-capable mobile handsets or terminals. Examples include "Find a Friend," fleet management, asset tracking, tracing of stolen property and vehicles, and person surveillance—monitoring children, prisoners, or people with disabilities. Friend Finder and fleet management applications have also been launched by several operators. Although so far it has not been an issue, it is expected that enhanced accuracy will be the biggest obstacle to the success of this type of application.

Information services leverage the user's location to provide her with more accurate, granular, and relevant information. Examples include yellow pages, telephone directories, traffic reports, news and weather updates, city guides, travel books, maps, and parking information. Yellow pages applications have also been launched by a number of operators. Again, increased accuracy will allow the enhancement of these services to provide the end-user with greater value and convenience.

Safety and emergency applications provide the user with timely and accurate advice. They can be of help even when the user can't provide his own location. Examples include emergency calls such as E911 (United States) and E112 (Europe), automotive towing and repair services, warnings about unsafe areas and nearest medical center information. Today, emergency calls in the United States can be located with cell-ID accuracy (within a kilometer) and some operators can locate mobile subscribers with much greater accuracy in selected locations. The FCC has mandated that all operators support enhanced accuracy (within tens of meters) by the end of 2002. Please see the Regulatory Issues section for more details on the ongoing E911 and E112 activities.

Entertainment applications are primarily location-based games. This category of applications has emerged as one of the most popular among young users and one of the first location-based applications deployed by operators. An example of a location-based game is BotFighters, which was developed by the Swedish company It's Alive. BotFighters is a location-based action game allowing mobile users to adopt a robot personality to play one another in their local area. It is an action game in which players locate and shoot their competitors. Automatic mobile positioning based on cell-ID is used to determine if players are close enough to shoot and hit each other.

Operator services help mobile operators to improve their customer services or capacity planning. Examples include location-based charging, location-based barring, location-based intelligent network (IN) services, traffic measurements, and network planning. With the initial focus being on consumer services, not many of these types of services have been deployed. An example of location-based charging would be the system that allows an operator to charge different rates for calls initiated at different locations. An example of a location-based barring service would prohibit calls initiated in a particular location.

Although these are the primary ways we envision LBS rolling out, it seems clear that location information can add value in just about any application. For example, instead of sending a short message to a friend asking where they are and giving directions to a restaurant, you could automatically locate your friend and send a map. Can you think of anyone who would not benefit from the intelligent use of location information? In the future, all applications should be able to utilize and transfer location-based information elements.

Critical Factors for Success

To be successful, LBS must become a mass-market business and address such issues as interoperability, security and privacy, and personalization. There are clear indications that the consumer segment will generate the vast majority of LBS revenues in the next few years, so industry focus must be on the quality of the end-user experience. Three factors that operators must consider will be especially relevant:

1. **Design of the service**—Ensuring personalized, localized services with good service access and reliable, up-to-date content. The service design should always reflect end-user perception and priorities, e.g., how many clicks it takes to get to the information sought, how long it takes to get an update, the refresh rate of information, etc.
2. **Robustness of the system**—Requiring a scalable, integrated 2G/3G system capable of delivering acceptable response times, high availability, smooth interoperability, smooth capacity planning (caching, load balancing), and a clear path for technology evolution.
3. **Location method quality and coverage**—Offering high-speed and accurate location calculations covering as much as possible of an operator's network. The integrity of the network data, as affected by such elements as antenna location and direction, is critical to providing accuracy.

The Location Interoperability Forum (LIF), an organization formed by Nokia, Motorola, and Ericsson to promote an interoperable location services solution that is open, simple, and secure, has defined three levels of location services based on network and handset support and location measurements within the network. The three are defined in Table 21.1.

LIF expects that 80 percent of location-based services can deployed at the basic service level, which is available today and is the least expensive option. It does not require consumers to buy new handsets in order to take advantage of the added value location brings to their service. Note that the same application can belong to all three service categories, with service quality improving from category to category. *Service quality* is defined as a function of three metrics: location detection accuracy, location data delivery time, and customer choice capability (on/off functions, privacy, billing etc.). All of this combined with relevant and personalized services builds up perceived service quality. In some cases such as a weather forecast, enhanced accuracy will not add any value to the service and is not needed. However, point of interest or tourist attraction services, although adding value for the subscriber at the basic service level, can be significantly enhanced with improved location accuracy.

TABLE 21.1 Service Categories		
Category 1 (Basic Service Level)	Location of all (including legacy) handsets with cell accuracy or improved cell accuracy (within a kilometer). Applications that can be deployed using this service level include news, weather, today's events, points of interest, etc.	
Category 2 (Enhanced Service Level)	Location of all new handsets with improved accuracy (within tens of meters). Applications requiring this improved accuracy include fleet management, nearest restaurant, and emergency services.	
Category 3 (Extended Service Level)	Location of selected new handsets with higher accuracy (within a meter). Applications requiring this level of accuracy are navigation/directions, navigation inside a building, and enhanced emergency services.	

An early success story in location-based services deployment comes out of Japan. The J-Navi service was launched by J-Phone on May 1, 2000. It gives the subscriber a color map that positions him in relation to a desired destination, plus some supporting text to provide routing information. The software platform was designed to accommodate 100,000 requests per day. On May 3, two days into its launch, the service received more than 1.5 million hits. Today it regularly averages 1 million hits per day. Key to its success are good network response time (an average of two seconds and a maximum of five seconds); ease of use (information can be accessed by the user in three to five clicks); and a low price of ¥ 20 (U.S. $0.18) per request. J-Phone claims that more than 4 million users have used the J-Navi service.

Technology Options

A variety of positioning methods are available for use in the mobile network, and these provide different levels of accuracy for different applications. Methods can be network based or handset based, or a combination of both. The figure below compares positioning methods and their relative accuracy. We'll go into more detailed for each of these methods in the following section.

Figure 21.3
Location method accuracies.

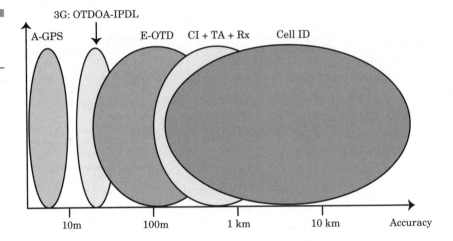

The *basic location method* uses a cell/sector identifier, cell ID (CI) for 2G networks and service area identifier (SAI) for 3G networks. The method may be enhanced with additional propagation time information such as timing advance (TA) or round-trip time (RTT) and with measured signal levels (RX) to locate handsets. This method can locate all legacy GSM handsets.

Location accuracy using this method is good enough for 80 to 90 percent of commercial applications such as weather forecast, point of interest, and tourist attraction services. Figure 21.4 illustrates the method.

Figure 21.4
Basic location method: Cell ID + TA.

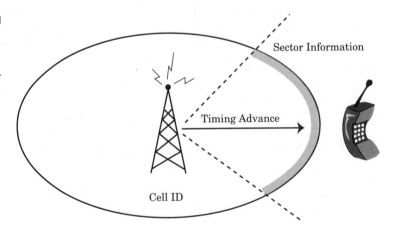

The *enhanced location method* uses handset-assisted enhanced observed time difference (E-OTD) in a 2G network or observed time difference of arrival–idle period downlink (OTDOA-IPDL) in a 3G network. It requires that the handset can hear at least three base stations for accurate triangulation. The handset measures the observed time difference (OTD) between the bursts from the serving base stations and the neighboring base stations, and reports these to the network when requested. Since GSM networks are not synchronized, the real time difference (RTD) between base stations must also be known by the network. The network calculates the location of the handset based on the OTD and RTD measurements and the locations of the base stations.

Figure 21.5
Enhanced location
method: E-OTD.

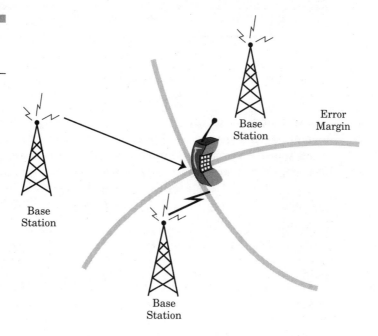

The *extended location method* uses a network-assisted global positioning system (A-GPS) to achieve the highest positioning accuracy for special applications such as detailed driving directions and navigation. A reference GPS network is established whose receivers are optimally located and have a clear line of sight to the GPS satellites. These reference receivers provide assistance data to the handsets to enable accurate positioning in dense urban and indoor environments.

Figure 21.6
Enhanced location
method: A-GPS.

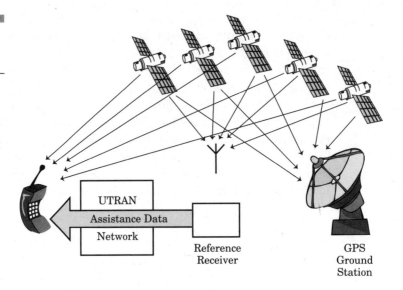

A combination of these methods will likely provide the best result, depending on the type of application and the accuracy required. GPS-based solutions lack good indoor coverage, while cell ID-based solutions lack a high degree of accuracy. Response time should also be considered—a cell-based solution can return a first location response within approximately 3 to 5 seconds while a GPS-based solution may take 10 to 20 seconds.

Customer Privacy and Security

The availability of mobile location-gased services has raised concerns about consumer privacy. Consumers are worried about "Big Brother"—the illicit surveillance of their mobile phones—and also about the extent to which data about their lives is gathered and aggregated without their consent. While this data may be gathered with the best of intentions for personalizing and localizing the user experience, the consumer is worried about how else it may be used and who can access it.

Privacy is undoubtedly the single biggest potential obstacle to the market take-up of location-based services. Both the United States and the European Union have introduced legislation to address this issue. The United States is likely to pursue legislation requiring authorization from the consumer before collecting, using, or retaining their location information. There will also be restrictions on the operator's ability to share location information with third parties. Check the FCC Web site (www.fcc.gov) for up-to-date information on the status of privacy legislation.

The European Union introduced a proposal for a Directive in December 2000, stipulating that location data may only be used with the consent of the subscriber and requiring that subscribers be provided with a simple way to temporarily deny processing of their location data, just as they do for calling line identification. The only exceptions to this consent would be for emergency services or criminal investigations. The European Union Web site (www.europa.eu.int) continuously posts the latest information on privacy in the European Union.

It is important to understand that the mere fact that a handset is in a certain location is not a privacy issue. The privacy issue comes in when this data is collated and aggregated to provide information about the movements of the mobile device and its user. Uncontrolled availability of this type of information can present a serious risk to individual privacy, and it is critical for the operator to ensure proper protections. Operators must provide the means to prevent any association of a subscriber's location and his identity.

Operators must also build trust with their customers—not only by allowing them control over their location information, but also by physically separating location data from the applications that use it. Customers should be given the opportunity and the means to decide who can use their location data and when it can be used (e.g., only during the week for the boss, only on the weekends for the kids). Operators should also educate their customers on the potential benefits of "being locatable"—benefits such as locating a lost child at the amusement park or personalized advertising that can provide them with coupons and discounts to the restaurant they have just passed or the shopping mall they have just entered.

Regulatory Issues

Regulations requiring carriers to be able to locate mobile subscribers prompted research and development into more sophisticated technology to perform this function. We are just starting to see the benefits of that work. In Europe, this technology is being used to provide value added services to subscribers, while the focus today in the United States is to meet the FCC mandates described below. It is expected that commercial services in the United States will start to take off at the end of 2002. Regulatory requirements have helped to make this technology available to enhance mobile services much sooner than it would have been otherwise.

United States

Since 1996, the Federal Communications Commission (FCC) in the United States has taken action to improve the quality and reliability of 911 emergency services for wireless phone users by adopting rules to govern the implementation of enhanced 911 (E911) for wireless services.

The FCC was driven by the concerns of the emergency service bureaus that were not able to respond to emergency calls originated by mobile phones. The increase in the number of these calls from basically none in 1985 to over 20 million in 1996 and over 43 million in 1999 drove the FCC to the actions they have taken.

The goal of the FCC's wireless 911 rules is to arm emergency services personnel with location information that will enable them to locate and assist wireless 911 callers much more quickly. With these goals in mind, the FCC has required wireless carriers to implement E911 services in two phases, subject to certain conditions and time schedules.

Phase I of the E911 implementation required carriers, as of April 1, 1998 (or within six months of a request by the designated Public Safety Answering Point) to supply the telephone number of the wireless 911 caller and the location of the cell site or base station servicing the call. This information enables timely emergency responses by identifying the general area from which the call emanates, and by permitting the emergency operator to re-establish a connection if the call is disconnected.

Phase II required wireless carriers to provide automatic location identification (ALI), which would provide much better accuracy, beginning October 1, 2001. The FCC recognized the differences between network-based and handset-based technologies and established separate accuracy requirements and deployment schedules. On October 5, 2001, the FCC agreed to push back this deadline after receiving waivers from every carrier, stating that equipment was not available to meet the deadline. As a result, each carrier is now required to submit quarterly status reports to the FCC and each carrier has a slightly different deployment schedule. For details of the accuracy requirements and current status on network operator deployment, please see the FCC Web site (www.fcc.gov).

European Union

The European Union was driven by some of the same pressures as the FCC—the increased number of emergency calls originating from mobile phones. However, it has taken a different approach to the development of regulatory requirements to address the problem.

European Union regulations for providing location information to emergency services were identified in July 2000 in the *Directive on Universal Services and Users' Rights*. This directive states simply that caller location information shall be sent to emergency service (E112) if it is technically feasible. There are currently no accuracy requirements. The European Commission has set up the Coordination Group on Access to Location Information by Emergency Services (CGALIES) as a partnership of public service and private sectors to find harmonized, timely, and financially sound solutions.

CGALIES will come to consensus between all players regarding the implementation of E112 services and recommend a time schedule for a coordinated introduction of these services. CGALIES will also address the liability of operators for the accuracy of location data, interoperability between different solutions, and support for international roaming. A research and development study commissioned by the group presented two scenarios—one market driven and one driven by strong regulations. The study proposed that the market-driven scenario be the starting point of the implementation, and continue through 2005. However, a review in 2004 will determine how successful the market-driven approach has been in providing satisfactory emergency services. If it falls short, a move towards a strong regulatory scenario will be initiated. It appears that CGALIES is leaning towards this type of recommendation.

CGALIES submitted its final report on January 28, 2002, recommending that member states should avoid too much regulation in meeting the proposed directive to provide location information. Based on information provided by network operators, it appears that all mobile operators plan to implement commercial location-based services with the accuracy of the location information available, improving with the development of technology and the penetration of location-enabled handsets. There are many differences within the European Union in the way E112 calls are processed and in how the public safety answering points (PSAPs) handle the calls. The aim should be to develop common standardized data formats and for the transfer of location information. If a de facto standard is developed, it will most likely develop a mass market for hardware and software, reducing the cost of implementation. CGALIES concluded that the biggest obstacle for implementing location information with emergency calls will be strained public budgets and recommended that responsible parties in the member states be notified of the importance of this issue and of the future public benefits that will be realized if the financing issues are solved. For the latest information on the CGALIES activities, please see their Web site (www.telematica.de/cgalies).

PART 5

Unless You Can Make a GPRS Toaster, You Need a New GPRS Terminal Device

The title of this section says it all. One of the biggest hurdles to the widespread adoption of GPRS is the need to equip customers with new mobiles or handies. Unless a terminal device is GPRS enabled, it cannot access GPRS services. Even WAP phones that can access the Internet today cannot access GPRS. That's because the existing WAP-enabled devices are circuit-switched and the GPRS network is packet-switched, and the two are not compatible. New phones and data cards for everyone!

If you are reading with a keen eye, you will have caught us using seven different terms so far for the thing with which we access the GPRS network, not including the toaster: mobiles, handies/handsets, terminals, stations, devices, phones, data cards. All used interchangeably to mean the same thing. Certainly it can be confusing, but it is benign and just reflects how young and vibrant the industry really is.

The term I'll use in this overview section is *terminal device*. Not that it is more correct, but rather because it best describes two distinct types of GPRS access mediums. A terminal is the term most closely associated with the products commonly used to communicate verbally, which are now morphing to incorporate data capabilities. They are, in other words, phones. Devices are other "things" not commonly associated with voice communication today, but which can nonetheless access GPRS networks. These include PDAs, MP3 players, PCMCIA data cards, data modules built into other devices, computers, telemetry devices, pagers, and others yet to be named or conceived. Typically they are not voice-enabled today and, when equipped to access the GPRS service, may or may not offer voice communications along with their data functionality. And just to be clear, Bluetooth-enabled devices by themselves cannot access GPRS services unless they are also equipped with a GPRS module. Bluetooth is a short-range wireless connectivity technology and does not, in itself, provide wireless access to the packet data network.

So what is the big deal about new product lines of GPRS terminal devices? Not much really, at least from the customer perspective. The early versions of GPRS devices will look, sound, and most likely even smell like their GSM predecessors. Most will resemble the voice phones which hundreds of millions of people use daily to communicate without cords or wires, with the exception that a GPRS module will be included to allow the device to communicate with the GPRS network as well as the GSM network. If they are tri-band, they will access GSM 900, 1800, and 1900 in the same way they do today. Adding another module shouldn't be that difficult, but then again, maybe it is....

Manufacturers of GPRS devices face a number of difficult challenges, not least of which is a customer expectation that the devices will not increase in size or weight. Batteries must also last as long as today's models, and longer if possible. And retail prices must remain similar to those purchasers find when browsing the high street or mall. It's quite a task to shoehorn a new, altogether different technology into the same amount of molded plastic. All this without considering how the device must evolve to make data browsing less complicated. Screen size and shape, number of pixels, monochrome or color screens, wheel mice and touch-screen capability all enter the picture. Oh, and let's not forget that the battery must now run two technologies at the same time without stepping back to forty-five minutes of use and four hours of standby time. And what about heat? More power consumption means more heat that must be eliminated in design. And so on. Terminal device engineers have never before had to navigate so many different objectives and their ramifications.

Still the industry cuts the manufactures little slack in this endeavor. GPRS devices are generally considered to be "late to market" with limited product lines and even fewer quantities of commercially available devices. And as the title of the chapter suggests, there are no alternatives to a new GPRS terminal device. So we wait. And wait. And wait, for these revolutionary GPRS devices to be designed, manufactured, tested, and offered for sale.

It is generally agreed that existing players in the mobile device manufacturing business are best positioned to enter the GPRS category initially. Companies such as Nokia, Sony, Ericsson, Motorola, Siemens, Sagem, Samsung, Alcatel, Philips, and Mitsubishi have the expertise to design and manufacture GSM devices in quantity. Adding GPRS data capacity is indeed a daunting task, but with lessons learned in more than ten years of wireless device manufacturing, these companies should be best positioned to morph their existing GSM technology. And since most are already building WAP-enabled devices running on circuit-switched GSM with browsers to access the Internet, transitioning to packet data with similar tools should be easier than starting from scratch.

On the other hand, PDA, pager, data card, and computer manufacturers entering the GPRS space face a struggle with an entirely new technology. They may or may not have the necessary expertise to bring these products to market. Should they decide to marry GPRS packet data capabilities with GSM voice service in a single device, two technologies must be brought together, one of which is likely to be unfamiliar. Outsourcing development to third parties will solve many problems on the design side for these manufacturers, but integration, testing, and the general learning curve may mean that their products follow after those of GSM device manufacturers already in the market space. So don't look for a plethora of new players delivering data devices with voice capability any time soon. GSM phone manufacturers should be able to establish their GPRS products before new entrants are able to gain a foothold in the marketplace with data only or data/voice devices.

Whatever the source of new GPRS terminal devices, environmental considerations will have a large impact on time to market. Because GPRS is a packet-based pulsing technology, more packets sent at ever-faster speeds means increased heat and possible absorption of heat and even low-level electromagnetic frequencies by the human body. These and other factors must be tested to ensure that national environmental and health regulations are met before products are offered for sale. GPRS devices are categorized into different classes based upon their ability to offer data, voice, and combinations of both. Heat and absorption rates will vary with the number of time slot transmission capabilities built into the device. Engineering design must take these requirements into consideration even as demand for faster data speeds drives the process in a contrary direction. The proper design and testing of GPRS devices will become more, not less, critical in the time-to-market calculation for new devices as their capabilities and functionality increase.

One way to ensure compliance with regulations is for the service provider to conduct random testing. There is no regulatory imperative for such testing, but purchasers do it anyway to find out if the devices meet their functionality and interoperability requirements. Always of foremost concern is the possibility that a particular model could inadvertently introduce compatibility problems into the network or cause interoperability problems for the customer when roaming. Therefore devices usually get properly tested both in the lab and in the field. Manufacturers also test extensively to uncover any bugs that may be lurking in their commercial products. Even so, many service providers require that new products pass field trials before placing thousands or even millions of devices on their network. Experience has shown that the upfront testing and conformance to requirements is less costly than a product recall.

Which brings us to the issue of standardized GPRS testing. It doesn't currently exist. There is no global agreement as yet about what types of laboratory testing will assure the interoperability of GPRS devices. GSM voice components can be tested against agreed standards, but GPRS components can't. Since there is no regulatory requirement for the launch of a GPRS device other than basic health and safety tests, which vary slightly around the world but are essentially the same, GPRS testing has thus far been left to the manufacturer's discretion.

The same lack of agreement on how to perform lab testing, or what to test for, means that almost all testing is conducted in the field on either test networks or commercial networks. Such field testing may be the best way to confirm that the device performs as designed, but it's also expensive and time consuming as geographies and GPRS infrastructures can differ significantly. And just because a device performs well in one commercial network on certain types of infrastructure hardware and software releases, there is no guarantee it will perform as well in another network with either different hardware or software.

One can infer that interoperability testing is complex and not scientifically reproducible. What appears one day may not reappear the next. What fixes a bug in one network may in fact introduce error in another. It is no wonder that device manufacturers have a difficult time improving time to market. They stake their reputations on the quality of the terminal devices they produce. This uneasy tension between new product introduction and interoperability causes consternation but little complaint among service providers awaiting shipment of new GPRS devices. Delays caused by interoperability testing are a curse, yet might be a blessing if the testing eliminates bugs that customers will ultimately pay the price for finding.

Until the GPRS industry agrees on a comprehensive set of tests which can be conducted on certified test equipment and verified through standard field trials on commercial networks, testing will continue to vary from manufacturer to manufacturer. Service providers will therefore continue to demand interoperability compliance results specific to their networks, and the process will go full circle again. Hence, the GSM Association has undertaken a program of compliance testing in coordination with device manufacturers. As of now, GPRS testing is not included in the GSMA's GSM Certification Forum activities, but many believe that the adoption of GPRS tests into the program will result in testing which is trustworthy enough to let them scale back or even eliminate internal testing.

In this section, experts will explore the types of GPRS devices that exist or are on the drawing boards. They'll review environmental issues faced by the industry as high-speed packet data devices appear, and they'll assess testing programs that can verify interoperability and compatibility with commercial GPRS networks.

We recognize that GPRS will never see commercial take-up without the introduction of new terminal devices. Today, the industry is plagued by a dearth of choices in available terminal devices. Service providers continue to insist that their networks and applications are ready for customers using GPRS to arrive en masse, but the lack of GPRS devices is hurting their business plans. It is hoped by all that this currently situation is drawing to a close and the mass production of GPRS terminal devices is just around the corner. If not, someone had better figure out how to make a wireless toaster transmit packets...and fast.

CHAPTER 22

Types of Devices and How They Grow

Ray Haughey

GSM Association

GPRS is already in the rollout phase and is a key element in bridging the gap between second generation (2G) voice-centric mobiles and the multimedia, content rich products of third generation (3G). Often referred to as 2.5G, GPRS represents the first real steps at providing 3G-like services within limited bandwidths.

Terminal availability in sufficient quantities for emerging technologies is one of the great challenges for manufacturers and operators alike. GPRS is no exception and, finding themselves in the proverbial chicken and egg situation, manufacturers are trying to evaluate both the market potential and level of investment required in the absence of reliable forecasts for either. Nevertheless, the leading manufacturers have committed to deploying GPRS equipment, with the U.S. telecom giant Motorola stealing an early market lead with Nokia, Ericsson, Siemens, and others in "hot" pursuit.

> *The availability of handsets is only one of the factors affecting GPRS deployment. Others which are as important are infrastructure stability, service availability, cost of service, and applications offered.* —Philips Semiconductors[1]

But what of those early forecasts for mass GPRS take-up? It was originally estimated that by the end of 2001 there would be in the region of 10 million GPRS handsets in circulation. Sadly, 2001 ended with far fewer GPRS subscribers, by some estimates as few as 500,000 worldwide. Still, by the end of 2006 it is expected that GPRS terminals will account for 76 percent of all 2.5G terminals (1xRTT, HSCSD, and EDGE making up the remainder).[2]

Overlaying a packet data network onto its legacy circuit-switched contemporary has posed some challenges for network operators, mitigated in part by the choice of similar infrastructure and terminal vendors. Economies of scale and network optimization provide an architecture that allows for a greater ratio of downloading capabilities (and therefore more down links or time slots from the network to the terminal) to network data uplinks. However, challenges remain for the GPRS terminal and internetwork security issues, not the least a threat from spamming, virus infections, and denial-of-service attacks or the inability to bar services in a roaming context. Early GPRS terminal underachieved data

---

[1]Curtis.M 2001.
[2]ARC Group 2001.

rates (13.4kbts in the uplink using one time slot) with multiples of that capacity depending upon on the number of allocated downlinks.

So, under current realities and growing expectations, what will GPRS mean to the customer?

Product Trends and Market Segments

The look and feel of a terminal will begin to evolve from an ever-smaller digitized assembly of plastic-coated circuitry capable of "occasional" voice service, a frustrating keypad-based SMS/data construction (only small fingers need apply), and a hard-to-read monochrome screen. That vision pushes us toward a new user experience characterized by a handset capable of personalization and no doubt similarly gratifying user experiences exploiting biometric advances to provide options for voice, finger, earprint....yes, you heard correctly...recognition and authentication techniques coupled with larger, colorful touch screens capable of carrying audio/visual services at data rates greater than a stopwatch missing a second hand. In this chapter we'll look at how realistic this happy vision is, and when we might expect to see GPRS and 3G terminals containing standardized universal functionality.

Not all things are created equal—ask any bald man disenchanted with doublesided sticking tape...

We are most likely to see handsets that are optimized for certain applications like games or location-based services. Market segmentation is expected to dictate that terminals be differentiated by service requirements, bandwidth, and quality of service. For example, games and entertainment services will rely on enhanced data speeds, good screen size and adequate battery life, while email and unified messaging will require improved keypad facilities. Downloading music and MP3 player/files will necessitate small, lightweight terminals with good sound quality.

As we'll see in the next chapter, there are several types of GPRS-capable terminals and devices that people will use for GPRS services.

GPRS "Regular Phone" Devices

This category of GPRS devices consists of "standard GSM" type phones with small screens and numeric keypads. These voice-centric devices offer GPRS, WAP, and Bluetooth functionality as an addition to its core purpose. Augmenting with simple data services running over GPRS will have limited appeal. Such devices are mostly for those entering at the low end of the market with voice calls as their primary intended use and limited data functionality needs. This segment of the device market will almost certainly be occupied by existing manufacturers such as Nokia, Ericsson, Siemens, Motorola, and others currently in the GSM space.

Figure 22.1
Phone devices.

Ericsson T68   Nokia 8310   Alcatel 502

Motorola T260   Sendo Z100   Samsung SGH-Q100   Panasonic GD96

Data Cards

Historically PCMCIA cards are credit card–sized hardware devices connected to the bottom of the mobile phone via serial ports. Data cards constitute a data-centric subcategory that allow for high-speed access to the Internet. GPRS will exploit them to offer data mobility for millions of portable computers around the world. Data cards increase speed of transmission over normal GPRS devices by employing more time slots (up to 8 time slots can be used without concern for heat build up or reduced battery life). Most analysts expect eight time slot cards to be available in quantity in the second half of 2002 from both traditional device manufacturers and new entrants in the wireless field, such as Intel.

Figure 22.2
Data cards.

DotSurfer, Merlin, Aircard, and Xircom units.

Card Phones

These devices are cards that slot into a personal computer. Card phones support voice via an earpiece and antenna. They provide all the functionality of a data card, including all nonvoice services, without the need to supply a separate phone. This accessory segment is expected to be somewhat limited except perhaps in the case of data cards with an earpiece for added limited functionality.

Figure 22.3
Card phones:
Globetrotter option
card.

Smart Phones

These are integrated computer and communication stations capable of carrying voice, nonvoice, and even Web browsing with its typical QWERTY and icon-driven user interfaces.

GPRS is itself an enabler; Nokia pioneered smart phones through its 9000 model series launched over HSCSD in the latter half of the 1990s. Moving smart phones away from HSCSD and toward GPRS will allow these devices to better flourish and maybe even to break out of the niche segment they currently occupy.

Other manufacturers such as Sony Ericsson, Alcatel, and Sharp launched smart phones towards the end of 1999. Smart phones are communications-centric; they can be PDAs with in-built wireless modems like the Palm Pilot or communication-specific devices like the Nokia 9110 or Ericsson R380. Morphing PDA-type devices towards wireless data and voice capability holds out potential for long term success. What are the obstacles? Short battery life and small screen size must be overcome to allow this category to realize its potential. Data entry and readability require significant simplification. However, if data entry and access, power management, and graphic capabilities can be successfully addressed in a user-friendly, inexpensive device, smart phones may turn out to be the largest market segment among GPRS device options.

This segment also holds out the greatest opportunity for new entrants to exploit as they enter the GSM/GPRS wireless device category. A question for the long term is whether it proves easier to add data capability to wireless devices or voice capability to data devices. New entrants, many of whom come from the data field, bet on the latter as they challenge the traditional manufacturers in the GSM category. Only time will tell whether traditional device manufacturers are able to retain their market dominance once the GPRS segment hits the consumer market. Either way, look for smart phones to be one of the benchmarks in the GPRS business.

Figure 22.4
Smart phones: Motorola Accompli 008 and 009; RIM Blackberry; Siemens SX-45; Trim Mondo; and Handspring units.

Multimedia Devices

These combine integrated and "add-on" devices offering sophisticated color cameras and remote screens for video purposes constrained only by available network bandwidth. Full integration of these services is already beginning with the advent of early 3G services. NTT DoCoMo launched their FOMA 3G service in October 2001 with a device that incorporates a built-in camera capable of still and limited video functionality. This segment too presents an opportunity for new entrants in the device manufacturing field to compete with entrenched GSM device manufacturers. Companies such as NEC have been successful in Asia precisely because of their experience in bringing full-motion video and color screens to the consumer. Others are sure to try to replicate their success such as Sony Ericsson who have already brought the T68i to market with a color screen and an add on camera.

Figure 22.5
The multimedia device: Nokia 7650 with integrated digital camera.

Black Boxes and Embedded Devices

Industrial devices consisting of stripped-down versions of standard terminals (minus the screen and keyboard) are used for applications such as remote meter reading and security alarms, and telematic systems. Siemens and Intel have produced many of these "black box" products. Further developments are underway for machine-to-machine applications using GPRS, which could become one of the largest GPRS product segments. There are literally millions of possibilities for machine-to-machine communications over secure, packet-based communications. While fixed-line access to some such devices is available at reasonable cost today, many other machine-to-machine applications are not possible today, for either technical or fiscal reasons, but may become opportunities for GPRS in the near future. The oft-quoted concept of the vending machine that sends data to a central monitoring location requesting new stock is but one example. The days of dispatching staff to the physical location of meters or machines to determine their status may soon be gone if GPRS black boxes are able to transmit vital information cost effectively on a near real-time basis. It's still not clear whether the full vision of GPRS embedded devices can be realized.

Figure 22.6
Embedded devices:
Nokia 30 M2M unit
(top left); Siemens
MC35 cellular engine
(top right); and
Wavecom units
(bottom).

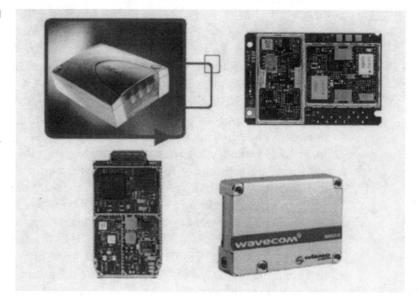

WLAN and Bluetooth

GPRS and Bluetooth are synergistic, creating in tandem whole new product and feature possibilities for terminal manufacturers. One trend to watch for is the mobile wireless hub that allows piconets and scatternets to form as needed wherever there is a concentration of appropriate devices. Piconets will also enable single-function device viability, giving industrial designers more scope to craft services, where form and function translate into an intuitive tool. Examples include the "pizza clicker" where one-click short code abbreviations or soft keys will be used to order your favorite fast food, the "taxi whistle" that signals you are in need of a taxi and gives your location to the driver, and the "half-full milk bottle" that initiates a product reordering process. Swiss Army Knife terminals, i.e., those that support many features, will no longer baffle users; instead modular solutions will let users pick and mix services at digital jewelry stores.

Handheld Computers

Increased interoperability between phones and handheld computers is a feature of product names such as Microsoft Pocket PC, Symbian (EPOC), PalmOS, and Sharp. Consumers may still encounter difficulties in configuring software and hardware to turn a computer into a messaging center (Mobile Lifestreams 2000), but as the line between mobile devices and handheld computers blurs, more and more similarities will be found between these two product groups. Some predict that the vast majority of GPRS devices will function like "handheld computers" while handheld computers begin to function more and more like "wireless devices." The blurring of these boundaries opens up new opportunities for the PC industry to access the GSM customer base without having to enter into agreements with existing device manufacturers. In one school of thought, GPRS will realize global success by means of a single feature: the capacity to add IP mobility to wireless handheld computers. The further addition of GSM voice capabilities then brings functionality full circle: PC, GPRS wireless data, and GSM voice service in a single solution. Much like smart phones, this type of device will succeed in attracting users based on simplicity of use, battery life, readability, and the access medium for data entry.

Although it's almost impossible to track who is doing what and when they're doing it in the GPRS device area, the GSM Association has undertaken a program to provide information as publicly as possible. Accessing www.gsmworld.com/technology/gprs_terminals.html will get you a high level overview of all GPRS devices made known by the manufacturer to the Association. The listing is updated on a regular basis and is free of charge. Other sources may provide similar information and readers are encouraged to verify all data with the manufacturer to confirm availability and technical specifications. This chapter has been written without such details in order to give you an overview of GPRS devices that won't go stale immediately, and to avoid burdening readers with information made irrelevant by new product launches.

Figure 22.7
Handheld computers: Compaq ipac and Sony Clie.

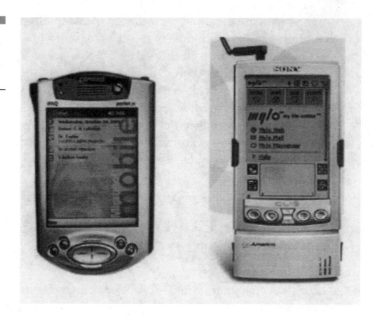

Is the Road to Success Paved Yet?

Will GPRS be the savior of WAP in its ability to provide a true mobile Internet experience to overly expectant customers (and those without child)? The short answer is probably but only if this technology is tested for adequacy, supported by sufficient terminal availability, and kept fresh with innovative customer services and applications.

> *The success of general packet radio service (GPRS) will hinge on whether operators can offer customers good data rates and reliability. After the negative publicity about WAP, the industry wants to make sure that GPRS users are satisfied.*[3]

Others have said that the key to GPRS take-up is predicated on two practicalities: global GPRS service with automatic or seamless roaming and simple-to-use devices. Just as the PC industry exploded with the availability of prepackaged applications that the consumer could install and operate without a PhD in computer science, GPRS will probably take hold through commercial consumer applications. However, those

---

[3]Wireless Europe, p31, 2001.

applications must run on a platform that's equally easy to use. It took a long time for the PC industry to bring together functionality, simplicity, and affordability in a package the mass market found appealing. The same can be said for GPRS. Devices must match the needs of the user, be compatible with commercial applications and services, and be cost effective both at purchase and on an operating basis. Once these goals are achieved, the addition of global GPRS roaming will cement the success of GPRS devices and allow them to become the norm throughout the world for mobile data services.

What the Customer Can Expect from the GPRS Terminal

GPRS lets customers access full-color remote services and a wide range of new features much faster than they can now. Initial data speeds could be as much as a dozen times faster than conventional GSM. The use of data-compression techniques enhances transmission for a myriad of multimedia services including image and video download so that they become desirable by the average consumer, and potentially offers seamless IP connectivity. The look and feel of terminals will evolve to ensure greatest possible screen size, thereby incorporating and embracing other forward-looking technologies such as voice recognition and activation to more activities which the customer can do, without supplying passwords and keyboard commands.

As video telephony establishes itself in an ever-expanding and ever-hungrier global consumer market, integrated digital cameras will become the norm, spurring on a host of new customer experiences, services, and supporting industries. Battery life will remain an issue for customers who always benefit from reduced consumption levels as data packets are transmitted in bursts as opposed to the full power consumption currently required for initiating a WAP session. The migration of SMS to GPRS frees up signaling resources for network operators (this benefit, however will largely be transparent to customers). Interoperability issues will emerge as newer terminals designed with network management control channel features are adopted, resulting in the necessity to recall and replace latent handsets.

Above all, GPRS represents to the customer an important milestone along the road to achieving third-generation–like services.

Conclusion

The advent of GPRS heralds an opportunity for the wireless industry to embrace and converge many existing sectors like financial, marketing, advertising, security, and telecommunications, to name but a few. New players with new viewpoints will generate valuable content and applications not seen before. For those involved, opportunities are rife to build value-added services that will convert readily to new revenue streams. And none of this will take place without risks to tested business models and long relationships.

From the customer's point of view (which includes those of us still trying to grapple with programming a 5-year-old video recorder to an equally out-dated TV!), GPRS terminals must provide non-techies with a "pleasant" user experience, complete with step-by-step tutorials and clear directions to engage essential functionality and ancillary applications. Customers have a right to expect this as a minimum.

If these goals are achieved, there still remains one important caveat. Reliability and quality of service in a data world will assume a significance all their own. Customers may not be so forgiving of incomplete data transmissions as they have been of the clunk-click, echo, and dropped calls so customary in early GSM.

Figure 22.8
Pogo.

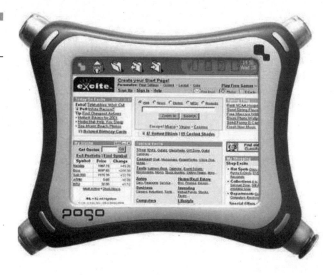

Figure 22.8 shows a hybrid device using data compression techniques to provide GPRS and I-Mode like services over existing GSM. It is aimed as a must-have device for early adopters of GPRS.

References

GSM Association—GSMWorld
Next Generation Mobile Terminals—EFT Conference Notes (2001)
Wireless Europe—(2001)
Yes2 GPRS—Mobile Lifestreams (2000)

CHAPTER 23

GPRS Devices

Rainer Lischetzki

Motorola

GPRS Classes

Before diving into the matter of GPRS handsets and devices, let's quickly review a bit of the GPRS network infrastructure and how it affects connectivity between handset and infrastructure. GPRS shares the base station with the GSM infrastructure and filters out data packets that then will be processed by the SGSN and the GGSN. In other words, GPRS is an overlay network to an existing GSM network. Despite the fact that they share resources and have some internal interconnection, it is useful to imagine GSM and GPRS as two different branches of a network served from a single access device. Each of these network branches offers specific services that must be supported in the handset.

Figure 23.1
Packet-switched and circuit-switched branch.

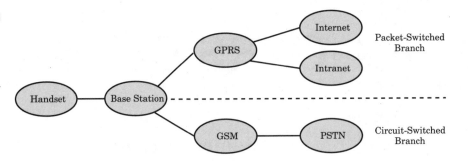

The functionality of voice and data services offered to users varies with the GPRS class. Three GPRS classes have been defined in the specification, each with its own distinct properties. In order for a handset to access GPRS services, at least one GPRS class needs to be supported; the actual implementation in the handset will determine whether this class of service is pre-assigned by the manufacturer or if users have privileges to change it. The three GPRS classes are known as A, B, and C.

Class A

A GPRS Class A device is able to register to the GPRS and GSM branches of the network simultaneously. Users can access voice and data services at the same time, using one timeslot per service. In total, a GPRS Class A handset has to support a minimum of two active timeslots in the uplink direction and downlink direction. Additional timeslots will increase maximum data transfer rate.

GPRS Class A allows users to transfer data over the GPRS branch of the network whilst engaged in a voice call over the GSM branch. The advantage of Class A is, in fact, to start and end GPRS data sessions and voice calls independently, so that the two services will not interfere with each other.

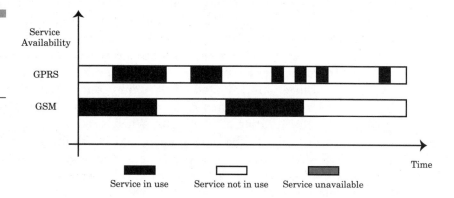

Figure 23.2
GPRS Class A, simultaneous use of GSM and GPRS services.

What is the user scenario behind Class A devices? I imagine a standard situation in the office: while speaking on the phone to a colleague, you are reminded of a spreadsheet or a PowerPoint presentation that could clarify the subject under discussion. In a cabled office environment the needed information is easily emailed, or a document link lets everyone access it. While the phone conversation continues, a data transfer is completed by the participants. The very same scenario is possible using a GPRS Class A-enabled PDA, permitting voice communication via headset/microphone combination and simultaneous download of data files: the office suddenly becomes mobile. And it's not just the office, of course—many consumer usage scenarios can benefit equally from GPRS Class A capabilities.

The implementation of GPRS Class A services is demanding, given that handsets use GSM and GPRS branches of the network simultaneously. Clearly, on the handset side, both GSM voice and GPRS functions need to work in parallel. From the network planning side, GPRS Class A functionality requires uninterrupted voice services. In addition, the quality of service (QoS) criteria for GPRS data services need to be matched in order to provide sufficient bandwidth for applications. A cell handoff essentially denotes the availability of two timeslots in the new base station to support handsets voice and data needs. It is therefore not too surprising that no GPRS Class A handsets are yet available in the marketplace at the time of writing.

Class B

A GPRS Class B handset is able to register to GSM and GPRS services simultaneously, just as GPRS Class A handsets do. However, voice calls and GPRS data sessions are allowed only sequentially, not in parallel.

When the user has registered the GPRS Class B handsets on the network, but has not started a GPRS data session yet, the handset behaves just like a standard voice handset: making and receiving voice calls works as in GSM. If the user has started a voice call, on the other hand, GPRS data services are not possible until the voice call has been terminated.

During an activated GPRS data session, the GPRS Class B handset uses the *suspend* and *resume* mechanism to "park" GPRS data sessions for the duration of voice calls. When a call comes in, the user can either continue with the data session, which will divert the voice call to voice mail, or can opt to suspend (park) the data session for the duration of the voice call. In the latter case, the user may resume the data session afterwards.

Figure 23.3
GPRS Class B, alternating use of GPRS and GSM services.

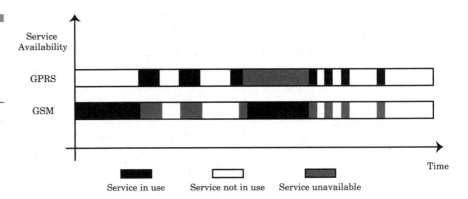

A GPRS Class B handset requires a minimum of one timeslot for uplink and downlink, to be used for either voice or data services. Additional timeslots will increase data transfer rates but will not change the GPRS Class.

Class B matches the needs of consumers running WAP applications over GPRS, or any other applications infrequently transferring smaller data amounts. It's well suited for playing a game or downloading Java applications. It is convenient to use for business applications in particular, and for accessing email over WAP and GPRS.

The mechanism for parking a GPRS data session during an incoming voice call naturally halts the user data flow. It is entirely up to the application how to control this interruption. WAP, for instance, is very forgiving, allowing the user to continue the GPRS data session at any time. In other cases, say a file transfer, once data flow has been cut off by an incoming call, the user may have no option but to start the file transfer again.

GPRS Class B handsets are commercially available on the marketplace.

Class C

As opposed to Classes A and B, GPRS Class C only allows registration in one of the network branches, either GPRS or GSM. In consequence, three flavors of GPRS Class C are used.

- GPRS Class CC devices will register to the GSM branch of the network, permitting circuit-switched GSM services only. It is worthwhile to note that all legacy GSM handsets will behave like GPRS Class CC devices.
- GPRS Class CG devices register on the GPRS branch of the network, allowing only GPRS services. Circuit-switched voice and data services are not accessible.
- The device manufacturer may allow users or applications to switch between Classes CC and CG either via man–machine interface on the handset display or via use of AT commands (when used as a wireless modem).

GPRS Class CC and CG require a minimum of one timeslot in both uplink and downlink directions. Additional available timeslots, when used as Class CG device for GPRS data services, will increase the data transfer rate.

When using a Class CG device, users may access GPRS services only. This restriction clearly has the advantage of providing uninterrupted data transfer, especially for file transfer and similar applications. On the other hand, it forces the user to track and, if required, change GPRS classes when placing or receiving a phone call. In other words, it's up to the user and the application whether Class CG behavior proves to be beneficial or limiting.

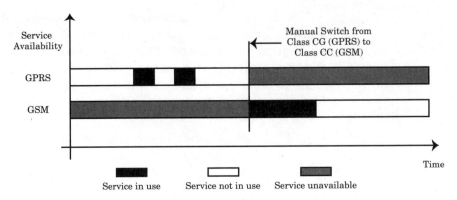

Figure 23.4
GPRS Class C, manual selection between GPRS or GSM service.

Typical usage scenarios for GPRS Class C include undisturbed file transfer using a handset as wireless modem. Class CG is an excellent choice for corporate intranet access, especially with purpose-built GPRS devices in the form of PCMCIA cards. It is also the class of choice for the telemetry and telematics environments, where applications are data-centric to a large extent.

GPRS Class C handsets are commercially available on the marketplace.

Multislot Classes

Apart from the GPRS classes described above, the Multislot Class of a GPRS device is the most distinguishing factor. It lets users determine maximum achievable data rates in uplink and downlink directions and is often used in the literature to differentiate a handset's GPRS capabilities.

The Multislot Class is characterized by the maximum number of timeslots a handset can use for communications in the uplink direction (handset transmitting) and the maximum number of timeslots in the downlink direction (handset receiving). The third main component of the Multislot Class is the total number of active timeslots a GPRS device can use simultaneously for uplink and downlink communications. Table 23.1 details the Multislot Classes.

Unfortunately, interpreting the Multislot Class requires some caution. Any Multislot Class 8 device will support a maximum of four timeslots in the downlink direction and one timeslot in the uplink direction. Therefore the same device will support Multislot Classes 1, 2, and 4 as well. Multislot Classes 3, 5, 6, and 7, on the contrary, are not supported because they require two or three timeslots in the uplink direction.

TABLE 23.1

Multislot Classes

Multislot Class	Downlink Slots	Uplink Slots	Active Slots
1	1	1	2
2	2	1	3
3	2	2	3
4	3	1	4
5	2	2	4
6	3	2	4
7	3	3	4
8	4	1	5
9	3	2	5
10	4	2	5
11	4	3	5
12	4	4	5

Please note that the Multislot Class only provides information on its maximum capabilities. In real life, the number of timeslots to be used for GPRS services will be governed by the user's subscription details, the GPRS network design, and the total number of GPRS users.

For practical purposes, the number of timeslots available for data transfer in a given direction multiplies the data rates achieved in single-slot operation. However, this generalization does not take into account the fact that applications will only benefit from an increased number of timeslots when data volumes are large enough, and not unnecessarily fragmented into small data packets.

APN and QoS

GPRS introduces two new concepts to wireless data communications. One is the access point name (APN), which in very rough terms represents the equivalent of a dial-in phone number for Internet/intranet services. The second, QoS settings, defines the quality of the data connection in terms of data rate, data integrity etc. APN and QoS are assigned by the network operator either on a per-subscriber basis or for groups of subscribers.

APN and QoS are, in most cases, programmed into the handset during the manufacturing process, according to network operator requirements. The APN can usually be accessed from the menu, and a user is able to modify it when an application needs to use an APN other than the default. QoS settings are typically not accessible from the menu, since setting them requires a lot of detailed knowledge beyond the capabilities of the vast majority of users.

For experienced and knowledgeable customers using the handset as a wireless modem, APN and QoS settings can be changed with GPRS-specific AT commands, to enable optimal performance.

Chip Sets

In today's GSM world, handsets and devices are constructed using "chip sets." These are integrated semiconductor circuits designed for handling specific tasks in a mobile environment. In combination with external components (keypad, display, microphone etc.) they make up the handset hardware. All chip sets require specific operating system software that will handle radio issues (timing of transmission and reception, channel selection, signaling, etc.) and drive external components (keypad, display). In addition, operating system software handles user interface and internal application clients. Some software components are available from third-party vendors and are integrated into the system operating software by the device manufacturer.

Two ways of bringing GPRS functionality into handsets are in use: (1) existing state-of-the-art GSM chips are modified to support GPRS routines, including major redesign of the operating system software, and (2) new chip sets are designed and new operating system software is developed to include GPRS routines right from the start. The first route proved to be the fastest way to market, enabling early availability of GPRS handsets. Meanwhile the first GPRS products using new chipset design are now available on a commercial basis.

The chip set employed for the design of the handset governs the maximum number of available timeslots for radio communications in uplink and downlink directions. Current GPRS-enabled chip sets offer Multislot Classes 1, 2, and 4, some Multislot Class 8. Other chip sets have been announced to support Multislot Class 10, offering two timeslots in the uplink direction.

When GPRS Becomes Widespread

As useful as it was to imagine GSM and GPRS services as two independent branches of a network in terms of end-to-end connectivity, all good analogies have their limits. In reality, voice and data services within the network are coordinated jointly. This is accomplished by the use of specific signaling information on signaling channels, which assist a network operator in administering network resources in the most economical way.

In GSM systems, signaling channels are used to set up, maintain, and end a circuit-switched call, with only a minimum of overhead created. GPRS introduces the concept of virtual connections, which increase the signaling information sent over the network. Instead of just signaling begin and end for a circuit-switched call, signaling information is sent over the network each time GPRS data is to be transferred. As the number of GPRS users grows and data traffic increases, traditional signaling channels will not be able to handle all signaling information without impact on the overall network throughput performance.

This situation has been addressed in the specifications and prompted the introduction of new GPRS signaling channels such as packet broadcast control channel (PBCCH) and packet common control channel (PCCCH). These signaling channels are mapped into the packet data channels (PDCH) carrying the actual data traffic. The cost is a hardly noticeable reduction of maximum user data rate, but the benefit is assurance that all GSM signaling channels are kept free for voice signaling, thus guaranteeing undisturbed voice service.

To cut a long and technical story even shorter (the above is about as short as possible, and there is much more to it): once GPRS becomes a success, it will be advisable for network operators to verify that handsets support the packet-signaling channels (PBCCH, PCCCH, etc.) in order to maximize spectrum efficiency. Packet-signaling channels are supported by a number of commercially available GPRS handsets.

External and Internal Applications

At first glance, it might seem like an academic question to ask where application software and clients reside. But at closer range, this turns out to be of vital interest for the design of the handset and its capabilities. In sum, internal applications and clients run in the handset, whereas external applications run on any computer or computing device connected to the handset.

All applications require runtime processing power and storage capacity. Today's desktop computing clearly shows to what an extent these resources are used (or, worded differently, absorbed) by newer applications. These factors frame any discussion of how well applications perform, and certainly the very same rules apply for GPRS devices. But when moving into the mobile environment, another factor is added: battery life. The more processing power and access to storage media an application requires, the more battery capacity is needed to fuel it. This reality sets limits on the applications that can be used in GPRS handsets.

External Connections...

The simplest way of running applications over GPRS is to use a handset as a serial modem, just the way a telephone modem is used. When connecting a handset with a laptop, you'll often find that the GPRS network provides Internet access as part of the standard subscription. A variety of standards have evolved in the past for serial communications. All promise seamless migration to GPRS and perfect results for GPRS. Here is a selection of serial connections supported by commercial GPRS devices:

- **RS232**—Serial cable connection, supported by most computers. Rugged and well understood.
- **IrDA**—Infrared connection, supported by a variety of computing devices including personal digital assistants (PDAs). IrDA requires line of sight between the two devices.
- **USB**—Universal serial bus, started replacing RS232 serial connections, allowing higher data rates.

Speaking a little more futuristically, Bluetooth has evolved as a new standard for short-range wireless communications, replacing serial cables. Currently, no Bluetooth-enabled GPRS handsets are available, but Bluetooth should be considered as the natural extension of GPRS (and 3G) data services when interconnecting GPRS handsets and modems with computing devices.

... Protocols ...

Serial connections only define the physical link between the GPRS handset and the connected computing device. The definition of protocols has been left to the GPRS specifications. Luckily the well-understood point-to-point protocol (PPP) is amongst the options provided in the GPRS specifications, and most handsets currently offering serial connection support it. PPP itself has been defined in RFC 1661 by the IETF, ensuring backwards compatibility with existing applications.

After the serial connection between handset and computer is made, the GPRS handset needs initialization. This is carried out by the application issuing the appropriate AT commands. AT commands are extensions to existing cable modem and GSM AT commands and are defined in GPRS specifications (GSM 07.07 and 07.060). Their implementation can vary slightly according to the implementation in the particular handset, however, the important commands all are mandatory. Once setup is concluded, the application invokes the PPP protocol stack, and an IP data session with a remote host can be activated.

The outstanding beauty of adopting this approach and implementing a PPP protocol stack in GPRS handsets lies in its backwards compatibility with the familiar wireline Internet standards. Any application able to use Internet and/or intranet services over a cable modem will be able to do the same over GPRS services *without changes.*

... and Additional Software

Having read the last paragraph, you may be thinking that GPRS usage seems perfectly easy. However, the user or application may be asked to submit a couple of parameters prior to activating a GPRS data session. Specifically, the APN and QoS parameters have to be entered, especially when the configuration of the GPRS data session is different from the network operator's defaults.

The configuration of GPRS devices for modem services requires intimate knowledge of the AT command structure and GPRS parameters. The entry either requires the applications to enter parameters or requires users to manually configure the handset. At best this is a cumbersome procedure, and in the worst case, it results in failure and frustrated users.

To the rescue of customers and users, handset manufacturers and independent software houses offer utility programs, which either are distributed with the GPRS handsets or are accessible via Internet. These utility programs can be installed on computers and in many cases already contain network- and/or customer-specific GPRS settings. In addition, they allow easy configuration of the handset, if this is required at all, by making minor modifications to a standard configuration. For a GPRS user, it's as simple as connecting the handset with the computer, running the utility program, and then starting the application.

Voice-evolved GPRS Devices

The first GPRS handsets to reach the market are what can best be described as voice-evolved devices. These handsets come in the traditional "candybar" form or in the more stylish folding design. They are operated from a numerical keypad and offer a 4 to 6-line display. The hardware architecture follows the "one processor" approach, where a manufacturer-specific operating system handles both radio communications and the applications embedded in the handset. As a consequence, the CPU time available for applications is limited and the handset will always put application processing needs behind the real-time requirements of the radio part.

Figure 23.5
Voice-evolved GPRS
handset.

Usually a manufacturer-specific operating system cannot be accessed by third-party application designers, and consequently all applications have to be factory installed during the manufacturing process. In addition, storage capacity for programs is limited, even though some recently announced GPRS handsets show quite a large capacity for applications. As a default application, a WAP browser is embedded in the majority of GPRS handsets.

Multislot Classes 1, 2, 4, and in some cases Multislot Class 8, are supported, allowing between one and four timeslots in the downlink direction and one timeslot in the uplink direction. Since the handsets are to be used for voice calls, they mostly support GPRS Class B operation, but they can very easily be used as wireless modems via the serial connection in order to access Internet/intranet services.

Data-evolved GPRS Devices

Inasmuch as the design of GPRS devices depends on the actual application to be used, a series of data-evolved GPRS devices have been launched or are about to launch. Among their ancestors are PDAs, so it is not a surprise that GPRS devices inherited large display and pen-based operation. Data entry can be done either by handwriting recognition or by pop-up keyboard on the display, allowing the user intuitive handling and operations without having to learn specific alphabets. Other designs include a built-in QWERTY keyboard for data entry and easy message editing.

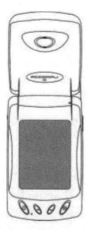

Figure 23.6
Data evolved GPRS device.

These PDA-type GPRS devices most often follow the "dual processor" approach, based on two logical sections with at least two dedicated processors. One handles all radio-related issues and the other runs applications and the user interface.

There are a variety of advantages of splitting the workload between two processors. First, the real-time requirements of the radio section are not impacted by application needs for CPU time. In turn, the operating system software on the adjunct processor handling applications can be made more accessible to third-party application developers. Having two processors usually relaxes storage capacity constraints, so applications can be quite complex before hardware limitations come into effect. Finally, applications do not necessarily need to be factory installed during manufacturing, thus enabling the user to change his personal portfolio of applications according to his needs and preferences.

A small shadow falls on the two-processor approach when it comes to battery consumption. Running two processors may not require more power per se, but as soon as the workload imposed by applications on the adjunct processor increases, current consumption increases too. This behavior certainly corresponds to that of PCs, but because of the much smaller form factors and the restricted battery sizes they mandate, the limits of processing power in GPRS devices are reached far earlier than in PCs.

On this type of device, internal applications range from email clients and calendar to WAP browsers and games. In order to give the user more data functionality, most devices provide either RS232 or IrDA connection for downloading user data and/or applications into the handset.

Although GPRS devices are sometimes larger than the typical PDA, they still fit in shirt or suit pockets. More importantly, they support full voice functionality and can be operated as standard handsets, either held against the ear or via earpiece and microphone combination. In terms of GPRS-specific functionality, these devices support GPRS Class B, Multislot Classes 1, 2, and 4, and some Multislot Class 8.

Making More Out of GPRS with Java

As the previous section highlights, restrictions apply when storing and running applications on GPRS handsets. A possible remedy for these limitations, and to prevent the ever-increasing complexity of the contemporary PC world from penetrating the handset space, is the introduction of Java.

Java is a twofold concept—that of a high-level programming language and that of a runtime environment interfacing to the operating system software and underlying hardware. The Java 2 Micro Edition standards (J2ME) in particular offer the environment required to run applications successfully when faced with limited processing power and memory. Java applications can be designed without intimate knowledge of the device on which they will eventually be executed. And probably more important yet, the installation process is quite simple in comparison with a PC: no re-boot is necessary. Once an application is outdated or no longer needed, it can simply be deleted.

Java is already implemented in some handsets, marking the first time applications can be downloaded over the air into the handset. Users can configure their own application portfolios on the fly, leveraging the advantages of GPRS, and at any time can adjust applications on the devices according individual needs—be it for professional use or personal entertainment purposes.

The first GPRS handsets supporting Java were made commercially available in 2001. They fall into the category of data-evolved devices. Java now is supported also in a large number of voice-evolved handsets.

Telemetry Devices and Data Modules

Telemetry and telematics are classic wireless data applications. These segments of the market call for specialized hardware designs in order to support the extended temperature ranges and mechanical robustness requirements for harsh environments.

The first GPRS modules commercially available were based on existing GPRS chip sets. In terms of the performance and functionality of GPRS classes and Multislot Classes, they closely resemble voice-evolved handsets. The obvious difference is the lack of display and keypad. A serial link is used to connect computing devices and allows control command exchanges and user data communications.

Some of these modules come with type approval and the EU declaration of conformity with the EMC directive (CE Mark). Still, some minimum integration is required in order to provide the necessary power source serial link adaptations (e.g., voltage level shifting for RS232 ports, mounting, etc.)

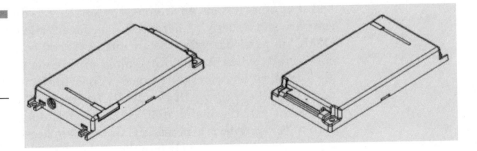

Figure 23.7
Sketch of GPRS data module, size approximately PCMCIA III.

Typically, these modules are distributed as OEM modules built into third-party software and then offered as part of complete solutions to corporate customers, automotive manufacturers, and others.

The first offerings that will incorporate data modules into rugged handheld terminals and vehicle-mounted wireless modems are now available from various manufacturers. These address primarily the courier marketplace, where collection and delivery information for parcels is scanned via the terminals integral barcode reader and then transferred via GPRS to a host computer.

PCMCIA Cards

The purest data-only devices announced to date (and by a number of manufacturers) are GPRS-enabled PCMCIA cards in the type II and type III formats. These cards provide the standard PCMCIA connector, allowing the card to be inserted in laptops with PCMCIA bays. They have an integrated onboard battery, which can be recharged from the hosting computer. A small antenna is sometimes attached to the outside of the PCMCIA card, or any standard GSM antenna can be connected via a connector to improve RF properties, e.g., inside a car when driving.

Once again, standard chip sets are used for the design of PCMCIA cards, with special care devoted to achieve the form factors set by PCMCIA specifications. In terms of GPRS classes, most PCMCIA cards will support Class B, enabling not only GPRS services but circuit-switched services as well. Manufacturers quote support of (up to) Multislot Class 10, offering at maximum two slots in the uplink and four slots in the downlink. Voice services supported by Class B PCMCIA cards can be used either via headset/microphone combination or by voice applications on the hosting computer.

PCMCIA cards are intended for the corporate marketplace, which typically uses GPRS as an access method for corporate intranets. Many corporate customers are prepared to go the next step and equip their mobile workforce with two devices, a handset for dedicated voice communication and a GPRS PCMCIA card exclusively for data services. These customers will need two SIM cards and therefore will receive two bills.

GPRS and Current Consumption

Aside from battery-consuming applications and clients installed in handsets, battery capacity is not really much of an issue with GPRS. For packet-switched services, the amount of data transmitted, rather than the time spent online, governs overall battery life. Terms like *talk* time and *standby* time, both of which play a major role in circuit-switched services, apply to a much smaller degree in GPRS. In short, the more data sent—especially in the uplink direction—the faster the battery will need a recharge. In this respect, GPRS does not differ from GSM voice services.

GPRS also makes intelligent use of signaling. Despite the fact that GSM and GPRS services are offered by different branches of the network, they use the same basic signaling. When not transferring GPRS data, signaling traffic will not increase and in turn will not have any adverse effect on battery life. This applies even for active GPRS data session, as long as no user data is transmitted. For instance, a WAP session over GPRS in which no user data is transferred can be maintained over an entire day without any impact on battery life.

Taking this concept a bit further, note that many applications require only relatively small amounts of data. To use the example of WAP once more, even heavy usage of WAP doesn't significantly impact battery life because total data volume is nominal.

In terms of current consumption, it does not matter whether a single timeslot or multiple timeslots have been used for data transmission. A data file can be downloaded using two timeslots instead of one. The total data transfer time then is cut in half compared to single timeslot operation, but the total keying time of the transmitter remains the same (assuming the same coding schemes and RF conditions). This results in the same current consumption for single and multislot operation.

Another interesting observation can be made in the case of asymmetric data transfer; i.e., more data is downloaded from the network than is sent from the device. A typical scenario is the downstreaming of audio or video files: only smaller data packets are sent in the uplink direction, with the vast majority of data transfer occurring in the downlink. Despite staying connected over long periods of time, the transmitter is keyed considerably less often than a voice call of the same duration. The resulting download time from the battery is much longer than the achievable talk time when engaged in an ordinary voice call.

In summary, in most scenarios, the inclusion of GPRS as data bearer server in the handsets will have no noticeable impact on battery performance. When we compare the battery performance of an application running over GPRS to the performance of the same application running over circuit-switched data services, GPRS even appears to offer an extended battery life.

CHAPTER 24

Environmental Issues

Jack Rowley, Ph.D.

GSM Association

The safety and environmental impact of mobile communications handsets and infrastructure is a topic of regular coverage in trade and general media outlets. On the safety question, the conclusion of expert bodies in Canada, the United States, the United Kingdom, France, Australia, the Netherlands, Singapore, Malaysia, Ireland, the European Union, other countries, and the World Health Organization (WHO) has been very consistent: the existing scientific evidence from more than 40 years of research on radio signals provides no convincing evidence of a public health risk from the use of mobile phones or living near a base station. From a research perspective there remain uncertainties in the scientific literature and the WHO has identified a program of work to support a formal health-risk assessment in 2004/2005. The global cellular industry has responded to this agenda by funding significant independent research programs in Europe, the United States, and the Asia-Pacific to ensure the information needed by the WHO is available. Both national and international safety guidelines exist for infrastructure and handsets. The metrics used to assess the compliance of handsets and base stations to safety guidelines will be discussed in further detail later.

The higher data rates available from GPRS handsets through the use of multiple timeslots raise questions about the possibility of increased levels of user exposure to radio frequency (RF) signals. Suppliers are aware of the need to ensure compliance with safety standards under intended usage conditions while operating in GPRS mode. To emphasize again, this is clearly a question of ensuring compliance with standards, not of intrinsic safety, as the consensus of expert groups is that there is no demonstrable evidence of a risk to human health from the use of compliant handsets.

In terms of other environmental impacts from GPRS, these mainly arise around questions of energy use, disposal/recycling, and visual impacts of infrastructure and have much in common with 2GSM technology.

Specific Absorption Rate

The compliance of GPRS handsets with human RF exposure standards is based on measurement of the *specific energy absorption rate* (SAR). This term quantifies the rate of absorption of RF energy (measured in watts [W]) per unit of body mass (measured in kilograms [kg]). Thus the units for SAR are W/kg (watts per kilogram) and in essence this is a measure of the *dose* of RF energy. As a point of reference, RF energy is converted to heat when absorbed in the body. If this heat load is greater than the thermal compensation mechanisms of the body, adverse health consequences may occur. While there has been much speculation about adverse biological effects at exposure levels too low to cause heating—so-called "athermal effects"—the experimental evidence is not convincing.

A GPRS device in use will normally be close to some part of the body (the head for a handset during a voice call) or perhaps the leg (if a GPRS data card is used in a laptop). In these usage situations, only a small proportion of the body is being exposed to the RF fields and hence we reference partial body exposure limits in the appropriate safety standards. As a side note, most Western RF exposure standards specify different allowable levels depending on whether the exposure is occupational or public. As GPRS handsets are generally considered a consumer device, the discussion below will address only public exposure limits unless otherwise stated.

There are two main partial-body SAR limits in use today, one derived by the Institute of Electrical and Electronic Engineers (IEEE) RF safety committee and one derived by the International Commission on Non-Ionizing Radiation Protection (ICNIRP), a WHO-recognized body. The rationales used by the two organizations in developing the exposure limits differ, thus leading to different final numbers.

In determining a partial-body exposure level of 1.6 W/kg, the IEEE committee considered both the threshold whole-body exposure level for adverse effects and the nonuniform nature of RF energy absorption in the body. A measurement mass of *phantom tissue* (more later) also has to be defined in order for the SAR measurement to make sense. The committee chose a volume of 1 gram—the practical minimum measurement volume at the time. For practical measurement purposes, this has been defined as a cube of 10 mm per side, rather than an arbitrarily shaped volume. The IEEE-based limit has been adopted by the Federal Communications Commission (FCC) in the United States and the Australian Communications Authority (ACA), and is being considered by other countries.

The approach taken by the ICNIRP was quite different and largely based on research conducted by the U.K. National Radiological Protection Board (NRPB). The ICNIRP took a more directly physiological approach by examining the ability of small volumes of tissue to dissipate extra heat. From this analysis they determined a level of 2 W/kg and, from a consideration of contiguous tissues, a measurement mass of 10 grams (about the mass of the eyeball). In this case the practical measurement volume is taken as a cube with a 21.5 mm side. In 1999 the European Council recommended adoption of ICNIRP-based limits by European Union member countries, and the ICNIRP guidelines have been adopted in Japan and many other Asian and African countries. After a review of the scientific literature, New Zealand changed from the IEEE to the ICNIRP level in 1999, and the 2002 Australian standard also aligned with ICNIRP.

The discussion in the preceding paragraphs applies only to the limits relevant to RF energy deposition in the head and torso. Both the IEEE and ICNIRP specify a higher (and identical) allowable exposure limit for the limbs of 4 W/kg measured in 10 grams. This limit would come into play for a GPRS device that was sitting on the user's leg or held in the hand.

There are obvious numerical differences in both the level and averaging mass between the two expert bodies. Numerical modeling and practical measurements show that the IEEE limit is effectively more restrictive than the ICNIRP guideline by about a factor of two. However, it should be noted that both limits are conservative relative to the threshold levels at which biological effects have been consistently observed, and that there are philosophical differences in the underlying rationale. As I write this, the IEEE RF safety committee is considering a proposal to adopt the ICNIRP partial-body exposure limit as part of its standard revision process. The WHO is promoting this move toward greater international harmonization.

Assessment of SAR Compliance

While handset manufacturers have been conducting SAR assessments for some time, it is only recently that internationally agreed testing standards have been formally adopted. The two main bodies active in this area are the European standardization body (CENELEC) and an IEEE sub-committee on RF product safety. Close cooperation between the two organizations has resulted in closely aligned standards. The CENELEC standard was adopted in July 2001 and the IEEE subcommittee expects to finalize their standard shortly.

Handset compliance with the SAR standard can be demonstrated in two ways: computational methods or direct measurement. In the former approach, a detailed mathematical model of a handset is constructed with specialist computer code and then the SAR is calculated in either a simple or complex model of the head. Uncertainties in modeling all of the important handset details and difficulties in verifying the accuracy of the results work against using this approach for final compliance measurements. However, it can be used in the design process to investigate the effect on SAR of locating the antenna in various locations, or as a research tool with an anatomically realistic multitissue model of the head to calculate the SAR in different structures of the brain.

The direct measurement method involves using a precision robotic arm to scan the volume of a phantom head with a small RF probe. The phantom consists of a plastic shell shaped to correspond to the outer contours of the head. The interior cavity is then filled with a semi-viscous liquid. This phantom tissue has the same electrical properties as composite brain tissue at the operating frequency of the handset to be tested—the tissue dielectric properties are different at 800/900/1800/1900 MHz. The shell of the phantom is based on an anthropomorphic survey of United States Army personnel and was chosen to correspond to the 90th percentile of head sizes (as a larger head model will tend to provide a conservative SAR assessment) with an insulating spacer to represent the ear. The handset being tested is placed with the earpiece over the phantom "ear" and the mouth-piece aligned with the phantom mouth, i.e., the intended use position. Under computer control, the handset is set to transmit at its maximum power, and the robot arm automatically searches for the peak level before calculating the volume average SAR relative to the applicable standard. Whole-body phantoms are available for measurement of SAR with the handset on other parts of the body. Overall accuracy of these systems is of the order of 30 percent. See Figure 24.1 for a typical set-up.

As noted above, the large phantom head is a conservative feature of this test method. Other factors designed to ensure that the measured SAR will tend to overestimate actual user exposure include the loss-less ear spacer and phantom shell, the choice of phantom tissue electrical properties, and the probe measurement algorithms. Additionally, the handset is tested at maximum transmit power and no allowance is made for the dynamic power control that can result in a GSM handset transmitting at less than 10 percent of the maximum for more than 75 percent of the time. Thus, the measured SAR figure should be *very* conservative for all possible users in practice.

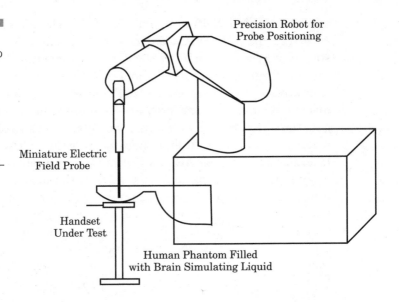

Figure 24.1
Precision robot set-up used to scan miniature measurement probe within the "brain" of a human phantom for SAR compliance testing.

Precision Robot for
Probe Positioning

Miniature Electric
Field Probe

Handset
Under Test

Human Phantom Filled
with Brain Simulating Liquid

GPRS Compliance Issues

At the present time, there are differences in the allowable SAR limit across regional and national boundaries. This has compliance implications not just for manufacturers who are designing product for specific markets, but also for GPRS handsets in an international roaming situation.

A number of studies and media stories have reported SAR figures for GSM handsets. The level of SAR can vary significantly between different phone models, with low values starting at around 0.1 W/kg in 10 g and high values up to 1.2 W/kg. Based on the publicly available data, most GSM phones will exhibit a SAR of less than 1 W/kg in 10 g when tested at 900 MHz with one timeslot in use. If multiple timeslots are used on the uplink, the SAR limit for some standards may potentially be exceeded if usage instructions are not followed. While GPRS is intended primarily for data services, and the user is unlikely to have the phone against his head, relevant exposure standards must be observed supposing, for example, that the phone is resting on the user's leg.

Initial GPRS handsets are designed for single timeslot transmission and multiple timeslot receive. For these handsets, compliance requirements will be similar to those of non-GPRS handsets. It is unlikely that GPRS handsets will support a significant number of uplink timeslots because the battery capacity of a phone will limit the available call time.

However, GPRS card phones, i.e., those designed to plug into a computer, will draw power from that source and may therefore be designed with capability for more transmission slots. For these handsets, the usage mode will mitigate compliance concerns with increased separation from the user.

Handset manufacturers are familiar with these issues and test their handsets for compliance in the markets and circumstances in which they are intended for use. The suppliers will therefore take into account the differences in national standards.

In addition, the major suppliers have proactively developed a program for reporting SAR information for handsets. It was launched in October 2001 for new devices. In essence, the scheme includes a declaration of compliance on the handset packaging material and the SAR test results, plus explanatory material in the handbook. It is important that consumers do not take these SAR figures as measures of relative safety—all handsets meeting the appropriate standards are regarded as being safe—but they may be considered like any other technical information (such as talk-time) in making a purchase decision.

RF Safety Compliance of GPRS Base Stations

There have been several published surveys of the level of RF signals from base stations in countries such as Australia, Hungary, Italy, New Zealand, and the United Kingdom. These surveys show that the average level of the GSM signals in the community is similar to or lower than radio and TV broadcast signals. Typical maximum levels in public areas are less than 1 percent of accepted scientifically based exposure standards.

As GPRS is essentially a way of getting greater data rates through improved signaling and multiple timeslots, the maximum RF signal levels in the community will not be significantly affected. The improved signaling efficiency of GPRS could even result in slightly lower average GSM signal levels as a result of less control-channel dialog between the handset and the base station. While it is true that the use of multiple timeslots would seem to reduce the network capacity available to service customers, the improved signaling of GPRS ensures that, at least initially, additional base station sites to serve customer needs probably won't be needed. This situation is aided by the overdimensioning of current GSM networks to deal with peak traffic levels. However, there may be exceptions where GPRS data traffic is high, e.g., in-building coverage in an airline lounge.

It is clear that community exposure to base station signals is at levels substantially below scientifically accepted standards. Despite this fact, some members of the public remain concerned about possible adverse health outcomes or effects on the amenity of property if a base station is located close to their home or place of work. This concern has made it difficult to acquire base station sites and led to the adoption of nonscientific "precautionary" policies by local authorities as a political response to manage the public perception of risk. Compounding the situation are poor community consultation practices employed in the past and the general distrust of corporations in surveys of community attitudes.

The WHO has warned that nonscientific policies can have the effect of undermining confidence in health standards and consequently these policies may cause unintended community alarm. Operators and their agents are becoming more sophisticated in managing community consultation, increasing their emphasis on two-way communication while relying at all times on the solid scientific base of evidence. Specific guidance may be found in the area of "risk communication" and the WHO has conducted two workshops on this theme. Responding to community concerns is also an opportunity to build improved relationships with key stakeholders, with direct effects on site acquisition and on the perception of the corporation. This trend may be seen as a specific expression of a more general acknowledgment of the operator's need to behave in a manner consistent with corporate social responsibility principles in order to enhance brand.

Interference by GPRS Handsets

Digital mobile phone technologies may cause electromagnetic interference (EMI) to other electronic devices. Interference may occur when the on-off nature of wireless digital signals is detected or *demodulated* by the electronic device, in much the same way an electric drill may interfere with a TV or radio. The level of any interference will depend on characteristics of the digital signal, proximity of the handset (or base station) to the electronic device, and the *immunity* or resistance of the electronic device to such radio interference.

The key signal characteristics for interference assessment are the peak signal level and the *frequency* or *repetition rate* of the digital signal; i.e., EMI is related to changes in the digital signal. Therefore, the signal peak rather than the average level will determine the extent of interference. Repetition rate is important because regular signals tend to be more potent interference sources. Electronic devices tend not to respond to higher repetition rates because of inherent bandwidth limitations. These last two points are especially important where the electronic device has an audible output, as do domestic audio equipment and hearing aids. With respect to GSM, the primary repetition rate is 217 Hz (Hertz) with additional components at 16 Hz and 2 Hz. As this is in the range of audible frequencies (about 20 Hz to 10,000 Hz), the interference may be readily perceived as an audible buzz on some devices.

For GSM, significant experimental work has been done to quantify the level of possible interference and to specify the immunity in electronic devices necessary to minimize interference effects. The table below summarizes the results of some of this work with a variety of digital wireless signals from handsets.

TABLE 24.1

Typical Ranges for Possible Interference by GSM Handsets to a Range of Electronic Devices

Type of Device	Interference Radius	Comments
Medical devices	<2 m	Based on advice from medical device administrations about safe usage distances.
Hearing aids	Range from 0.0 m to ~3.2 m	Experimental results on wide range of hearing aids. Older devices have poor immunity; modern aids may be usable with GSM handsets or simple hands-free accessories.
Domestic electronics	<2 m	Typically have lower immunity than medical devices. Effects include audible buzz or interference to remote controlled devices.
Pacemakers	<0.15 m	Based on medical device administrations' recommendations about safe usage distances.
Automotive electronics	<0.50 m	No evidence of interference to ABS or airbag systems. Limited interference with audio and central locking systems.
Aircraft electronics	Unclear	Limited evidence from UK Civil Aviation Authority tests during 2000. Present practice is to ban use of all types of radio transmitters while aircraft are in the air.

With respect to GSM infrastructure, typical design guidelines posit that interference will not occur in medical institutions if the GSM signal strength is below 1 V/m and in domestic residences for signal strengths below 3 V/m.

As GPRS does not change the peak GSM signal level, it can generally be expected that the level of interference will be similar to that of existing GSM handsets and base stations. One issue that is still unclear in relation to this conclusion is the effect of multiple timeslots on the repetition frequency. For example, if two adjacent timeslots are combined, the repetition frequency will continue to be 217 Hz—the characteristic buzz of an operating GSM handset brought near a speaker. If the timeslots are not adjacent, then the fundamental repetition rate will differ from this frequency and may vary as timeslots are assigned to different users. In summary, the level and radius of interference by a GPRS device is likely to be similar to GSM, but the audible frequency may vary.

Other Issues (Surface Temperatures, Energy Efficiency, etc.)

Users of mobile phone handsets will be aware that the surface temperature of some handsets may perceptibly increase during use. This is primarily a result of energy produced by chemical reactions in the battery, the electronic processing within the handset, and the limited efficiency (about 55 percent) of RF power amplifiers. To put it in some context, a modern GSM handset has the processing power of several PCs that require fans for cooling. When the handset is placed against the face there is less opportunity for air circulation to dissipate the heat.

Measurements of a variety of analog and digital hand-held mobile phones that compare the temperature difference between the side of the face with the phone and without (after six minutes use) show a difference ranging from 1.7°C to 4.5°C. Elsewhere, it has been reported that maximum surface temperatures for GSM phones in use varied between 36°C and 39°C. This is well below the level for thermal injury to the skin but is likely to be perceptible. Note that the temperature rise was caused by component heating with no measurable contribution from the radiated RF signal.

Because GPRS uses the same modulation scheme as GSM, the RF power amplifier will offer similar efficiency. Therefore, using additional timeslots on the uplink creates a clear potential for increased surface temperature. It would also mean a direct reduction in transmit time. As noted previously, and given the consumer expectation of small handset size, it seems likely that GPRS terminals with the capability for significant numbers of uplink slots are likely to draw their power from external sources.

The significantly increased processing demands of GPRS over GSM (by a factor of more than two) implies increased heat output due to signal processing, though this may be mitigated by improvements in the design of the phone electronics. As excess heat production from processing demands is directly related to reductions in usable battery life, it is a clear driver for suppliers to design to minimize temperature rises.

On the infrastructure side, the increased processing required may mean greater heat dissipation by base station and allied equipment. While this can usually be easily managed with existing air conditioning systems, it may run counter to efforts to increase the overall efficiency of base stations by the use of fans and convection cabinets. These efforts are motivated both by economic considerations (cost savings and reduced maintenance) and by the environmental benefit of reduced greenhouse gas production.

Obviously, mobile phone use requires energy, and much of that electrical energy is derived from burning fossil fuels. One U.K. operator, in its 2000 environment report, estimated that communications for each customer generated approximately 9.5 kg of CO_2 per year, or about 7.4 grams per minute of conversation using GSM. This is only the network energy usage, not the energy used to operate the phone. To put these figures in context, the average annual production of greenhouse gases by an individual in the United States from home energy use and transportation is 8279 kg. So, if the United Kingdom to United States proportions hold true, mobile phone usage contributes on the order of 0.1 percent. Achieving greater efficiency in handsets and infrastructure will directly reduce this greenhouse gas impact. On the positive side, improved communications can mean a reduction in travel requirements through greater e-working. A similar calculation for transportation shows that a medium-sized car produces about 0.22 kg of CO_2 per kilometer. For air transportation the figure is 0.18 kg of CO_2 per kilometer. That means talking for about 30 minutes uses the same amount of energy as 1 kilometer of road or air travel. To put it another way, a 60-minute mobile phone call produces less than 7 percent the CO_2 of a

30 km round-trip drive and only about 0.04 percent the CO_2 of a 1000 km round-trip flight.

Recycling of Non-GPRS Handsets and Sustainable Design

It is somewhat of a truism in the mobile phone industry that customers will tend to change their mobile phone every two or so years as part of an upgrade process or to follow design trends. Add to this the significant changeover impetus of a technology change, e.g., analog to GSM or now to GPRS, and potential for a significant environmental challenge emerges, especially if the prediction of 800 million GPRS users by 2005 is validated.

Mobile phone handsets contain a variety of materials that may need to be managed when the phone is discarded, either because of potential for recycling or potential toxicity. Excluding the battery, a typical handset is composed of about 50 percent plastic, 25 percent metal, and 16 percent glass/ceramics by weight. The remaining materials include components, the display, and flame-retardants. Copper compounds are the major metal components with other metals including iron, nickel compounds, zinc compounds, silver compounds, and a small lead content (<1 percent). Older mobile phone batteries were based on nickel-cadmium but these have largely been replaced with nickel-metal hydride or lithium-ion batteries, which perform better and don't have the toxic nature of cadmium compounds. Network infrastructure subsystems tend to have a higher proportion of metals than plastics. It is easier to recover metals, and so their greatest environmental impact may be through energy consumption.

It may prove possible to recover useful quantities of precious metals or other compounds from the metal recycling process. However, the achievable efficiency of recycling is limited by the need to minimize contamination by other materials and to control waste products from the recycling process. Recycled materials may be used in the production of stainless steel (nickel), new batteries (cadmium) or in furniture (plastics).

Increasingly, the electronics industry, like many other sectors, is being tasked with management of the whole product lifecycle (see Figure 24.2). Therefore the industry is exploring ways to incorporate energy efficiency, nontoxic materials, and recycling strategies at the design stage. Some companies are becoming partners in schemes to recover

Figure 24.2
Diagrammatic representation of a life-cycle assessment approach to the production, design ,and recovery of mobile phone technology.

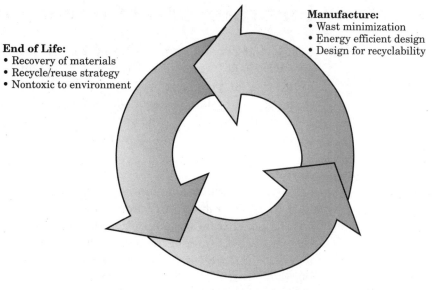

End of Life:
• Recovery of materials
• Recycle/reuse strategy
• Nontoxic to environment

Manufacture:
• Wast minimization
• Energy efficient design
• Design for recyclability

Production:
• Raw material extaction
• Sustainable resources
• Social impacts

redundant product for reuse or recycling. This general philosophy may be termed *sustainable development* or *product stewardship*. In design terms, it includes appropriate choice of materials; energy efficiency in manufacture and use; ease of assembly/disassembly; takeback and recycling offers; and waste management strategies (at all stages including raw material production). Examples of supplier initiatives for sustainable product development include the use of shape memory polymers to allow "automatic" disassembly of handsets and biodegradable polymer clip-on covers that could be composted.

The mobile communications industry has also implemented takeback or recycling schemes in a number of countries. In Ireland, a network operator provided pre-paid envelopes to recycle handsets when the analog system was closed. An Australian program based around collection boxes in over 700 retail outlets has collected approximately 30 tons of mobile phones, batteries, and accessories since 1999. The transition to GPRS handsets can also trigger the development of suitable programs in other countries, and means that the programs will be in place before the advent of 3G handsets. It seems to be a golden opportunity for the cellular industry to demonstrate the same leadership in the environmental sphere that it has in technological development.

Environmental Conclusions

GPRS handsets and infrastructure are being designed to ensure compliance with safety standards. There is no substantive evidence of a public health risk for compliant products or infrastructure. New environmental challenges are related to energy efficiency and recycling. Responding to those challenges will provide another opportunity for the wireless industry to demonstrate leadership and provide community benefits.

Acronyms

ACA	Australian Communications Authority
EMI	Electromagnetic Interference
FCC	Federal Communications Commission
Hz	Hertz (unit of frequency, also referred to as cycles per second)
ICNIRP	International Commission on Non-Ionizing Radiation Protection
IEEE	Institute of Electrical and Electronic Engineers
kg	kilogram
RF	Radiofrequency
SAR	Specific energy Absorption Rate
W	Watt
WHO	World Health Organization

Tomorrow: Just Like Yesterday with a Byte of the Unknown?

The question of what will happen and when, keeps people awake at night. Tomorrow can sometimes be envisioned by looking back, not necessarily for specific answers, but for guidance on what has worked and what shouldn't be tried again. GSM gives us such guidance.

According to the GSM Association, the GSM family consists of GSM, GPRS, EDGE (Enhanced Data Rates for Global Evolution), and 3GSM (a.k.a., Wideband CDMA) as a collection of unified, standardized, global wireless technologies providing seamless services for customers across 168 countries of the world. This family of technologies was developed through an unparalleled exercise in technological co-operation between standards development organizations, operators, and equipment manufacturers. Its cooperative origins have invested the GSM family with a clear evolutionary path through GPRS and (in some markets) EDGE, to 3GSM or Wideband CDMA. This graceful, well-marked path enables the world's wireless carriers to evolve and roll out GPRS, EDGE, or 3GSM services, based wholly on their individual network capabilities, market demand, and development around the world. The vast majority of the world's 3G operators have chosen the third generation of the GSM family or 3GSM (W-CDMA) as their ultimate 3G technology, in part because it provides:

- A clear upgrade path with graceful customer services evolution
- A way to leverage existing network investments
- Unprecedented global economies of scale
- An open architecture with technological superiority
- Global roaming and interoperability
- Choice and competitive availability of infrastructure and equipment

It is a great benefit for operators not to have to guess what path will prevail for continued technology enhancement and feature growth. That doesn't imply that technology options are closed; operators still enjoy business plan flexibility regarding timing, implementation options, service-level offerings, and investment spending. But all these are made more comfortable by the knowledge that the GSM family is designed to be evolutionary, forward- and backward-compatible, with global economies of scale.

Design for growth makes a compelling story, but what about the specifics? Is GPRS to be installed, developed to some degree of capacity, and then replaced with EDGE or 3GSM? Not likely. In fact, while the original concept is thought to have hard and fast timelines for each evolutionary step, today most operators believe that GPRS will be around for many years to come, with EDGE or 3GSM built as overlay, and with capacity enhancements where service demands are the greatest or regulatory requirements come into play. The rather grandiose vision of global ubiquity for 3GSM is no longer pervasive, because cost and investment return potential do not justify it.

Just as GSM was the best choice for the wireless voice business over the past decade (and will more than likely continue for at least the near term), GPRS is the best technology choice for GSM operators entering the wireless packet-data business sector. Over time, EDGE may prove to be the GPRS enhancement choice for some, maybe even many operators around the world. Others will elect to bypass the EDGE upgrade and move directly to implementation of 3GSM WCDMA. There will be no rules prescribing which option best serves an operator's business case. North and South America is likely to be where EDGE is first installed and operated, with other regions embracing the technology based upon customer demand, infrastructure, and operational cost considerations. Other business case variables for operators to determine individually include terminal device availability, regulatory requirements, and spectrum availability. The fork in the road which operators will face post-GPRS may not prove to be decisive when the global evolution picture is considered. Both prongs—the path from GPRS to 3GSM by way of EDGE, or the path from GPRS to 3GSM bypassing EDGE—have been carefully laid out.

Not all technology hurdles have been cleared. There are literally thousands of issues yet to be sorted out for GPRS service, not counting those yet to be encountered for EDGE and 3GSM. Time will tell if the standards evolve in a straightforward manner or are pushed by catastrophic events in the marketplace. Either way, the evolution roadmap has been published and is being implemented. Nuances are sure to arise as these technologies move from paper to the R&D labs and then to the field. So, is it inspired or foolish to make such detailed plans about the future of a technology? Tomorrow we can all relive the events which will answer that question, but today what we know for certain is that nobody has tried this before.

CHAPTER 25

Testing and Approvals

Darren Thompson

VoiceStream

With distinct anticipation you go to the store to buy your new GPRS mobile. You can hardly wait to have that device in your hands, all charged up and waiting to be connected. When you turn it on, you expect it to work; in fact you envision it working pretty well, and you even harbor a hope that it will amaze you. Will it? A lot of things happened to that device before you encounter it at the store, or even before it was gently placed upon the shelf, and some of them bear review before it can be considered market ready.

If you build it, they will come. If it doesn't work, they will leave...and complain to everyone they know. That's why testing and approvals are so important.

It's natural to think that the way to find out whether a GPRS terminal works is to try it out on a network. Plug it in and see what happens. WAIT JUST A SECOND! Would you want to see new car prototypes on the freeway, or would you prefer to have them tested in a lab by a qualified mechanic? If you test it on the open road (the equivalent of the operator's network), a prototype could cause havoc with the existing traffic (paying customers). Besides, the network is not really a convenient environment for testing device limitations and capabilities; it's too hard to separate what's wrong with your prototype from what's wrong with a great many other variables regularly encountered on a network. Eventually, we will "take her out and see what she can do" on the network, but first we need to do some controlled testing.

In a test lab, you control the environment in which the GPRS device will function and, as a result, you can explore device behavior under various conditions. You can test features and even create errors just to see how the device reacts. In theory, if the GPRS terminal was designed to comply with the GPRS specification, and the network also complies with the spec, then everything will work. Unfortunately that is not reality; throughout all the best efforts of the individuals who work to create specifications and communicate them with clarity, there is always a level of interpretation involved. And that's not a bad thing; that's how unique implementations are inspired.

The challenge for testers is to create ways to verify all the features and functionality of the GPRS device. Manufacturers, test houses, operators, and industry experts all assist in the development of GPRS tests. The principle of the tests is to ensure that any lab adhering to the test procedure will get exactly the same results. Hence a *validated* test is one that has been successfully implemented by an approved test lab to produce an accurate and repeatable outcome.

In a perfect world, these tests would be written as the specifications are being written, but at last look the world was not perfect and tests are still being validated for GPRS.

The next step after validation is to incorporate the newly validated tests into product approval. Unapproved products will not be able to roam from network to network, not be distributed for sale, or at the very least not be able to compete successfully against approved equivalents. So whom are you going to ask for approval—the GPRS Lords? The answer depends on some variables like where in the world you are and whom you're trying to assure. In the next few pages I'll explain the kinds of approvals available before delving into the formal world of directives and standards.

Regulatory Testing

These tests are mandated by government. Every terminal must pass such tests before it can be sold in that government's country. Regulatory testing determines that:

- **The unit will not cause injury to the user**—Ruling out overheating, sparks, or unusual emissions that could change the channel on your television set, pacemaker, or produce virtual reality.
- **The unit will not cause harm to the network**—You only use the network resources that you need, and you will not cause interference to other phone users (remember using a CB and hearing three or four other conversations at the same time?).
- **The unit will not cause harm to other networks**—This proviso is like the previous one, except it's tested against networks for which the device was not designed (such as FM radio broadcast).

In the United States there are additional regulatory requirements:

- **E-911 Location Technology**—The ability to determine caller location in an emergency voice call.
- **TTY**—Text telephony for the hearing impaired.

In short, regulatory testing establishes that callers can neither do harm nor come to harm by using their phones. It's a bare minimum, meant to protect tax-paying citizens. Naturally, operators of GSM networks will always ask for more than just the regulatory testing. In North America, the operators mandate additional testing through the PTCRB. PTCRB, for those not in the know (which accounts for all but 200 or so people who are intimately involved), is the PCS Type Certification Review Board.

PTCRB

The PTCRB consists of the GSM operators in North and South America, the manufacturers of GSM phones, the authorized PTCRB labs that perform the testing, GSM test equipment manufacturers, and anyone else who wishes to participate.

The process. Where do GPRS tests come from? They are selected from 3GPP TS 51.010 by the operators and presented to the PTCRB members for comment. A test laboratory must then validate the tests. Test equipment must be commercially available before it can be used in the required tests.

All PTCRB test labs must be members of the PCS Validation Group (PVG). They are responsible for ensuring that all laboratories comply with the requirements and for resolving disputes in tests or procedures. PVG also assists in the validation of test cases requested by the PTCRB operator members.

Once a manufacturer of a GPRS terminal successfully completes the PTCRB tests, the terminal is assigned its international mobile electronic identification (IMEI) and announced as approved.

No terminal with an approved IMEI may be refused service within PTCRB networks.

In Europe, a similar procedure is administered on behalf of the operators by the GSM Certification Forum (GCF). Before I explain what that is, it's relevant to look at a little of their history.

Voluntary Testing

Until April 2000, the European Union operated under the telecommunications terminal equipment (TTE) directive, which required mobile manufacturers to test their phones against GSM 11.11 specifications in a laboratory not directly under that manufacturer's influence to ensure independent results. After April 2000 the requirement to perform these test was removed; manufacturers had lobbied their respective governments to waive GSM 11.10 tests in the interest of faster time to market. Instead, manufacturers were *authorized to declare themselves compliant*. Surprisingly, even this very modest hurdle was officially voluntary for the manufacturers. GSM operators look for GSM Certification Forum approval on the phones they purchase as a matter of course, but anticompetitive laws within the EU prevent them from making it a requirement. (Read more about GCF later in this chapter.)

Self-declaration

Asking for self-declaration, in my opinion, is a little like asking a new parent to decide whose child is the cleverest in the nursery. "Oh yes, he is a little darling isn't he? Takes after his father, I'd say!" Simply put, there are potential barriers to objectivity in this practice. Global Certification Forum (GCF) is based on self-declaration by the manufacturer(s). When a manufacturer believes he is compliant with all the standards that govern his product design, he tells the purchaser (the operator) that the product is approved. Today, standards continue to be tested and test scripts continue to be approved, but the major difference is that tests do not have to be conducted by an independent laboratory.

GFC does require that the tests carried out by manufacturers are an agreed-upon suite acceptable to both manufacturers and operators. The validated tests are taken from 3GPP TS 51.010.

Most of the major manufacturers and GSM operators around the world (excluding the Americas, Australia, and Asia) participate in GCF. GCF observers are the test labs, test equipment manufacturers, and any other interested parties.

Field Trials

Eventually the operators in Europe stepped in and said, "OK, you've declared your product design to be fully compliant through the GCF, but how do we know that your mobile will work in OUR networks?" Not even the most rigorous of specifications can guarantee that phones will work perfectly under all circumstances. We've already noted that specifications all leave room for interpretations, sometimes intentionally and sometimes not. To find out where an interpretation may have gone wrong, the terminal must be tested in different networks. In this case "different" means not just different operators but also different manufacturers of network equipment. Repeating the same series of tests across these carriers and platforms is necessary to ensure consistent operation.

The operator's goal is to find the problems with a phone by using it as a subscriber would, trying to do as many things in as many ways as possible. The tests, in other words, have to make sense from the standpoint of real-world usage patterns, and so the operator will try to send SMS messages in the midst of a call, or make a series of voice calls in a row, or conference in a third party, placing him on hold while reading the phone book on the screen. I should note that this kind of testing is done by all operators and not just those within GCF.

Interoperability Testing

It's important to remember that each operator has a unique network. Even among operators with the same manufacturer's infrastructure, individual choices about how to deploy and configure network elements are almost guaranteed to result in different implementations. Ultimately the only assurance the operator wants is the assurance that a terminal will work in "my" network. The customer, on the other hand, wants assurance that "my" mobile will work on any network. To establish the extent to which this is true, interoperability testing is essential.

The phone is tested in as many networks as possible, simulating months of service in a very short period of time. Where field trials establish that terminals function well in the operator's network, interoperability tests verify that it functions in networks other than the intended operator's, and in roaming networks.

TTE Directive

Until fairly recently, telecommunications equipment such as mobile telephones were required to meet the Telecommunications Terminal Equipment Directive (91/263/EEC). For a GSM product, the TTE Directive laid down *common technical regulations* (CTRs) defining the essential technical and regulatory requirements that terminal equipment must satisfy before being granted type approval. The technical basis for these regulations was developed by European Telecommunications Standards Institute (ETSI). A notified authority issues a certificate against these CTRs. Compliance with the CTRs also allows GSM manufacturers to place the certification mark on their products.

TBRs refer to the specific test requirements laid down in the ETSI document ETS 300 607-1, also referred to as GSM 11.10/3GPP TS 51.010. The TBRs specified for phase 2 type approval used version 4 of GSM 11.10 /3GPP TS 51.010. The CTRs and TBRs for GSM terminals are 19 and 20, respectively, for GSM 900 phase 2, and 31 and 32, respectively, for GSM 1800 phase 2 and dual band.

The TTE Directive (as defined in Annexes 1–4) provides three common routes to compliance for GSM terminal manufacturers:

- **Annexes 1 and 2**—The terminal is type-examined, and the manufacturer undertakes a product check agreement.
- **Annexes 1 and 3**—The terminal is type-examined, and the manufacturer maintains a production quality assurance approval (PQAA).
- **Annex 4**—The terminal manufacturer performs a full quality assurance and makes a declaration of conformity.

Under the TTE Directive, manufacturers are also obliged to meet other essential requirements, such as the EMC, Machinery, and Low-Voltage Directives, in order to brand their products with the CE mark.

Under the TTE regime, however, compliance testing was not necessarily the whole story for GSM terminal manufacturers. Typically, manufacturers would perform many additional in-house tests and conduct extensive field trials. These field trials would often be conducted in areas of known weak signal strength or complex cell structures to evaluate and improve the real-world performance of handsets. Further testing might also be warranted for products to earn approvals from specific network operators who conducted their own acceptance testing.

R&TTE Directive

The Radio & Telecommunications Terminal Equipment (R&TTE) Directive appears to be one of the few pieces of legislation emanating from the

European Commission in Brussels that is actually aimed at reducing both the financial burden and the time to market for manufacturers. It was formulated by the commission to facilitate free trade and free movement of goods, and to speed up the emergence of new technologies throughout European Union (EU) member states. The *European Economic Area* includes the fifteen member states of the European Union: Austria, Belgium, Denmark, Finland, France, Germany, Greece, Ireland, Italy, Luxembourg, Netherlands, Portugal, Spain, Sweden, the United Kingdom, and three of the four states of the *European Free Trade Association*: Iceland, Norway, and Liechtenstein.

Under the R&TTE Directive, GSM terminal equipment must satisfy the relevant harmonized standards to demonstrate compliance. Essential requirements include electrical safety (73/23/EEC), electromagnetic compatibility (89/336/EEC), prevention of harm to both the network and its users, and efficient use of the spectrum. (Some additional requirements are yet to be decided by the commission, including interworking, information protection, antifraud, emergency services, and features for the disabled.)

The directive defines conformity assessment procedures in a series of annexes, including:

- **Annex I**—Equipment not covered by this directive
- **Annex II**—Internal production control
- **Annex III**—Internal production control plus specific tests
- **Annex IV**—Technical construction file
- **Annex V**—Full quality assurance

Because a GSM terminal contains a transmitter and a receiver, manufacturers have the choice of using the conformity assessment procedures outlined in Annex III, IV, or V. Regardless of which route is taken to demonstrate compliance, it's the manufacturer's responsibility to test the product adequately and keep the necessary test records.

The *Official Journal of the European Communities* has listed a number of TBRs as harmonized standards for the R&TTE Directive. GSM terminal manufacturers are required to apply TBR 19 Edition 3 (October 1996) for GSM 900 terminals and TBR 31 Edition 2 (March 1998).

The R&TTE Directive will have two major implications for the GSM world: terminal manufacturers, by declaring compliance, will take on increased product liability, and network operators will get less protection from regulation. An operator cannot bar access to the network for a terminal that is compliant.

Network Operators' Response

The GSM Association, which includes many network operators in its membership, was concerned about the risk of reduced regulation and initiated a voluntary certification scheme. (Please note that this scheme is not mandatory even now, and network operators are free to purchase GSM terminals that have not participated.) The current name for the scheme is the Global Certification Forum (GCF), and it was expected to be fully operational by the time the R&TTE Directive had become law. Development of the GCF has been a cooperative effort of the GSM Association representing the GSM network operator community and GSM terminal manufacturers, to ensure the interoperability of GSM terminals worldwide.

As of July 2002, the GCF is still mainly European in terms of participation. The Americas have the PTCRB and Asia has their own process, but work is ongoing to minimize duplication of testing.

The GCF provides GSM terminal manufacturers with a process for the verification of terminals against specified technical requirements. This certification aims for global recognition and is designed to create cost efficiencies. The GCF scheme does not provide any commercial or quality assurances from a terminal manufacturer to a customer. New requirements and tests may be incorporated into the forum once a standards authority has published them and validated test equipment is available. The GCF is evolving as GSM technology evolves so that it can encompass such future technologies as high-speed data and third-generation wireless air interface specifications.

The GCF program is a three-step process:

1. The first step is to become *quality qualified*. Terminal manufacturers are required to self-declare that they have a recognized quality assurance program in place. The program should cover all aspects of the design, development, and manufacture of the GSM terminal.

2. The next step is to ensure a *means of test*. The means of test should satisfy the test requirements detailed in the GCF requirement tables (currently 3GPP TS 51.010 tests) and also further requirements that are published through a relevant standards authority. The testing phase should also include field trials. It is recommended that testing should be conducted on at least five networks, which should represent infrastructure implementations from all major suppliers.

3. Finally, each GSM terminal manufacturer must determine whether its GSM product meets all the necessary certification criteria. Evidence of this must be recorded in a compliance folder.

The creation and development of the GCF has been a difficult endeavour over the past couple of years. There have been some particularly contentious issues between the network operators and terminal manufacturers. One such issue is that the GCF requirement for terminal manufacturers demands that GSM products satisfy not only test criteria (TS 51.010), but also all related core specifications. The latter requirement is potentially troublesome for a terminal manufacturer because test cases do not exist to prove all of the core specifications.

Another issue has been the decision on which version of GSM specification is to be applied. As mentioned earlier, under the TTE Directive, version 4 (phase 2) of the specifications was used. Under the GCF scheme, terminals are required to meet version 5 (which includes phase 2+ features) in the short term, and then newer versions later in the year. This does not mean that terminal manufacturers must design in all the phase 2+ features, but it does mean that special care should be taken wherever changes have been made to existing phase 2 features. Manufacturers will have to ensure that their projects are in line with this progression of GSM versions.

Incrementing and using the newer versions of GSM specification should encourage the development and implementation of GSM phase 2+ features. The GCF scheme has highlighted a number of features for which terminal manufacturers, test equipment manufacturers, and test laboratories are now preparing. Some of these include additional short message service (SMS) phase 2 features, multislot phase 2+ features, and supplementary service phase 2 features (e.g., multiparty, call forwarding, and call waiting).

The current process of developing, issuing, and validating test cases will not look very different in the near term. Accredited laboratories will continue to validate test cases against the test specifications, TS 51.010, and core specifications, before they can be categorized as applicable. However, the business case for independent test labs has been greatly reduced by the adoption of self-declaration, and test labs are not even members of GCF.

Regional and National Approval

GSM terminal manufacturers who intend to sell products outside Europe must meet the requirements of national regulators. This is an increasingly important consideration in light of the dramatic growth of GSM and its adoption across the globe. Recent figures show that 215 million subscribers in 129 countries, 72 million of whom are *not* in Europe, use GSM terminals.

Schemes for approval are largely based on ETSI specifications and European TBRs. It is unclear exactly what regional and national regulators will do, if anything, in response to changes in the European requirements, or whether they will recognize results from schemes such as the GSM Certification Forum or the PTCRB.

In Australia, the Australian Communications Authority (ACA) uses Technical Standard 018 to assess compliance. The main requirement of a GSM terminal is to comply with the ETSI GSM standards, which manufacturers can demonstrate by applying the European TBRs. In addition to the TBRs, TS 018 states that the terminal shall also satisfy the following requirements:

- Comply with the provisions of the radio-frequency radiation requirement AS2772.1.
- Maintain International Mobile Equipment Identification (IMEI) integrity and adequate protection against change.
- Support "000" and "112" emergency calls.
- Comply with "prevention of inadvertent ignition of flammable atmospheres by radio-frequency radiation."

Manufacturers wishing to have products approved in North America are required to apply to the Cellular Telecommunications Industry Association (CTIA) for GSM-1900 Type Certification. The requirements specified by the CTIA and a PCS-specific review board include compliance with PCS 11.10, a modified version of the European standard GSM 11.10. Most of the differences are changes to the frequency bands.

One significant difference between the North American (PTCRB) and European (GCF) approval systems is that the North American system uses features from several standards releases as defined by ETSI—release 97, release 99, and version 4 (phase 2)—and in practice uses higher versions of TS 51 .010 for specific features and test requirements. North America already implements the subscriber identity module (SIM) toolkit and leak-tolerant earpiece testing as applicable test requirements, as well as enhanced general packet radio system (EGPRS) also referred to as enhanced data for GSM evolution (EDGE).

A GSM 1900 terminal manufacturer may choose to apply for additional CTIA certification by demonstrating compliance with such requirements as hearing aid compatibility and audioport accessibility, but CTIA certification is only applicable for the United States.

Certification by Notified Bodies

The notified bodies are laboratories able to provide an "expert opinion" on conformance to the R&TTE directive. Many labs have recently made a certification scheme for equipment available to manufacturers, the *Certified for Network Connection.* This scheme will be based on TTE practices and will provide a means for suppliers to ensure both customers and network operators of independent verification of compatibility with telecom networks. Basically, it amounts to doing testing in an independent laboratory in the old-fashioned third-party way!

Strategies for Manufacturers

All GSM manufacturers wishing to sell products in Europe must meet the R&TTE Directive's essential requirements. The R&TTE Directive has been a hot topic of conversation, and because harmonized standards or GSM certification schemes have not been formalized for very long, many manufacturers are still unsure as to the best route to take.

For a complex technology such as GSM, manufacturers will have to test their products against the standards as *part of the development process.* Manufacturers with large budgets may choose to equip their own laboratories or subcontract the test and evaluation to a third-party laboratory. However, although some test equipment is within the financial reach of medium- or smaller-sized manufacturers, a full system simulator for GSM can be cost prohibitive for all but the largest manufacturers. Many manufacturers will opt to continue using accredited third-party laboratories so that they can obtain a full independent report demonstrating compliance. Test reports like these can add value and credibility to GSM terminal products and can act as powerful marketing and sales tools. A common strategy will most likely be the application of the TBRs as a minimum requirement.

Whatever strategy is employed, it is clear that manufacturers will have to maintain or even improve their test efforts for GPRS, especially with the growing complexity of terminals, to ensure a future for global roaming on the GSM communications system.

CHAPTER 26

Enhanced Data Rates for GSM Evolution

Clif Campbell

Cingular Wireless

This chapter provides an overview of the emerging radio technology called enhanced data rates for GSM evolution (EDGE), which will be used to provide higher speed data and third generation wireless services by a number of GSM carriers. We'll consider the technology, its uses, the driving forces behind its implementation, its evolution, and the current standardization process.

What Is EDGE?

EDGE was originally developed by the European Telecommunications Standard Institute (ETSI) as a GSM technology to provide much higher data rates than either GPRS or circuit switched data (CSD). This higher-speed GPRS, based on the EDGE technology, is referred to as enhanced GPRS (EGPRS). The Universal Wireless Communications Consortium (UWCC) adopted EDGE as its third generation wireless radio technology for the ANSI-136 systems.

Several factors are contributing to the push by GSM operators who want to deploy the EDGE radio technology:

- The need for higher bit rates over the GPRS core network in order to provide increased user throughput/performance in wireless packet data applications
- In the United States, the lack of available 3G spectrum in which to deploy such radio technologies as wideband CDMA
- The opportunity to utilize existing spectrum and infrastructure to introduce 3G service capabilities
- The need to ensure that GPRS remains competitive as other higher-speed wireless data technologies are launched
- Low incremental infrastructure cost
- The uncertainty in market demand for wireless applications requiring 2 Mb/s data rates

The objective of the EDGE specification was to provide a significant GPRS throughput improvement with very little change to the infrastructure, while meeting the IMT-2000 requirements for wide area coverage at 384 kbps. This will enable EDGE-equipped networks to support third generation services and capabilities.

EDGE Technology

EDGE uses the existing GSM radio frame structure consisting of eight timeslots per radio carrier. Each timeslot may be used to transport packet data traffic to and from the EGPRS core network.

As we've said, EDGE is designed to provide a much higher data rate than that provided by GPRS. To achieve the higher data rate, a new modulation is adopted in addition to the existing GMSK modulation. The modulation scheme selected is eight-phase shift keying (8-PSK). The lower EDGE data rates are provided with GMSK modulation, while the higher data rates are based on the 8-PSK scheme.

TABLE 26.1

EDGE Coding Schemes and Data Rates

Coding Scheme	Data Rate Kbit/s	Modulation	Family
MCS-1	8.8	GMSK	C
MCS-2	11.2	GMSK	B
MCS-3	14.8	GMSK	A
MCS-4	17.6	GMSK	C
MCS-5	22.4	8-PSK	B
MCS-6	29.6	8-PSK	A
MCS-7	44.8	8-PSK	B
MCS-8	54.4	8-PSK	A
MCS-9	59.2	8-PSK	A

The higher data rates result from the fact that, with 8-PSK modulation, three data bits are transmitted for every symbol, compared with one bit/symbol for GMSK. In GPRS there are four channel coding schemes defined as coding scheme 1 through 4 (CS-1, CS-2, CS-3, and CS-4). With EDGE, that number increases to nine. The various coding schemes and associated bit rates for EDGE are shown in Table 26.1.

Based on the supported coding schemes, EDGE is capable of providing a peak data rate of 59.2 kbps (MCS-9) per radio timeslot. Based on the bit rate of MCS-9, EDGE has a theoretical limiting bit rate of 473.6 kbps, assuming that all eight timeslots are available on the terminal device. A more likely scenario is that the terminals will support a maximum of four downlink timeslots, giving a maximum accessible bit rate of 236.8 kbps. The data rates defined in Table 26.1 are measured above the RLC/MAC layer in the air interface protocol stack.

Mobile stations supporting EGPRS are required to support all coding schemes, MCS-1 through MCS-9, while the networks supporting EGPRS may elect to support only certain coding schemes (by family designation as shown in Table 26.1).

The EDGE radio parameters are summarized in Table 26.2.

TABLE 26.2

EDGE Radio Parameters

Carrier spacing	200 kHz	
Modulation rate	270.1 symbols/s	
Frame length	4.615 ms	
No. of slots/frame	8	
Modulation	GMSK	8-PSK
Radio interface data rate		
■ 1 slot	22.8 kbps	69.6 kbps
■ 8 slots (1 frame)	182.4 kbps	556.8 kbps

Spectrum

The primary push for the deployment of EDGE has come from United States-based wireless operators, stemming largely from the fact that the FCC has not been able to provide the industry with radio spectrum specifically assigned for third generation services. This is particularly significant to TDMA and GSM operators since competing operators with CDMA systems are able to support higher-speed data services within their existing 2G radio bands. Today, several operators in South America have also committed to deploy EDGE and others around the world are carefully evaluating the technology and may soon commit to EDGE as an upgrade to their GPRS service.

EDGE was designed for deployment in a 4/12 reuse pattern with minimum spectrum requirements of 2.4 MHz plus guard bands. This design allows it to be deployed in the existing 2G frequency bands without addition of spectrum, which is particularly advantageous for network operators wishing to provide 3G services but lacking the spectrum to do so.

If 3G services are to achieve significant penetration in the United States, additional spectrum (over and above the existing 2G spectrum) must be allocated. Any such allocation of US spectrum will also have to be harmonized with the frequency bands being used internationally in order to facilitate roaming and provide handset economy of scale. In the meantime, however, EDGE provides GSM and TDMA operators with the opportunity to support the same services as the CDMA operators.

In 2000, the World Radio Conference (WRC) proposed the allocation of two bands for next-generation wireless services, namely the 1710 to 1850 MHz band and the 2500 to 2690 MHz band. In addition, the WRC recognized the need for an additional 160 MHz of spectrum to meet the future needs of IMT-2000. Of the two bands proposed, the 1710 to 1850 MHz is clearly favored by the international wireless community. Both bands are currently in use within the United States and are unlikely to become available anytime soon for commercial 3G wireless services.

In summary, the allocation and subsequent clearing of globally harmonized spectrum for 3G services is not going to happen soon. In the absence of an alternative, U.S. operators will undoubtedly be at a disadvantage in offering such services. EDGE provides the GSM and UMTS operators with the opportunity to start offering 3G services within the existing 2G spectrum as early as 2002.

EDGE Performance

While most of the focus on the technology has been on the peak data rates achievable, the real measure of usefulness is the data rate observed and experienced by end users of the EGPRS system. This measure is determined by the actual system performance. Throughput, which is a measure of how much error-free data is delivered in a given time interval, is probably the most significant measure of performance from a data user's perspective.

Several factors interact to determine throughput performance for EGPRS. In addition to the number of timeslots available on the terminals and the assigned radio timeslots on the infrastructure, throughput performance will be directly affected by the carrier-to-interference ratio (C/I) on the air interface. The C/I is a measure of the quality of the air interface (i.e., the radio) link. In general, it can be assumed that throughput rate decreases as the quality of the link degrades. To ensure that throughput over the air interface is maximized for the end user, a quality control mechanism known as link quality control (LQC) has been devised and incorporated into the design of the EDGE technology. LQC compensates as well as possible for changing environmental conditions without impinging on user throughput rate. The two link quality control techniques adopted in the EDGE specifications are link adaptation (LA) and incremental redundancy (IR).

With *link adaptation*, a hybrid automatic repeat query (ARQ) is used, with coding rates and modulation varying with the quality of the radio link. Channel quality is continuously monitored and the modulation, and coding scheme (MCS) selected will provide the maximum link bit rate for the link condition detected. The hybrid ARQ uses a forward error correcting (FEC) code to correct errors at the receiver as well as an ARQ mechanism to detect remaining errors through a frame check sequence (FCS) and to request retransmission of the blocks in error. This approach to error detection and correction is referred to as a type I hybrid ARQ scheme.

With *incremental redundancy*, a block of data is first encoded with a low rate FEC code. Only a part of this codeword (or sub-block) is transmitted initially. When the frame check sequence finds blocks that are in error at the receiver, transmission of additional sub-blocks from the same codeword is requested. They are combined at the receiver with the previous sub-block. This process is repeated until there is a successful decoding of the initial data. This is an incremental addition of redundancy to the transmitted data. The primary drawback of this approach is that the receiver needs memory to provide temporary storage for the various sub-blocks received until the decoding of the data is completed.

Link adaptation is intended to provide performance enhancements over the air link through design enhancements at the RLC/MAC sub-layer.

The nine modulation and coding schemes (MCS) defined for EGPRS can be used in both LA and IR mode.

Compatibility

Upgrading GSM cell-sites with EDGE will mean sharing radio resources among GSM voice, and GPRS and EGPRS data. As previously indicated, EDGE will share the same carrier bandwidth and timeslot structure with the existing GMSK radio. In addition, EDGE deployment will have an impact on the BTS, Abis links, and the BSC. There is no impact on the GPRS core network elements (the SGSN and GGSN) or on the MSC; the latter should only require software addition.

In today's network, the Abis links provide for a maximum data rate of 16 kbps per traffic channel. With the deployment of EDGE, which provides greater per-timeslot bit rates, Abis links must be upgraded as well. Infrastructure manufacturers have the option of either using multiple links to each traffic channel or the upgrading the links to support 64 kbps data rates.

Operators will deploy EDGE in existing GSM cells to preclude the coexistence of both GSM and EDGE in some cells and frequency bands. Therefore, no extensive modification of cell or radio frequency planning is expected for EDGE deployment.

Evolution

EDGE was developed to support GSM services in the 850, 900, 1800, and 1900 MHz bands. In addition, a peak data rate objective of 384 kbps (48 kbps per timeslot) was established for indoor/low-range outdoor (up to 10 km/hr speed) as well as for urban/suburban outdoor (up to 100 km/hr speed). In rural outdoor applications, the peak data rate was established at 144 kbps for speeds up to 250 km/hr.

EDGE provides an enhanced GSM Phase 2+ physical radio layer, with the new specifications incorporated in the existing GSM specifications. Standardization of EDGE is being developed in two phases as follows:

- **Phase 1**—Providing single and multislot packet-switched services and circuit-switched services
- **Phase 2**—Providing real-time services not included in Phase 1

Phase 1 of EDGE standardization is now complete and available as enhancements for GPRS (EGPRS) and circuit-switched data (ECSD). Phase 2 is scheduled for introduction in the 3GPP Release 4 with enhancement provided in subsequent 3GPP releases. The Phase 2 development has been incorporated in the work of the GSM/EDGE Radio Access Network (GERAN) Technical Specification Group (TSG) of the 3GPP.

The GERAN Reference Architecture is shown in Figure 26.1. GERAN supports several interfaces to the core network: the Iu-ps, Iu-cs, A, and Gb. Iu-ps and Iu-cs provide connections between the base station subsystem and the 3G SGSN and the 3G MSC respectively. An additional interface, Iur-g provides for the direct interconnection of two GERAN base station subsystems as well as between a GERAN BSS and the

UTRAN RNC. The Iu-cs interface is used to support circuit-switched services over the UMTS core network, while the A interface will continue to support circuit-switched services over the GSM network. Iu-ps supports packet services over the 3G SGSN node. The Gb interface allows for backward compatibility with the existing 2G GPRS network (SGSN).

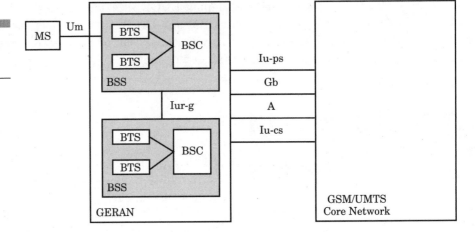

Figure 26.1
GERAN reference architecture.

Two terminal operating modes have also been defined:

1. **A/Gb mode**—For terminals connected to a core network with no Iu interface
2. **Iu mode**—For terminals connected to a GERAN with Iu interfaces towards the core network

With these two operating modes defined, the issue of backward compatibility with legacy terminals as the network evolves is assured.

Summary

EDGE allows operators of GPRS-equipped networks to support much higher data rates than are possible with the basic GPRS technology. No changes are required to the GPRS core network and no new spectrum has to be acquired in order to deploy EDGE.

With the development of the GERAN specifications in 3GPP, EDGE-equipped networks are evolved towards support for real-time and 3G services, strengthening the business case for adopting the EDGE standards.

CHAPTER 27

GPRS Evolution

Carol Politi
Megisto Systems

The year 2001 was a great year for working out the kinks in a very complex new network deployment. GPRS brought with it new radio technology, new core networking technology, new network services, new applications, new mobile content, and an array of middleware and OSS systems to assist with the delivery and billing of these new services. It was also the first time truly mobile data communications became possible on a broad scale at affordable prices.

However, while early GPRS services provide high value, they still only hint at the real potential for the wireless Internet. The full potential of GPRS will be realized as more network capacity is deployed through base station upgrades worldwide, as an array of diverse terminals becomes available, and as optimized mobile applications and content are more fully deployed. Key to a successful future will be delivery of a broader range of services, increased access capacity (aggregate and per user), and expanded accessibility—without impact to terminal size, battery life, or terminal form factors.

Standards

GPRS standards development began in 1994 and the primary GPRS standards were approved by 1997. Network equipment has been manufactured, and network deployment has begun. Standards for GPRS radio access networks are relatively fixed, with only incremental evolution in order to incorporate the results of real-world experience.

The GPRS core network is a strategic resource that will develop over time, and will support many flavors of network access in addition to GPRS—including EDGE, UMTS, and even wireless LANs. This core network is actively evolving to support enhanced capacity, quality, and more intelligent network services—and will continue to do so apart from any evolution in the GPRS radio access network.

In both segments of the network, as advanced packet data services are more broadly deployed, the industry will work to assure interoperability between handsets, radio allocation schemes, quality of service implementations, and advanced billing and mediation agreements. Such interoperability challenges will drive incremental changes to standards; however, no fundamental changes to GPRS specifications are expected at this point.

Radio Access

The future of radio access is inherently tied to the air link—and it is clear that air link technology will evolve even over the very near future. In addition to GPRS, EDGE, and UMTS, hotspot technologies such as Bluetooth and 802.11b will be expected to support mobile data terminals (Figure 27.1).

Figure 27.1
The future—diverse radio technologies supporting mobile communications.

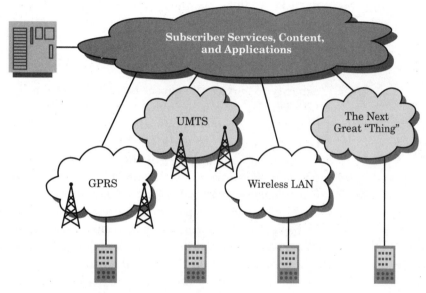

Source: Megisto Systems.

There are also efforts underway both within standards organizations and within the industry to come up with new airlinks that offer higher speeds and higher quality.

GPRS offers access to radio carriers operating at 115 kbps. As previously described, this carrier is divided into eight timeslots (each operating supporting data transfer at about 13.4 kbps) that are shared on an on-demand basis among multiple terminals. The evolution of networks supporting GPRS is closely associated with new radio technology—carriers are already moving to deploy EDGE (with carriers of 384 kbps) and UMTS (with carriers up to 2 Mbps) within islands (for example, urban areas with a higher density of subscribers) to increase the capacity of a single radio carrier. Terminals will increasingly support multiband and multimode capabilities in order to access carrier networks through a multitude of radio access technologies.

Within GPRS, the data rate available to any single subscriber will depend on the amount of radio capacity deployed by the operator for GPRS services, the quality of the subscriber connection, and the ability of the subscriber terminals to access that capacity.

The initial upgrade to GPRS is a relatively simple one involving limited investment in both infrastructure and spectrum capacity. Most GPRS radio infrastructure upgrades are primarily software upgrades to existing access networks, and data is transmitted in the same spectrum as is used for today's 2G voice services. Early GPRS networks have made only limited upgrades to the radio capacity available for GPRS service in advance of customer demand. However, as is true with today's 2G networks, in order for an operator to add subscribers (whether voice only or voice and data), the operator must size the network to ensure sufficient capacity to service the subscribers. Therefore, as GPRS succeeds, operators will be driven to invest in their network infrastructures to add capacity. If not upgrading radio technology, operators will upgrade their capacity through the addition of more radios per sector and through further cell-site subdivision (in the same way operators upgrade capacity on their 2G networks today as they add subscribers to the network). Such capacity upgrades allow for enhanced support for both voice and data by increasing capacity for both services.

Typical GPRS terminals will support transmission rates of up to 50 kbps, but the actual data rate available to the terminal at any instant will be dependent on the overall capacity available in the cell site (i.e., that allocated to GPRS by the carrier to GPRS), the congestion in the network (due to real-time demand from other terminals), and the quality of the access link. Subscribers will most likely see instantaneous data rates of between 10 and 50 kbps. Will this be sufficient to support subscriber expectations? The answer seems to be yes. The key to subscriber satisfaction appears to be the convenience of always-on service—not the raw speeds available to an individual terminal. Figure 27.2 illustrates services that appear to be of interest to users along with an indication of the capacity required to effectively support the service.

It is also useful to consider several vertically integrated wireless data services that have achieved high levels of success already—these include the Blackberry service in the United States and the I-Mode service in Japan. Today's Blackberry service supports email and messaging applications at less than 20 kbps. The initial I-Mode service supported per-subscriber data rates of less than 10 kbps and achieved an extremely high level of success (data rates have subsequently been increased to 28.8 kbps).

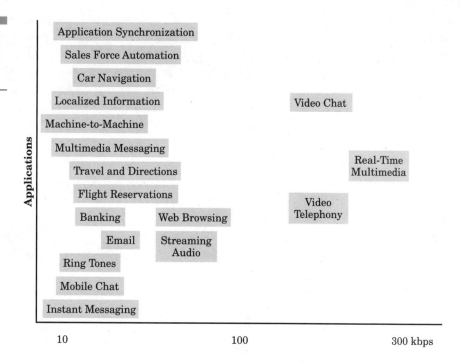

Figure 27.2
Data rates required for typical mobile applications.

Mobile Core Infrastructure

The GPRS mobile core sits at a critical intersection of two rapidly growing markets—the Internet and mobile wireless. IP networks were not originally designed to track mobile devices. Likewise, wireless networks were not originally built to carry packet data or handle IP services. Yet the birth of the mobile Internet is forcing the two worlds to collide for the first time.

The GPRS mobile core we have discussed will not only service the GPRS access network, but will also serve as the framework for EDGE and UMTS networks as these are deployed. And, as the GPRS core evolves, it is expected to form a single, unifying architecture for mobile Internet roaming—not only for mobile users accessing their home networks through licensed spectrum but also through unlicensed spectrum such as wireless LANs.

The evolution of the mobile core is independent from that of the mobile access network. The core is made up not only of the actual packet data network components required to manage the basic network roaming services (e.g., the SGSNs and GGSNs), but also of the Internet serv-

Figure 27.3

The GPRS core forms the evolutionary framework for future radio access networks and mobile network services.

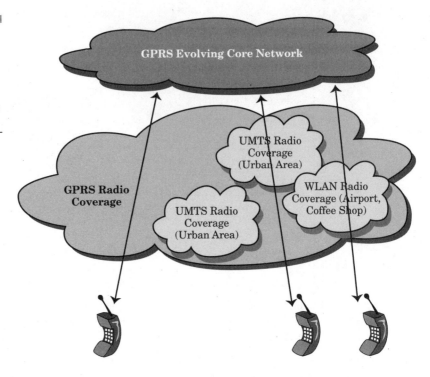

ice gateways (if not incorporated within the GSN nodes), and a variety of operational and support systems required to authenticate and bill subscribers for services.

While many early GPRS networks have already been launched, most of the system issues that have been ironed out at this point are related strictly to basic network connectivity. The mobile core brings with it challenges that become obvious only as services become broadly deployed. These challenges are not driven by a requirement for high aggregate throughput, but instead are related to the design of the mobile network, the scale of subscribers expected, and the type of personalized services that will be deployed.

Early GPRS network models envisioned the delivery of intelligent data services within the visited network. As network services are deployed, however, it has become clear that most operators are managing subscriber services from the core network. This network design drives some interesting challenges—one of which is scale. Since subscribers are terminated and managed from their home networks, the *network service nodes* required to provide mobile data services for a subscriber (a combination of GGSNs, Internet service gateways, address translation devices, and firewalls) are deployed within core nodes in the home network (often called "points of presence" or POPs). As a result, a network with 15 million subscribers will have over a million subscribers to manage per POP. And as GPRS evolves, each of these subscribers will access 2 to 3 services—meaning that in this scenario each POP has to manage between 3 and 4.5 million service contexts. This translates into a significant amount of state that must be maintained in real time by mobile core POPs in order to deliver always-on mobile data services. Each piece of state is a piece of information relating to the subscriber, the current visited network serving area, and the range of services and billing information associated with that subscriber.

Figure 27.4
Advanced services core.

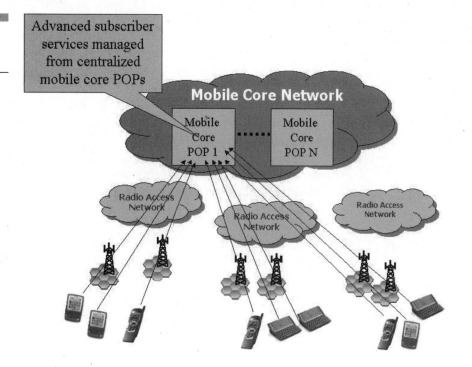

The mobile core will evolve over the next few years to manage increasing numbers of subscribers and their associated services. This means technical evolution in the form of supporting more always-on subscriber connections as well as new mechanisms for subscriber addressing, and service evolution to enable a broader array of partnerships between mobile operators and value-added service providers. Changes that can be expected include:

- Support for termination of an increased number of roaming subscriber connections, and an increased number of services per subscriber. As GPRS gets more broadly deployed, and as mobile subscribers accessing the Internet outnumber fixed subscribers, new core equipment will be deployed that can manage far more subscribers than available with today's core nodes. Today's equipment can typically support management of only tens of thousands of mobile subscribers. Tomorrow's core networking equipment will support delivery of service to up to a million mobile subscribers.

- Migration from an all-IP version 4 (IPv4) addressing scheme to an IP version 6 (IPv6) addressing scheme. Even early mobile networks will outstrip the IP version 4 addresses available, so mobile operators are providing terminals with "private addresses" (which are hidden from Internet-connected devices outside the mobile operator's network). These addresses are translated into public addresses before the mobile data packets are sent to the Internet. However, it is more desirable to have an open connection between terminals and the Internet and to provide the mobile terminals with persistent addresses, allowing Internet-attached servers to use the address to find the mobile terminal (simplifying their ability to push content to the mobile). Mobile data networks are likely to be the first data networks to have a driving need to change today's standard scheme of addressing for Internet-attached devices.

- Delivery of services from many network service and content partners. Mobile network operators traditionally managed not only the subscriber relationship and access to the network, but also the bulk of the services delivered to the subscriber. Mobile networks are currently evolving from these walled-garden environments to more open environments where not only applications, but also network services and service relationships, are often managed by third parties. As network capacity becomes more fully deployed, more of these relationships are expected to form. The mobile core network will then end up being a resource shared by a number of service providers but managed by one—typically the mobile network operator but in some cases an independent IP network operator (see Figure 27.5).

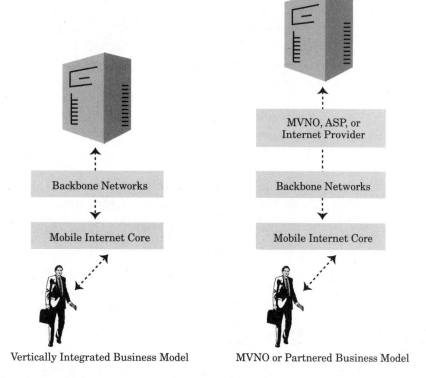

Figure 27.5
Multiple service
delivery models.

Vertically Integrated Business Model   MVNO or Partnered Business Model

Source: Megisto Systems

- Enhanced, end-to-end, traffic management. Today's GPRS standards support delivery of services with a variety of quality levels, but early networks are unlikely to deploy much in the way of guaranteed service levels. As network capacity is more fully deployed, and as access network equipment becomes more sophisticated, it will be realistic to deliver (and charge a premium for) enhanced services (e.g., lower latency or higher bandwidth).
- Broad support for usage- and content-based billing schemes. Today's mobile cores are only beginning to integrate service provisioning and billing systems so that operators can easily manage and bill for delivery of voice and data services using a variety of flexible billing models. IP networks have traditionally not billed based on usage or the value of the content delivered, so equipment being used in today's early core networks has little in the way of capacity to accurately collect charging data records. Pre-paid mobile services are increasingly popular, and heavily dependent upon accurate collection of charging

data records, as well as the capability to alert the user in real time to a pre-paid balance. Over the next few years, the mobile IP core network will increasingly integrate within operator billing environments, enabling not only seamless support of both pre- and post-paid billing services but also billing based on a number of more sophisticated usage parameters (per transaction and by content value). Operators may also bill for content on behalf of the content or application provider, and may provide free services for content providers looking to provide incentives for mobile subscribers to use their services (a data "800 number" service).

In addition to the evolution expected to support delivery of sophisticated and scalable GPRS services, the mobile core will evolve to support a broad range of additional services that become realistic as higher-capacity radio access techniques are implemented. These include "all-IP" networks carrying voice traffic, as well as IP multimedia subsystems delivering an array of integrated voice and data services.

Services

Services, applications, and content are often discussed without any level of differentiation. To the network operator, services are the options made available to subscribers when they sign up for their GSM accounts. Users will typically pay their network operators for services—but they may also make payments to their network operators or other providers for applications (e.g., the ability to execute stock transactions), and for content (e.g., for the ability to gain instantaneous access to a used-car history).

Mobile network operators have traditionally delivered highly vertically integrated services, providing everything from network access to voice services, to SMS and WAP services and applications. And historically, the difference between an application and a network service has never been clear. The term *SMS* applies to both a bearer service (the control channel used to deliver SMS) as well as the short message application itself. Operators have typically been dependent on their network infrastructure providers to modify pieces of the core network to support these advanced services. 800-number dialing, SMS, call waiting, etc. are all services that were dependent evolutions in the core network.

With GPRS, for the first time subscribers have access to a standard, high-performance bearer service that is independent of the applications and content being delivered to the subscriber. GPRS uses IP as the common bearer, enabling network operators, applications developers, and content providers to independently deploy new applications and provide access to new content without a dependency on their core infrastructure provider. As discussed in the previous section, the mobile core will certainly evolve. However, this evolution will enable classes of applications rather than specific applications (Figure 27.6).

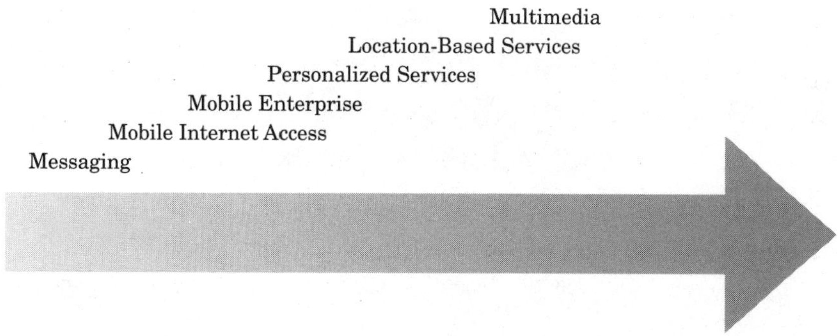

Figure 27.6
Services enabled or enhanced by GPRS.

Multimedia
Location-Based Services
Personalized Services
Mobile Enterprise
Mobile Internet Access
Messaging

Within each of these service categories, applications will be deployed. The first GPRS applications are simply higher-performing variations of applications available today via mobile circuit switched and CDPD bearer services. Delivery over the GPRS bearer means that the user will gain instantaneous access to these services via an always-on connection. GPRS will provide access to more efficient messaging and email services, mobile-optimized IP portals, sales force automation and vehicle tracking, machine-to-machine, and telematics applications. For example, GRPS is a bearer that will enable messaging services, although it will be up to applications providers to deliver instant messaging, email, real-time chat services, short message, and multimedia messaging applications over the GPRS bearer. We will discuss the evolution of "mobile Internet" applications rather than "GPRS applications," as the applications evolution should be independent of the network bearer services (i.e., of whether delivered over GPRS, EDGE, UMTS, or WLAN). The expected evolution of mobile Internet applications is highlighted in Figure 27.7.

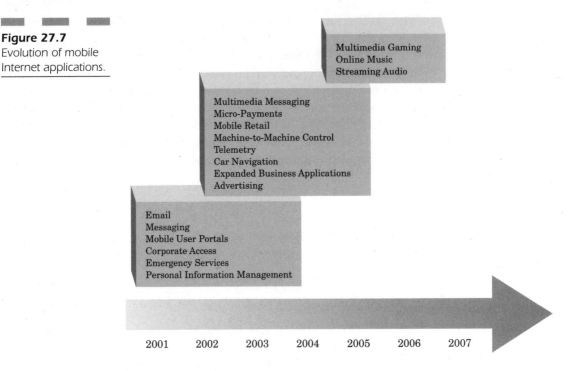

Figure 27.7
Evolution of mobile
Internet applications.

The majority of the first mobile Internet applications are being delivered as "pull" applications—where the user pulls information or content from the network. The primary advantage delivered by GPRS during this time is efficient, always-on access to data while on the move—subscribers get instantaneous access to content whenever and wherever they need it.

A second wave of mobile Internet applications is making expanded use of the ability to push intelligent content to the user, taking advantage of the capability of applications to initiate contact with the user (i.e., to find the user's "network" location), as well as the capability to determine the user's physical location. Such applications can include delivery of instant messaging, loyalty services, traffic updates, etc. on a scale broader than is possible with today's SMS services.

Future mobile Internet applications will take advantage of the increased capacity available from bearer services such as EDGE and UMTS. These applications will increasingly make use of multimedia and rich voice capabilities. However, even after full deployment of 3G bearer services, it is considered doubtful that the user will have access to the type of applications available via other broadband bearer services (such as DSL and cable). Mobile services are unique in that they provide the subscriber with instantaneous access to applications and content—but given the complexities of delivering service in a fully mobile environment, it is not likely that mobile data rates will compete with fixed data rates over the next decade.

Key to the evolution of the Mobile Internet is the fact that the mobile operator may or may not be involved with delivery of these applications. The mobile operator will first and foremost be involved with the delivery of network services—providing network access, the ability to pull data from the network, push data to the user, and to obtain location information. Many network operators will also implementing mobile portals—providing the user with access to a variety of optimized mobile applications (internally developed or developed with partners) as well as to their own OSS services. Other mobile operators will focus only on providing network access. Table 27.1 provides an illustration of the classes of services that may be provided by different mobile operators.

TABLE 27.1

Mobile Classes of Service

Operator Type	Business Models
End-to-end mobile service provider (mobile access, IP core, portal)	Consumer mobile Internet Enterprise mobile VPN Wholesale mobile access (roaming) Wholesale IP core for facility-less MVNOs
Facility-based MVNO (IP core, portal)	Consumer mobile Internet (via roaming) Enterprise mobile VPN Wholesale IP core for facility-less MVNOs
Facility-less MVNO (portal only)	Consumer mobile Internat (via roaming)

Source: Megisto Systems

These different business models have a significant impact on the mobile services available to the subscriber. One of the key competitive changes with the deployment of GPRS is that the landscape moves from one where competition is based on network coverage, to one where competition is based on access to the services and applications of interest to the subscriber. In the early stages of network deployment, when bandwidth resources are scarce, we see operators primarily focused on providing these resources for their direct subscriber base. But new operators looking to increase market share, and operators with sufficient bandwidth resources, are now looking to gain further competitive advantage through comprehensive applications and service partnerships.

Compatibility

A key challenge to the delivery of *uniform* services (presenting the same service to the user regardless of the network in which the user is roaming) will be the compatibility of visited networks to deliver these services. It was originally believed that most services would be delivered from the visited network, but initial GPRS deployments have made it clear that it is far easier to manage and bill for sophisticated services from the user's home network. In order to deliver sophisticated services via a visited network, the home network operator must be assured that the visited network infrastructure can support not only the capacity and quality of service required for delivery, but also the appropriate levels of authentication, billing, security, and policy-based forwarding capabilities. An operator may have agreed to send all data from a particular subscriber through a parental content filter. In order to provide such a service out of the visited network, the visited network operator must have an identically configured content filter and knowledge of that particular consumer, or the traffic must be sent back to the home network for security processing. Corporate enterprises will typically use a specific security gateway and have a secure "tunnel" between the mobile operator and their enterprises. If services are to be provided via the visited network, the corporate enterprise must support a tunnel to each visited network, or the home network must provide some kind of aggregation for tunnels from multiple visited networks.

The downside could be an issue called "tromboning," the fact that the user's data must be brought back home prior to forwarding it to its final destination. This means that a subscriber from Italy who is looking for simple Internet access when roaming into the United States may have to wait for his traffic to be sent back to Italy prior to gaining access. This is an issue to the user, if he is accessing a server actually located in the United States.

While tromboning carries with it highly negative connotations, there are two reasons why the delivery of sophisticated services from the home network should not be considered a fundamental problem. First, many subscribers do want to roam back to their home countries to get access to their home network services in their local language. They will likely have customized a portal and a set of services that they want to access—independent of their location at any point in time. Second, it is not necessary to choose to deliver 100 percent of a subscriber's services from either the home or the visited network. The beauty of APNs is that the operator can elect to have some services provided from the visited network and some provided from the home network (Figure 27.8). This means latency-sensitive voice or multimedia services can be delivered effectively from within a visited network environment.

Figure 27.8
Mobile service
delivery points.

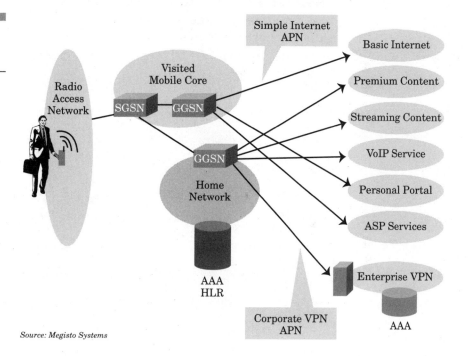

Source: Megisto Systems

Risk Factors

GPRS networks have been upgraded worldwide, and a number of key applications and services are now being made available. The risk factors are well known—GPRS brings with it new terminals, new radio and capacity allocation schemes, a new core network, and entirely new business models. However, operators have now had over a year to work out the issues with these new technologies. Handsets are becoming broadly available, and the initial handset–network compatibility issues are successfully being worked out. Basic services have been deployed and new services such as multimedia messaging with picture phones are becoming more broadly available.

The biggest risks remaining are whether the services delivered by operators will meet users' expectations, and whether operators will be able to make up with high-revenue data services for the decreasing margins they are experiencing in their maturing voice services. Key to the success of future, even higher-performing radio technologies is the demonstration that data services can help operators make money, and that deployment of these technologies will provide the operators with sufficient payback on their investment.

CHAPTER 28

All-IP
Networks

Simon Cavenett

Mondo Techno, LLC.

Before we examine the subject of all-IP networks, let's take a moment to establish what a network is—at least for this discussion. A network can be defined simply as a collection of *nodes* that, when interconnected and operated in collaboration, perform a set of functions or enable a service. A node can be either logical or physical, or both. Confused? Well, think in terms of your classic high school physics—where the world is wholly modular and composed of lots of atoms. In that simple world the recipe for any matter was overly simple—make up a bunch of different atoms using different ratios of electrons, neutrons, and protons. With these different atom types, you can compose any type of matter that exists.

OK, so let's ignore the scientific oversimplifications of those school day physics lessons but take the simplistic model of the nature of matter and use it to develop a simple model for telecommunications and computing networks. Starting at our initial definition that a network is a collection of nodes, we can further define a node as being the basic building block of a network. Like our atomic model, there are different types of nodes and each node is composed of smaller particles, or elements.

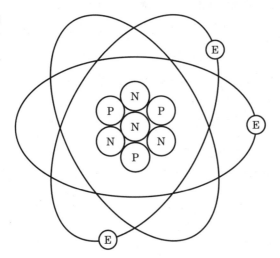

Figure 28.1
Simple atomic model.

To avoid overcomplicating this model for networks, let's just propose that a network is made of more than one node and a node is made of one or more *elements*. A node in this network model is comparable to an atom in our simple model for matter. At least right now it is. Although the comparison is a little too broad to really justify under close examination, it greatly simplifies the discussion. And so, by extension, we can assume that we can have different types of nodes. Now what defines a node as being unique for our model is that it performs a unique function. A network is made of nodes connected to one another. Matter is made of atoms bonded together.

What Is an "All-IP Network"?

As a core protocol underlying the Internet, the Internet protocol (IP) is the default networking protocol for most computing networks today. But keep in mind that IP is a protocol defined and used for the purposes of transmitting data between two or more nodes. In data networking, the nodes are more commonly referred to as hosts. For example, if I access a Web site via my Internet service provider from my personal computer (PC), I consider my PC to be a local host and the Web server a remote host. Both hosts communicate via their respective ISPs and the Internet.

Figure 28.2
Accessing a Web site
via the Internet.

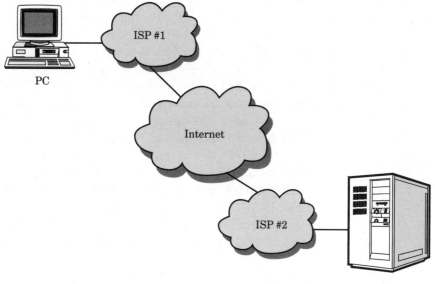

PC

ISP #1

Internet

ISP #2

Web Server

The IP protocol is exclusively for enabling the transmission of digital binary-encoded data. Further, the IP protocol operates on discrete chunks of data rather than a continuous stream of digital data. Chunk switched data never really caught on as a technical term, so the discrete chunks are more commonly referred to as *packets*. And so the term packet switched data (PSD) is common today. A packet of data is simply a segment of a data stream. The data stream is segmented so that it can be encoded and labeled for transmission between hosts. This segmentation, encoding, and labeling process is referred to as *encapsulation*. (For more discussion on packets, see Chapter 9.)

What is important to realize is that the IP protocol in itself does not provide a service directly to a user. By itself, IP is useless to the average person; it is only useful for the transmission of data. With no data to transmit between nodes or hosts, IP is unnecessary. A non-networked PC doesn't need any IP capability. To illustrate the point, the Microsoft Windows 95 and Windows 98 operating systems do not have IP capability installed by default. It is only required if the PC is to be networked using a dialup modem to an Internet service provider or an Ethernet-based connection to a LAN. And this often requires the user to hunt for the Windows disc when installing a networking device on his PC.

The Internet protocol allows the exchange of data between nodes that are networked. A host can thus be considered a particular class of node. Simply put, a host is either receiving or generating the data being exchanged, and typically the data is exchanged between hosts by being transmitted across a network, or multiple networks, linking the hosts. A person using a PC at home to browse Web sites on the Internet requires the services of at least three networks. Dialing up to the ISP, the user encounters the first network—the ISP. The user's ISP may also have the Web server on its network and in that case there is only one distinct network linking the user's PC and the web server. But more typically the Web server is connected to the Internet via another ISP.

The Internet is a common network (a backbone) to which multiple networks are linked. Many media pundits use their own atomic model for the Internet, defining it as comprising any network that connects to it and exchanges data over it. But more accurately, the Internet is the backbone through which different networks exchange data using the IP protocol. The Internet literally is an inter-network.

The history of the Internet and its rise in popularity is a whole topic in itself—full of sex, sin, intrigue, mystery, suspense, short-sleeved white shirts, and pocket protectors. Well, at least the shirts and pocket protectors. However, we don't need history lessons to recognize that one of the most important factors leading to the rapid growth and acceptance of the Internet was the early definition and implementation of a set of common standards. The Internet protocol is one such standard, and it was the result of a concerted effort to develop such standards.

So What Is an All-IP Network?

Well, the Internet is an all-IP network. "Duh!" you may say. But the Internet is not unique in this. The local area networks (LANs) that exist in most offices are usually all-IP networks as well. At least that's true today. IP was not a frontrunner as a network protocol two decades ago when the LAN was an emerging technology in the marketplace. CP/M once ruled as the operating system of choice for microcomputers. What made IP the winner was that it better suited internetworking. Why? Simply put, IP is deliberately a common and open standard. And when connecting different networks together, whether between office buildings or across the globe, it made obvious sense to use a nonproprietary solution—at least for the data protocols. IP was designed purposely for internetworking between networks on a national, and later a global, scale. An influential factor was also the fact that the Internet as it existed then (ARPANET) was already standardized on the IP protocol. After all, it is easier to plug your lamp into the wall if you use the plug that matches the outlet socket.

And (literally) closer to home, if you decide to network the computers in your house using one of the common networking solutions—such as a wireless 11Mb/s 802.11b network or a wired 10/100BaseT Ethernet, then most likely you too will have an all-IP network. The point to consider here is that all-IP networks are very common today. As a technology solution it is mature and it is ubiquitous globally—for computer networking. For telecommunications networks, however, it is a different story.

Circuit-Switched Networks for Telecommunications

Most telecommunications networks that exist and operate today are not IP based. Let me qualify that statement further—most telecommunications networks whose primary purpose is the provision of voice telephony services, and associated value-added services, are not IP based. Instead they're circuit-switched. Most computer networks today are packet-switched. So convergence in telecommunications refers to the evolution from circuit switch-based networks to packet-switched networks. What's a packet-switched network? An all-IP network, such as the Internet, is an example of one.

Let's explore circuit-switched networks a little further. A circuit-switched network is composed of nodes that allocate some of their resources to voice and associated services on a dedicated basis for the duration of the service. What does that mean? The system resources necessary during the time a service session exists are allocated exclusively to that service for the duration. To explain, consider a telephone call on a plain old telephone system (POTS). Your local telephone company typically deploys its telephone network across its service area by having multiple local telephone exchange buildings situated across the major population areas.

It is no coincidence that the local telephone exchange exists so often near the center of the town or suburb. I digress, but the reason is simple: most telephone networks were planned and installed using copper wire cables that were run between the telephone exchange building and the subscribers' premises. Copper cable has a resistive loss—the longer the cable, the weaker the signal power. So it makes sense that if the cable is too long, the signal becomes too weak for adequate service. For the gauge of cable commonly used, the maximum cable distance is around 20,000 feet. So to provide the best service to the most subscribers, telephone exchanges were normally located at what was estimated to be the *copper center* for the local area. The copper center was mathematically the geographic centroid such that the total copper cable distance for the entire area was minimized. In other words, the exchange building was located in the least bad position so that the cable distance to the most number of subscribers was the shortest possible and therefore the quality of the telephone lines were a best average for most subscribers. Confused? Well, in North America, the local telephone exchange is also commonly referred to as the central office (CO)—although a large enough town or city may have many COs distributed across it and networked together.

To define a circuit-switched network, let's use a simple network plan. The town of Smallville has one central office operated by the local telephone company. So does the nearby town of Springfield. Homer lives in Springfield. Clark lives in Smallville. If Homer decides to call Clark, his call will be routed (trunked) between the respective central offices in each town.

The telephone call requires resources from the equipment at each central office and also from the transmission equipment between the two central offices. For the call, a circuit is set up and maintained between Homer and Clark. This circuit is a *bearer* to transmit their conversation. Importantly, this bearer between Homer and Clark is tempo-

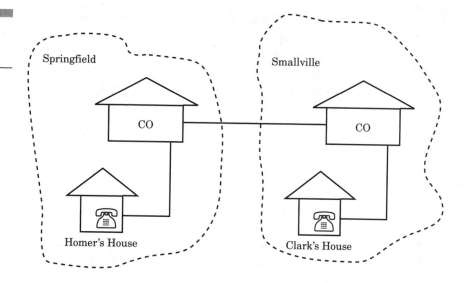

Figure 28.3
Simple telephone network example.

rary—in other words it is established (constructed) only for the duration of the call and will be "torn down" (deconstructed) at the end of the call when both parties hang up. During the existence of the call, the resources used from the necessary network nodes are dedicated to the provision of the bearer for the call. For example, at the Springfield central office, a voice circuit is established between Homer's house and the transmission equipment connecting the Springfield and Smallville central offices. Since the call is for a subscriber in Smallville, the call is routed over a transmission link between the towns and during the call it occupies resources for a voice-grade circuit on it. Likewise at the Smallville exchange, resources are allocated to connect to Clark's house. During the call, the circuit established end-to-end across the entire network is exclusive to Homer and Clark's conversation—regardless of the actual amount of information exchanged between the two.

Is this end-to-end temporary circuit all-digital as well? For voice-grade telephony, the answer is maybe. It does not have to be. Most telephony systems and networks in North America are digital *and* circuit-switched. Huh? It comes down to the encoding method used for voice. Somewhere at the central office end, or between the subscriber's premises and the central office, the analog voice signal is digitized. For telephony, the standard encoding scheme used for digitizing voice is pulse code modulation (PCM). The default for a standard 3100 Hz bandwidth analog voice-grade circuit is to encode using 64 kbps PCM. Thus in our example, the call between Homer and Clark may have been all-digital—at least from central office to central office—meaning that a temporary 64 kbps PCM circuit was set up and switched through the network for the call. But circuit-switched networking theory was developed for all-analog networks. (Factoid: one of the inventors of the technology was an undertaker.) Digital telephone exchanges and digital transmission equipment were, in general, developed to be architecturally and functionally compatible with analog. If Smallville still relies on older analog equipment in its central office, the telephone call in our example may involve both analog and digital voice circuits with part of the call being handled by one sort of equipment and the rest being handled by the other.

Circuit-switched equipment, and therefore circuit-switched networks, can be all analog or all digital or a combination of both. Normally, the majority of nodes in a telephone network are either all analog or all digital. Thus, in circuit switching the importance of analog versus digital primarily affects the interfacing of nodes within a network. Before any telecommunications engineers start warming up the barrels of tar and plucking the chickens, it should be noted that such a statement is a very broad oversimplification. But for the purposes of this discussion, it is a reasonable assumption to use. Signaling System #7 (SS7)-enabled telephone networks, the ones that separate the call setup and control signaling information from the voice switching and transmission, can be operated equally well with digital or analog switching and transmission equipment.

Our discussion here of circuit-switched networking has posited voice telephony as the service provided. This is because voice has been, and still is, the primary service controlled and carried by circuit-switched networks. Whether those networks are for wireline or wireless telephony, circuit-switched voice (CSV) is the dominant type of traffic carried by circuit switched networks today. But data services can also exist on these networks using circuit-switched data (CSD) technologies.

For wireline networks, CSD is typically routed off the network at the earliest opportunity onto a packet-switched network, because carrying data over circuit-switched networks is not as efficient as carrying data over packet-switched networks. For example, an ISP may have a modem bank or access switch co-located at the telephone company's CO. Incoming dialup calls to the ISP are switched to this modem bank to move them off the telephone carrier's "voice call-optimized" network. From the modem bank forward, the ISP traffic is carried on a packet-switched basis through to the Internet.

For wireless networks, the story is pretty much the same. Resource efficiency is a very significant issue for wireless networks since the comparable cost of network resources considerably exceeds that of a wireline network. Unfortunately for most second-generation (2G) wireless networks, including GSM, a CSD session requires more network resources than a regular CSV session. In a GSM wireless network, the first opportunity to route data off the circuit-switched network is at the mobile switching center (MSC). The MSC routes the data to an inter-working function (IWF) for delivery to a packet-switched network; in other cases the IWF routes data back to the MSC for delivery to the public switched telephone network (PSTN).

Actually the CSD session requires *at least* the same set of system resources on the wireless network. A circuit must be established and maintained end to end between the user's mobile and the IWF. If the user is connecting to his ISP of choice and using his mobile as a data modem, the circuit must be routed back through the MSC and out to the PSTN. Indeed, in this example, the CSD session requires twice as many circuit resources from the MSC node than a mobile-to-PSTN voice call. Further, the session to the ISP is typically a local call and so the operator does not gain a sizable share of roaming revenue from it. (For circuit-switched voice, roaming and toll revenue can be a significant percentage of the operator's income.)

For service plans that offer a monthly quota, or *bucket*, of call minutes with national roaming included (as commonly offered in North America), this philosophy still holds. The hook to the subscriber is the offer of free long distance, and the operator is relying on the fact that the average subscriber only uses a fraction of his monthly allocation. On this model, circuit-switched data does not present a favorable business case to most mobile network operators. The predominant demand by users is for voice telephony. Coincidentally, this is why many mobile operators choose to market circuit-switched data services "gently."

Nevertheless, circuit-switched data is an important service to offer for competitive marketing purposes. At present, the relative amount of data services carried by existing mobile networks is low compared to future potential. Most mobile operators, however, seek profitable methods of exploiting this potential for wireless data service demand. It is one of the key drivers for the introduction of more resource-efficient data transmission technologies using wireless networks—such as GPRS.

Drivers for All-IP
Networking in Wireless

Second-generation wireless network technologies such as GSM are circuit-switched for both voice and data. As we have discussed, circuit-switched data services for wireless networks are inefficient resource-wise and have low potential to generate revenue—at least compared to providing voice services over the same network. So the casual reader may inquire, "Why bother with wireless data at all? Why not simply focus on voice services?" Good question. Many startup companies have come and gone in the last two years that could have benefited by asking themselves those questions. And many industry analysts, too.

The answer to both is revenue—the revenue a mobile operator can generate on a per-user basis. With many developed countries considered to have mature competitive mobile telephony markets, defined as a penetration of mobile users exceeding 30 percent of the adult population, the potential for increasing revenues from voice telephony services is approaching saturation. While the gross number of mobile users continues to increase, the average revenue per user (ARPU) in most markets continues to decline as a result of competitive pressure between operators as well as the general trend of commoditizing mobile telephony services—in much the same way that wireline telephony was commoditized.

Compounding the weakness in the business model for mobile operators is the common practice of bundling services with a monthly bucket of minutes that typically includes any long distance charges. The benefit of this trend is that low-usage subscribers in these bundled plans usually overpay for what they actually use. The penalty of this trend is that high-usage subscribers may actually use all of what they have paid for and more. As we've seen, the profit/loss margin is based on average usage. So, while the revenue per user is made more predictable when operators promote bundled service plans, the tradeoff is the increased effort necessary to raise average revenues; in such plans we have to overcome a much greater subscriber inertia in order to increase a monthly subscription commitment.

Unfortunately, the subscriber population is composed of people, and most people have a threshold for the maximum amount of time they use or do anything (teenagers excepted). Particularly when it costs money to do. In the United States, most dialup service plans from ISPs offer in excess of 150 hours per month of Internet access for a flat monthly rate. One major ISP offered 1,000 hours for a one-month free trial. That comes in handy for those 42-day months, but otherwise the potential value of the bargain reaches the sublime. The average user of the Internet in North America logs between 20 and 30 hours of online time per month. For an ISP offering 150 hours of usage per month with a flat rate service plan, the revenue potential to be gained by encouraging subscribers to increase their monthly online time is nil. Less than nil in fact, given that it costs the ISP more in infrastructure and operating costs to support higher average usage. The alternative for such an ISP is to attract more revenue by offering additional services that users are willing to pay for.

For mobile operators of voice-centric networks, the potential to squeeze additional revenue from traditional voice telephony services is evaporating—particularly in highly competitive and highly subscribed markets. In the short term, pre-paid voice for wireless networks opened up a new market segment by providing a new means of service payment for those unwilling or unable to meet the credit requirements of a monthly commitment under an annual contract. However, pre-pay also has its saturation level.

For voice, ultimately it comes down to how many minutes per month a user is willing to spend talking on the phone. And to whom, and from where. The mother-in-law market segment aside, it can be quickly concluded that many markets are reaching the saturation point for voice telephony usage—a reality against which further discounting of service charges becomes ineffective. Look at it this way. If it cost *nothing* to make a voice call on a mobile network, how often would you talk on your mobile phone? You may make more calls on your mobile than you currently make on your wireline service—but probably only for convenience's sake. Considering what you currently make today across your wireless and wireline services combined, the number of calls would likely be about the same. Voice telephony is a service commonly used for communication between people, and here's the point: if the amount of information to be communicated remains the same, the amount of a service utilized to communicate it likewise remains the same.

So What About Convergence?

Earlier we noted that circuit-switched data over wireless networks is currently the default, and that CSD over wireless in itself has little-to-negative revenue potential. It's a vise on the industry that we increasingly look to convergence to loosen. A common basis for computing and networking would go far toward stabilizing the business model.

So what are the drivers for all-IP networking to be adopted by wireless industry? Considering voice telephony as a service that has a usage saturation level for the average user in any market and mobile operators who are beginning to sight it, the need exists to develop and offer new types of revenue-generating services that leverage the existing market and existing network infrastructure.

The toe in this particular bathtub, so to speak, was WAP over CSD: wireless application protocol over circuit-switched data. Unfortunately, WAP always promised so much and often delivered so little. The comedy of errors that WAP largely became for many mobile operators was caused by many factors, but hype tops the list. Nevertheless, WAP did provide some useful real-world experimentation of nonvoice services for wireless networks across a broad subscriber base in many markets. The final conclusion on this global experiment appears to be that the technology wasn't quite ready to deliver and the service was slow, awkward to use, and unreliable. In general, it was easier to use an alternative method to perform the task.

If the technology solves a problem well, the solution has a good chance of success. Consider that for a moment. Now consider circuit-switched data. It suits some applications but it has severe limitations:

- It is slow to connect since the end-to-end data circuit must be set up each time a session is initiated.
- It requires dedicated network resources devoted to each session.
- It does not mesh well with packet-switched network techniques.

Data is generally bursty by nature, meaning that it occurs as the application generates it. For example, IP traffic is exchanged when loading a Web page into your browser but little or no traffic is exchanged while you read the page. Most data applications generate traffic that is well suited to multiplexing, i.e., sharing a common transmission channel by interleaving data from different sources. Following this reasoning, it can be seen that a packet-switched network best suits data traffic.

"But wait a minute!" you may say. If voice is the prevalent service used today on wireless networks, and circuit-switched network architecture suits this purpose, why convert the network to an all-IP packet-switched architecture? Well, what if voice was no longer the most significant service carried on the wireless network? What if packet-switched data was? It's feasible both technologically and economically as long as the supported services are marketable. But to this I will add a caveat: the collective consciousness of the marketplace will determine what becomes popular and what does not. The marketplace is littered with the remains of technologies that never achieved enough popularity thrust to escape the gravitational pull of financial disaster.

Advantages of an All-IP Network

With wireless networks evolving toward an all-IP network architecture, it is relevant to discuss the advantages and disadvantages. For computing networks, IP networking is a mature technology that underpins much of the common networking infrastructure that exists. For wireless telecommunications, data services and data networking are still in their infancy. Putting technological arguments aside, let's consider the simple law of market forces as it applies to the convergence of telecommunications and computing. IP networking technology underpins computing networking today. Circuit-switched networking underpins telecommunications today—but it was primarily developed for voice services. Wireless telecommunications seeks to add substantial nonvoice services and furthermore seeks to internetwork significantly with existing IP networks such as the Internet. It follows that wireless networks should align with the IP networking technology. Alignment allows seamless integration. Seamless integration allows convergence. QED?

We can also prove the point by focusing on some key advantages of all-IP in a wireless network. An end-to-end IP network architecture allows multiple operations to be performed on the packet data as it traverses the network. For example, certain data can be tagged for classification and prioritization in the aggregate data stream. Packet data can also be rerouted through the network—for example to avoid congested paths. This can be achieved dynamically, unlike circuit-switched data routes. Significantly, the packet-switched domain allows the end-to-end session to be isolated from the physical transmission links in between. Thus, a session can be maintained by dynamic rerouting even if a physical transmission link is removed or lost. This is not possible with circuit-switched data links—lose the physical link between the mobile and the network and the entire session is lost. Experienced users of WAP over any of the current 2G wireless networks should be familiar with that problem.

Packet-switched data is, by design, ideally suited to multiplexing. That is, data generated by multiple service applications can be carried on the same transmission link while still being identifiable in the aggregate stream (by inspection of higher layer protocols).

Disadvantages of an All-IP Network

Existing second-generation wireless networks, which are not based on an all-IP architecture, were very expensive to build. The accountants are not going to let them be ripped out anytime soon. Besides, there are over 900 million people in the world who would not be very happy if they were forced to buy a new GSM mobile handset overnight. So we must consider evolution, not revolution, as the viable path for the wireless industry from circuit-switched networks to all-IP networks. How long is evolution? Let's allocate a decade as a rule of thumb for most major mobile operators.

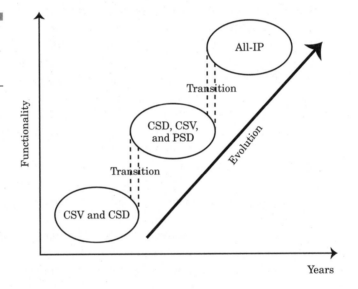

Figure 28.4
Evolution path from all circuit-switched to all-IP.

Evolution brings with it some tradeoffs. One such tradeoff arises in conjunction with the air interface standards for second-generation networks that were primarily developed for circuit-switched voice transmission. At the mobile and the network ends, voice is encoded using compression schemes such as EFRC and EVRC. The details of these vocoder techniques are not important here except to note that they are optimal *only for circuit-switched voice links.* Radio links inherently encounter variable transmission conditions—introducing bit errors, frame errors, and lost data. Accordingly, current transmission schemes implement interleaving of the vocoded voice data for radio link transmission, so that the loss of one or a few data frames only slightly reduces the subsequent voice quality.

Being optimized for voice transmission, existing 2G radio link protocols are designed to allow a residual (continuous) error rate—typically 1 to 2 percent for GSM or CDMA systems. Tolerating this residual error rate allows the radio link to operate at much lower carrier-to-interference ratios. This results in more efficient use of the available radio spectrum and a higher total network capacity for traffic. In other words, allowing some of the information transmitted to be received in error, or lost, lets the network carry much more traffic, which in turn makes for a more cost-effective network to build and operate. Error rates up to 2 percent introduce only a minor degradation in subjective voice quality. However, voice is a unique type of service in that it is interpreted subjectively by the listener. The encoding data is discarded as soon as it has been transmitted, decoded, and the information regenerated as an analog audio signal for the listener.

Humans, unlike most software applications, are fairly forgiving data processors who can tolerate reasonable error levels in voice transmission. However, most data generated by nonvoice applications is intended for errorless transmission end to end. Thus, the effect of errors across the radio link on IP packet data can be significant for the application at the receiving end. Typically, transport of IP packet data is managed by the transmission control protocol (TCP), a higher-layer protocol that is encapsulated by IP protocol data packets. TCP was developed for wireline transmission networks that typically do not encounter a residual error rate. Also TCP was developed to react to network congestion conditions and to throttle packet transmission rates by the sending end until the congestion is perceived to have cleared. Unfortunately, TCP can misinterpret packet loss across a radio link for network congestion and thus react incorrectly by throttling the transmission rate rather than simply resending the lost packet(s). The effect of this is to slow the throughput of data across the network, even though the network is operating under reasonable conditions; stalling data transmissions, or an error condition occurring within the receiving application which interrupts the service.

Voice over IP (VoIP) is a packet-switched data protocol often proposed to enable voice services for packet-switched data-capable wireless networks. For all-IP networks, VoIP has potential to become the default voice service protocol—particularly since the core network can be efficiently architected to use H323- or SIP-enabled media gateways for service control and management of the services. For all-IP networks, VoIP potentially offers an alternative to circuit-switched methods for voice services. In particular, VOIP offers greater flexibility for selecting and achieving the desired quality of service of the voice transmission.

However, for existing GSM networks that have a circuit switch-optimized architecture, the justification for VOIP is much less compelling because existing vocoding techniques are efficient for voice, and by definition, all existing GSM handsets are compatible with them.

To support VoIP, a mobile device would have to run VoIP applications, which implies more complex devices than those commercially available today. As with TCP, VoIP performance may prove problematic when encountering radio link errors, and the existing GSM circuit-switched network architecture was optimally designed for voice services. With the newer vocoding techniques for GSM such as adaptive multi-rate (AMR), there already exists an upgrade path for voice transmission that is compatible with existing network operations. And not to be ignored, VoIP may represent a means to bypass the mobile operator's circuit switched voice network and thus pose a threat to voice service revenues. This possibility is not viewed kindly by operators, who may try to block unauthorized VoIP usage on their networks to prevent just this outcome.

Finally, the requirements for the mobile device must be factored into any accounting of disadvantages. Currently mobile devices capable of providing voice and data services on a circuit switched basis require only a few custom DSP chips as the core of their hardware. To implement applications that will take advantage of IP network protocols, on the other hand, will require more processing horsepower, added memory, better and larger displays, more complex radio circuitry, and higher battery capacities to power it all (see Chapter 23). Add to that the requirements on size, form, functionality, and thermal and RF emission limits. The development of mobile devices leads to a nearly paradoxical situation: in order to justify the development of new mobile devices, the networks must exist and the demand for their services must exist, but in order to justify network deployments, the devices must exist (in commercial quantities). As one executive of a mobile operator commented during the initial rollout and marketing of GSM, "God Send Mobiles!"

The Evolutionary Path to an All-IP Network—and Beyond

Circuit-switched wireless networks are currently configured and operated as homogeneous networks; i.e., the network is composed of multiple nodes but is only functional with a minimum set of specific node types. The functions performed by these nodes are neither divisible nor relocatable to an external network or even to another node type within the network. To illustrate this, consider a basic GSM network. The basic network consists of at least one base transceiver station (BTS), one base station controller (BSC), and one mobile switching center (MSC).

The BSC is primarily responsible for managing and maintaining the radio link between network and mobile device. The BSC controls all BTS, handles most mobility management, and circuit switches the user data to the MSC. The MSC handles all service authorization, service control, and interfacing to all external networks. While it is possible to physically combine two nodes or even all three (BTS, BSC, and MSC), they must still exist as functionally complete in a logical sense at a minimum. In other words, it is not possible to remove some functionality from a BSC node and insert it into, say, a BTS node.

To use the high school physics model analogy, the BTS, BSC, and MSC are atoms. They may be composed of smaller elements but they are essentially the primary building blocks.

For GSM, the evolution to all-IP networks begins with GPRS. GPRS introduces new network nodes specifically for packet-switched data. These GPRS nodes include the SGSN, GGSN, BG, and PCU. A GPRS-capable GSM network thus becomes a circuit-switched and packet-switched network hybrid. The "attachment" of the packet-switched network nodes is achieved with the packet control unit (PCU)—which acts a gateway between the circuit-switched network connecting at the BSC and the packet-switched network connecting at the SGSN.

In this hybrid configuration, a GSM network is often referred to as 2.5G or "second generation plus." The use of this term is debatable since the network is still second generation but with enhanced functionality. Unfortunately, the term 2.5G is often incorrectly interpreted to imply "next generation."

The wireless network evolution continues with the first release of architecture specifications for the third generation of the UMTS terrestrial radio access network (UTRAN), more commonly referred to as *3G*. Specifically, these are a set of specifications defined by the 3rd Generation Partnership Project (3GPP) that can be used for a 3G network. Got that? As this is the last chapter, I hope you have it by now. Anyway, the 3GPP specifications have been planned to include multiple releases—each release being a further step toward an all-IP network. There are currently four releases identified and scoped, and all are being developed in parallel. The first release, Release 99, builds from the existing GSM/GPRS specifications—and so is a hybrid network architecture combining both packet-switched and circuit-switched nodes. This is unashamedly intended to be for evolutionary purposes for existing 2G and 2.5G networks.

All-IP network architecture debuts in Release 4 of the 3GPP specifications. The significant change from Release 99 is the removal of all circuit-switched nodes from the architecture. The homogenous nature of what was a GSM network is also replaced by the concept of heterogeneous networking. That is, the radio access-specific nodes can be allocated to a dedicated access network. The access network interfaces to a core network—where nodes such as the SGSN, GGSN, and MSC server reside. Interfacing to non–packet-switched networks is achieved via media gateways.

The removal of circuit-switched nodes from the network architecture enables the convergence of telecommunications and computing network architectures to be fully achieved. An important benefit of convergence is the ability to separate network transport functions from control functions, such that the functions can be allocated, or aggregated, into the specific-purpose nodes. This allocation relates directly to the separation of control planes and transport planes for packet-switched data. Interfacing functions can also be allocated to gateway nodes. Thus the combined functionality that was integrally bound and indivisible in a circuit-switched node, such as an MSC implemented with a Class 5 circuit switch, can be unbundled, redistributed, and regrouped across the network on other nodes.

In the long term, the concept of separate control planes and transport planes for all-IP networks leads to distributed network architectures. The application of distributed network architecture principles to wireless networks remains an active area of study. Further ahead on the evolutionary path, actual implementation of these principles has "limited visibility" at the moment. In particular, it is unclear whether the principles of distributed architecture will be fully embraced by mobile operators.

A Simple Network Model Proposed

All-IP architecture for wireless networks, such as that defined by the 3GPP Release 4 Specifications, can be perhaps be better interpreted using our high school physics analogy of atoms and matter. If we construct the overall network as a collection of homogenous networks, the latter will be analogous to our simple atoms—they are composed of smaller elements but are irreducibly basic building blocks. The homogenous networks can be clustered into heterogeneous or homogenous environments. Further, this networking model can be expanded to include wireline networks as well.

In this model, examples of homogenous networks include access networks—such as the radio access network for 3GPP, the radio access network for an 802.11 fixed-wireless network, and (on the wireline side) a DSL access network or a dialup ISP access network. These networks are considered unique and homogenous in nature. They are homogenous in the sense that they are not divisible while still remaining a functional access network, and they are heterogeneous in that they can be clustered into a heterogeneous environment.

Figure 28.5

A proposed simple model for all-IP networks.

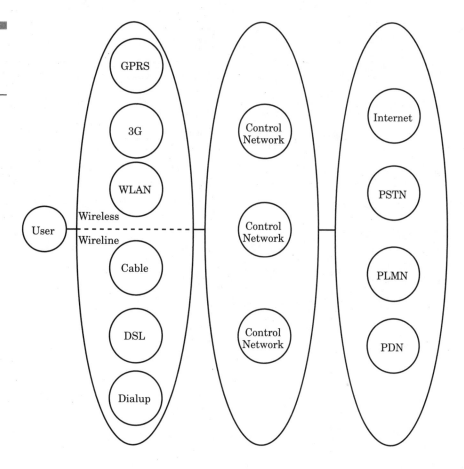

The core network of an all-IP network can alternatively be labeled as a *control network*. This term is selected since "core" can be ambiguous in the context of networks like those specified in the 3GPP Release 4 specifications. Also, the Internet is commonly considered a core network. The third class of network building blocks is defined as the *Service network*. The service network can contain nodes that provide or enable the various services available to the subscribers—such as media gateways, Web servers, and application servers. A service network typically connects to a control network, and an access network typically connects to a control network.

As the names of building block classes suggest, the access network provides the link between a user and a network; the control network performs primary service authorization and control functions for the user; and the service network provides the network end of the requested service.

As a guide to applying this simple model to all-IP networks, a basic network topology is defined for the model. This classification scheme allows any homogenous network to be correctly classified as an access, control, or service network, simply achieved by classifying an access network as primarily managing layers 1 though 3 (according to the OSI Reference Model), a control network as primarily managing layers 3 through 5, and a service network as primarily managing layers 5 through 7. Layers other than those allocated to each network class will also normally be implemented—for internetworking functions for example. However, as they relate to a particular IP-based service, the layers identified are primarily managed and controlled by the nominated network class.

Figure 28.6
Simple network classification scheme.

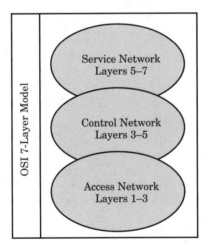

Like the high school physics model of atoms and matter, this simple network model for all-IP networks is presented as a tool for learning concepts in a general sense only. Its purpose is to clearly and graphically convey the fundamentals of all-IP network architecture. The model makes some gross assumptions and generalizations, so caution is advised when applying it.

Conclusion

The evolution of wireless networks from circuit-switched architectures, optimized to provide voice services, to all-IP packet-switched networks, optimized to provide multiple services, is primarily being driven by the need to diversify the types of services that can economically be offered to mass populations via terrestrial radio networks.

Figure 28.7
A paradigm shift for subscriber and billing management with all-IP networks.

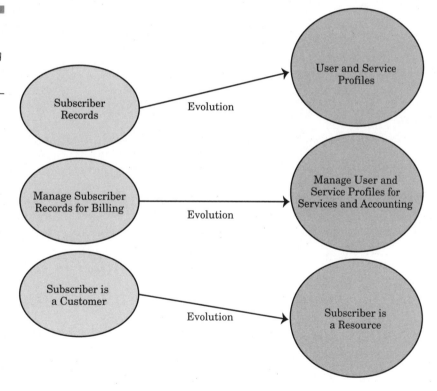

The circuit-switched infrastructure dominating most wireless network operations today was installed on the basis of significant capital investment. With circuit-switched voice currently the dominant revenue-producing service on these networks, the requirement for evolution away from circuit-switched architecture toward all-IP packet-switched network infrastructure is rooted in strategic business needs rather than purely technological considerations. The pace of this evolution will be determined by how long it takes operators and subscribers to recognize its revenue-enabling and service-enabling advantages, and by how swiftly the required technologies become commercially available in sufficient quantity, and in a complementary manner.

Now we know why the succesful implementation of GPRS is so critical. Getting it right from the beginning sets the path to a smooth evolution to an all IP wireless technology. Get it wrong and we have to sort out the mess before the evolution to an all IP GSM service offering moves forward. Let's all work together to get GPRS right from the start and once again change the landscape of wireless communication as we did with the launch of GSM in 1992.

John Hoffman

APPENDIX

ACRONYMS AND ABBREVIATIONS

The following is a list of abbreviations that you may encounter in one or more chapters of this book. While not as a good as a magic decoder ring, it may help you find your way through the alphabet soup that is our industry.

2G	2nd Generation
2.5G	2nd Generation Plus
3GPP	3rd Generation Partnership Project
3G	3rd Generation
AAC	Advanced Audio Coding
AAL5	ATM Adaptation Layer type 5
ACA	Australian Communications Authority
A-GPS	Assisted Global Position System
AMR	Adaptive Multi-Rate
APN	Access Point Name
APNIC	Asia Pacific Network Information Centre
ARIN	American Registry for Internet Numbers
ARPANET	Advanced Research Projects Agency Network
ARPU	Average Revenue Per User
ATM	Asynchronous Transfer Mode
AUTN	Authentication Token
BG	Border Gateway
BSC	Base Station Controller
BSSAP+	Base Station System Application Part +
BSSGP	Base Station System GPRS Protocol
BTS	Base Transceiver Station
BVCI	BSSGP Virtual Connection Identifier
CCU	Channel Codec Unit
CDMA	Code Division Multiple Access

CDR	Call Detail Record
CE	Customer Edge (the IP VPN terminology for a CPE)
CGALIES	Coordination Group on Access to Location Information by Emergency Services
CGF	Charging Gateway Functionality
CGI	Cell Global Identification
CI	Cell Identification
CK	Cipher Key
CLNP	Connectionless Network Protocol
CLNS	Connectionless Network Service
CMM	Circuit Mobility Management
CO	Central Office
CPE	Customer Premises Equipment (see CE)
CP/M	Control Program for Microcomputers
CSD	Circuit-Switched Data
CSS	Cascading Style Sheets
CSV	Circuit-Switched Voice
DHCP	Dynamic Host Configuration Protocol
DNS	Domain Name System
DSL	Digital Subscriber Line
DSP	Digital Signal Processor
DTM	Discontinuous Transfer Mode
EFRC	Enhanced Frame Rate Coder
EGPRS	Enhanced GPRS
EMI	Electromagnetic Interference
EMS	Enhanced Messaging Service
E-OTD	Enhanced Observation Time Difference
ESP	Encapsulating Security Payload
EVRC	Enhanced Variable Rate Code
FCC	Federal Communications Commission (US)
FFS	For further study (smile)
Ga	Charging data collection interface between a CDR transmitting unit (e.g. an SGSN or a GGSN) and a CDR receiving functionality (a CGF).
Gb	Interface between an SGSN and a BSS
Gc	Interface between a GGSN and an HLR
Gd	Interface between an SMS-GMSC and an SGSN, and between an SMS-IWMSC and an SGSN
Gf	Interface between an SGSN and an EIR
Gi	Reference point between GPRS and an external packet data network.

Gn	Interface between two GSNs within the same PLMN
Gp	Interface between two GSNs in different PLMNs. The Gp interface allows support of GPRS network services across areas served by co-operating GPRS PLMNs.
Gr	Interface between an SGSN and an HLR
Gs	Interface between an SGSN and an MSC/VLR
GEA	GPRS Encryption Algorithm
GGSN	Gateway GPRS Support Node
GIF	Graphics Interchange Format
GMM/SM	GPRS Mobility Management and Session Management
GPRS	General Packet Radio Service
GPRS-SSF	GPRS Service Switching Function
GPRS-CSI	GPRS CAMEL Subscription Information
GPS	Global Positioning System
GRX	GPRS Roaming eXchange
GSM	Global System for Mobiles (or Mobile Communications)
GSMA	GSM Association
GSM-SCF	GSM Service Control Function
GSIM	GSM Service Identity Module
GSN	GPRS Support Node
GTP	GPRS Tunneling Protocol
GTP-C	GTP Control Plane
GTP-U	GTP User Plane
HLR	Home Location Register
HPLMN	Home PLMN
HTTP	HyperText Transport Protocol
Hz	Hertz (unit of frequency; also referred to as cycles per second)
ICMP	Internet Control Message Protocol
ICNIRP	International Commission on Non-Ionizing Radiation Protection
IEEE	Institute of Electrical and Electronic Engineers
IETF	Internet Engineering Task Force
IK	Integrity Key
IMAP4	Internet Message Access Protocol
IN	Intelligent Network
IP	Internet Protocol
IP-M	Internet Protocol Multicast
IPv4	Internet Protocol version 4
IPv6	Internet Protocol version 6
IPX	Internet Packet eXchange

IREG	International Roaming Experts Group
ISO	International Standards Organization
ISP	Internet Service Provider
ITU	International Telecommunications Union
Iu	Interface between the RNS and the core network; also considered a reference point.
IWF	Inter-Working Function
kbit/s	Kilobits per second
Kg	Kilogram
KSI	Key Set Identifier
L2TP	Layer-2 Tunnelling Protocol
LAN	Local Area Network
LIF	Location Interoperability Forum
LL-PDU	Logical Link Control Protocol Data Unit
LLC	Logical Link Control
MAC	Medium Access Control
Mbit/s	Megabits per second; 1 Mbit/s = 1 million bits per second
MIME	Multipurpose Internet Mail Extensions
MIP	Mobile IP
MMS	Multimedia Messaging Service
MNRF	Mobile station Not Reachable Flag
MNRG	Mobile station Not Reachable for GPRS flag
MNRR	Mobile station Not Reachable Reason
mp3	MPEG 1 – Layer 3
MSC	Mobile Switching Center
MSIG	M-Services Interest Group of the GSM Association
MTP2	Message Transfer Part layer 2
MTP3	Message Transfer Part layer 3
NAT	Network Address Translation
NGAF	Non-GPRS Alert Flag
N-PDU	Network Protocol Data Unit
NS	Network Service
NSDU	Network Service Data Unit
NSAPI	Network layer Service Access Point Identifier
NSS	Network SubSystem
ODB	Operator-Determined Barring
OTA	Over the Air
OTD	Observed Time Difference
OTDOA-IPDL	Observed Time Difference of Arrival—Idle Period Downlink
PC	Personal Computer

PCM	Pulse Code Modulation
PCU	Packet Control Unit
PDA	Personal Digital Assistant
PDCH	Packet Data Channel
PDCP	Packet Data Convergence Protocol
PDN	Packet Data Network
PDP	Packet Data Protocol, e.g. IP
PDU	Protocol Data Unit
PIM	Personal Information Management
PLMN	Public Land Mobile Network: the terrestrial part of a mobile operator's network
PMM	Packet Mobility Management
POP3	Post Office Protocol version 3
PPF	Paging Proceed Flag
PPP	Point-to-Point Protocol
PSAP	Public Safety Answering Point
PSD	Packet Switched Data
PSTN	Public-Switched Telephone Network
PSV	Packet-Switched Voice
PTM	Point to Multipoint
PTP	Point to Point
P-TMSI	Packet TMSI
PVC	Permanent Virtual Circuit
QoS	Quality of Service
RA	Routing Area
RAB	Radio Access Bearer
RAC	Routing Area Code
RAI	Routing Area Identity
RANAP	Radio Access Network Application Protocol
RAU	Routing Area Update
RF	Radio Frequency
RFC	Request for Comments
RIPE NCC	Réseaux IP Européens Network Coordination Centre
RLC	Radio Link Control
RNC	Radio Network Controller
RNS	Radio Network Subsystem
RNTI	Radio Network Temporary Identity
RRC	Radio Resource Control
RTD	Real Time Difference
RTT	Round Trip Time
SAI	Service Area Identifier

SAP	Service Access Point
SAR	Segmentation and Reassembly
SAR	Specific Energy Absorption Rate
SDU	Service Data Unit
SerG	Services Group of the GSM Association
SGSN	Serving GPRS Support Node
SIP	Session Initiation Protocol
SMIL	Synchronized Multimedia Integration Language
SMS	Short Message Service
SM-SC	Short Message service Service Center
SMS-GMSC	Short Message Service Gateway Mobile Switching Center
SMS-IWMSC	Short Message Service Interworking Mobile Switching Center
SMTP	Simple Mail Transfer Protocol
SNDCP	SubNetwork-Dependent Control Protocol: the protocol used on air interfaces instead of GTP
SN-PDU	SubNetwork Dependent Convergence Protocol-Protocol Data Unit
SNDC	SubNetwork Dependent Convergence
SNDCP	SubNetwork Dependent Convergence Protocol
SPI	Security Parameter Index
SRNC	Serving Radio Network Controller
SRNS	Serving Radio Network Subsystem
SS7	Signaling System #7
SVC	Switched Virtual Circuit
TA	Timing Advance
TCAP	Transaction Capabilities Application Part
TCP	Transmission Control Protocol
TFT	Traffic Flow Template
TEID	Tunnel Endpoint IDentifier
TLLI	Temporary Logical Link Identity
TOM	Tunnelling Of Messages
ToS	Type of Service
TRAU	Transcoder and Rate Adaptor Unit
TWG	Terminal Working Group of the GSM Association
UAProf	User Agent Profile
UDP	User Datagram Protocol
UEA	Universal Mobile Telephone System Encryption Algorithm
UIA	Universal Mobile Telehone System Integrity Algorithm

UMTS	Universal Mobile Telephone System
URA	UTRAN Registration Area
USIM	User Service Identity Module
UTRAN	Universal Mobile Telephone System Terrestrial Radio Access Network
Um	Interface between the mobile station (MS) and the GSM fixed network part. The Um interface is the GSM network interface for providing GPRS services over the radio to the MS. The MT part of the MS is used to access the GPRS services in A/Gb mode through this interface.
Uu	Interface between the mobile station (MS) and the UMTS fixed network part. The Uu interface is the UMTS network interface for providing GPRS services over the radio to the MS. The MT part of the MS is used to access the GPRS services in Iu mode through this interface.
VoIP	Voice over Internet Protocol
VLSM	Variable Length Subnet Mask
VPLMN	Visited PLMN
VPN	Virtual Private Network
W	Watts
W3C	WWW Consortium
WAP	Wireless Access Protocol
WBMP	Wireless Bit Maps
WHO	World Health Organization
WEP	Wired Equivalent Privacy
WIM	WAP Identity Module
WML	Wireless Markup Language
WSP	WAP Session Protocol
WTLS	Wireless Transport Layer Security
WTP	Wireless Transaction Protocol

CONTRIBUTORS

Laurent Bernard
An engineer, physicist and optimisation specialist, Laurent Bernard has worked on the Internet from its commercial beginnings in France in 1994. After creating the Internet helpdesk for one of France's major ISPs, he became interested in communities and government markets. During this period, Laurent served in the French Internet Authority (AFNIC) in the capacity of French-Zone Internet Naming Commission Member and French ISPs Consulting Committee Member.

In 2000, Laurent Bernard joined the Carriers teams at France Telecom Long Distance. As International Project Manager he oversaw the creation of France Telecom's GRX product line and took an active part in the GRX Task Force, where he chaired the early DNS Working Group.

Charles Brookson
Charles of CEng FIEE AFRIN has worked in the telecommunications industry for over 25 years, specialising in security issues. He is Chairman of the GSM Association Security Group and participated in the definition of many mobile radio standards such as GSM, DECT, and 3G.

Clif Campbell
Clif Campbell is Director, Technology in the Cingular Wireless Strategic Planning organization. In this position he is responsible for Cingular's activities relating to the development and adoption of wireless network standards and specifications. Over his career he has been involved in the development of wireline and wireless services, packet data networks, intelligent networks and voice messaging systems. He is also the chair of the Data Working Group of the GSM North America organization, one of the Regional Interest Groups of the GSM Association.

Simon Cavenett
Simon Cavenett, Principal, Mondo Techo LLC, is a respected senior telecommunications consultant and industry analyst with particular focus on wireless telecommunications and related technologies. He has fifteen years experience across the telecommunication industry spanning network operators, equipment vendors, and engineering consultants. His career started with Telstra in Australia where he worked on numerous key projects including the rollout of the

national analog cellular (AMPS) network, the rollout of the national digital cellular (GSM) network, and other key projects including Australia's first major digital fixed wireless system deployment. Over the last six years, Simon has lived in the United States and has worked on numerous national and international telecommunications projects with LCC, Nortel, TWS International, and Blue Sky PCS. Prior to founding Mondo Techo in August of this year, he served as CTO of Avian Communications, a startup wireless network infrastructure vendor.

Robert Conway

Rob Conway is the GSM Association's Chief Executive Officer. He has a wealth of industry knowledge and experience, including the management of global wireless telecommunications operations. Conway previously held key executive positions with Motorola Inc.'s international cellular operations including CEO, Director, and chairman of the steering committees for major cellular joint ventures. Most recently, he headed up global business development for Motorola's International Network Ventures Group. Of interest, he was instrumental in Motorola's decision to invest with Entel in Chile to introduce the first GSM network in Latin America and in Motorola's investment in Egypt's GSM network.

He is also a lawyer with industry merger and acquisition experience and served as General Counsel for Motorola's subscriber terminals group and Deputy General Counsel for one of its largest international telecom equipment joint ventures.

Rob Conway represents the GSM Association externally and internally and is responsible for the overall management of the operations of the GSM Association reporting to the Executive Committee. In addition he works closely with the Executive Committee on strategic directions for the organisation.

Carolyn Davies

Carolyn Davies is a research analyst with Baskerville: Part of the Informa Telecoms Group, where she researches and writes market reports. Prior to joining Baskerville, Carolyn was a research analyst with the ARC Group specialising in the mobile Internet market, and was also a conference researcher with IBC Global Conferences, where she produced mobile commerce, handset technology and IP events.

Axel Doerner

Dr. Doerner earned his doctorate in theoretical mathematics before joining the IT department of Vodafone in Germany. His primary responsibility is the management of IT projects that implement support for new services like CAMEL, GPRS in the billing infrastructure. Since 1994 he has been active in standardisation efforts related to charging and accounting. As part of this service he has also acted as a chair person for several working parties within the GSM Association.

Scott Fox

Scott Fox is a 23-year veteran of the wireless industry and currently Group President—Wireless Facilities, Inc. Scott was instrumental in developing and driving the company's international expansion. The global leader in telecommunications outsourcing, Wireless Facilities, Inc. designs, deploys and manages wireless networks for many of the largest cellular, PCS, and broadband wireless carriers and equipment suppliers worldwide.

Mr. Fox is Chairman Emeritus and a member of the Board of the global GSM Association. He is also Chairman of Mobileum—a privately held, VC-backed company providing enhanced roaming and enterprise mobility services to wireless carriers globally.

Prior to joining WFI, Scott was Vice President Strategy and Chief Technology Officer for BellSouth. He has held a broad variety of executive positions within the wireless industry at MCI Corp, MobilMedia, McCaw Cellular, Metromedia, Southwestern Bell, and Radio Telephone Company.

Scott has published numerous articles, is a frequent invited speaker at industry forums and conferences, and serves as an Advisor and Director for numerous companies within the Telecommunications and Venture Capital industries.

Kim Fullbrook

Kim Fullbrook is engaged in activities associated with the design of the 3G IP network architecture for BT Cellnet, now O2. Kim has an MA in Engineering and has worked in the IT and telecommunications industries for almost 20 years. He first became involved with GPRS in early 1998 as designer and workstream leader for IP networking activities and is now the Technical Architect for IP aspects of the BT Cellnet 3G and GPRS networks. In mid-2000 he authored a white paper on IP addressing for GPRS mobiles that was widely circulated within the industry. Kim is a member of the GSM Association and, with coauthor Jarnail Malra, has led the way on IP addressing for all GPRS operators by producing the GSM-A IR.40 document on GPRS IP addressing policy for network infrastructure and mobile terminals.

David Gordon

As Manager of the International Services Department, Mr. Gordon has been responsible for the establishment and operation of international services and international relations for Partner Communications Company from day one of the company, which operates the Orange network in Israel. He is a company delegate to the GSM Association work groups and plenary, served as Chairman of the High Speed Data Interest Group within the GSM Association, and is a member of the SMS Steering Committee.

Prior to Joining Partner Communications Company Mr. Gordon served as a senior editor at Ma'ariv Daily Newspaper, one of Israel's major daily newspapers, publishing issues related to computer hardware, software, Internet and on-line services, telecommunications and hi-tech industries. He also conducted market research of the Israeli telecommunications market in collaboration with International Data Corporation (IDC).

Conchi Gutiérrez

Conchi Gutiérrez has a six-year Degree in Telecommunications Engineering. She joined Telefónica Móviles in Spain in 1996, where she was a project manager in the mobile services development division with special focus on the international roaming business. She has represented Telefónica Móviles at many international fora, including the GSM Association, 3GPP-SA and the UMTS Forum. Currently she is responsible at the corporate level for coordinating Telefónica

Móviles global positions in the various international organizations to which the company contributes.

Telefónica Móviles was created in 2000 as a new business line to manage Telefónica S.A. mobile services worldwide. It is present in 14 countries worldwide and currently manages 27.5 million customers, of which 12 million are in Latin America. Telefónica Móviles also takes part in two joint ventures: Terra Mobile (a mobile portal provider with over 5 million customers worldwide) and Mobipay International (a mobile payment solution supported by all the main banks and mobile operators in Spain).

Ray Haughey

Ray has spent 30 years in the telecommunications industry, 15 of them in security, fraud and risk management. In 1996 he graduated from Loughborough University with a Masters Degree in Security and Risk Management. He has held the position of GSM Association Fraud & Security Manager with responsibility for coordinating and developing the Association's Fraud (FF) and Security Groups (SG) worldwide, and more recently the position of Head of Security and Risk Management with a mobile operator in Ireland. Today he is a managing partner with a risk management firm in the UK.

In 2000, Ray formed his own consultancy business and has worked with the GSM Association in a number of Working Group and Task Force director positions including; IMT 2000 Steering Group (ISG), Services (SerG), Terminals (TWG), International Roaming (IREG), Roaming Task Force (RTF), M-Commerce Interest Group (MCIG), End to End MMS (E2E) and Minimum Performances (MPR) Task Forces.

Gerhard Heinzel

In 1991 Gerhard Heinzel earned a Master Degree in Computer Science from the Friedrich-Alexander University of Erlangen-Nuremberg in Germany. His subsequent career with the startup company SIGOS (Nuremberg) included several assignments with network equipment vendors as well as fixed and mobile operators. With SIGOS Gerhard become a specialist in the design and development of automated test systems.

In 1998 Heinzel joined Swisscom's business development branch for which he managed projects in the Network Evolution and Services area. In 2000 he was appointed Head of Service Validation and Testing of Swisscom Mobile's newly formed Service Creation & Integration Department.

As a result of his involvement in Swisscom's GPRS project, Gerhard took an active role in the GSM Association's GPRS working party and was elected the first chairman of the GSM Association's GRX Task Force, which he chaired until mid 2001.

Babak Jafarian

Dr. Jafarian heads the Network Solutions Group at Wireless Facilities Inc. (WFI), whose main activities are design and performances analysis of 2.5G/3G mobile and broadband fixed wireless networks. Services include technology assessment and market analysis for 3G mobile systems (cdma2000 and UMTS). In this capacity Dr. Jafarian has supervised projects for a number of operators to identify technical issues and carry out performance analysis for 3G migrations.

Prior to joining WFI, Dr Jafarian was an assistant professor in the Centre for Telecommunications Research, King's College London, where he established a Mobile Networking group and was one of the coordinators of BRAIN (broadband radio access for IP based networks)—an important European project working toward 4th generation mobile systems. As a research fellow at BT, he was involved in ETSI for 2.5G standardization (GPRS). His fields of expertise include mobile and cellular telecommunications (GSM, GPRS/EDGE and 3G), mobile networking (wireless mobile ATM, mobile and cellular IP) and wireless services.

Stephan Keuneke

Stephan Keuneke works for T-Mobile International, the group that includes among others T-Mobil of Germany, Radiomobil in the Czech Republic and max.mobil in Austria, all of which commercially launched GPRS in their networks early on. A psychologist by education, marketing specialist by profession and engineer by heart, he has worked in the telecommunication and multimedia industry for more than five years.

Joerg Kramer

Joerg Kramer is GSM Association's Services Group (SerG) Vice Chairman and was the editor of the M-Services Guidelines document (Phase I). After more than 6 years of activities in the GSM Association, ITU, ETSI, 3GPP and the WAP Forum representing D2 Vodafone, Joerg now works for the global Vodafone organization in Global Product Management and is responsible for 3G Launch Planning, Media Streaming and Java activities.

Rainer Lischetzki

Rainer Lischetzki studied at the Technical University Darmstadt obtaining a Dipl. Ing. in Electrical Engineering (Masters Degree equivalent) and now resides in the UK. His interest in mobile data arose during his years at university when designing satellite receivers. In 1992 he joined Motorola as a Data Account Manager for wireless packet data handsets. Later roles included Technical Support Manager and Sales and Support Manager for Europe.

Rainer is currently the Technical Marketing Manager for Motorola's Personal Communications Sector (PCS) in Europe, Middle East and Africa. He is primarily responsible for providing marketing and sales-focused technical expertise for mobile handset solutions involving technologies such as GPRS, Java and MMS. He plays a significant role in product commercialisation, facilitated by ongoing liaison with network operators, representation at industry conferences and exhibitions, and public speaking.

Philippe Lucas

Philippe is currently Head of Standardisation and Architecture for Orange. He is a member of the GSM Association Executive Committee and sits on the Board of Directors of Open Mobile alliance (OMA).

Previously he worked as an independent consultant and a GSM Association representative for MTN South Africa. He has also been the adviser for TIW in charge of the 3G licence in the UK & France and developed the mobile Internet strategy for Dolphin Europe. With SFR, Philippe was in charge of R&D activities for GSM and UMTS. In that role he participated in the definition of the UMTS and in the first WAP services on a GSM network.

Philippe has been actively involved in the GSM and UMTS/3G standardisation bodies in GSMA/3GPP/WAP Forum/ETSI (GSM—Tiphon) for the last decade. He has participated in the GSM Association framework since 1995, becoming chair of the Services Group in 1999.

Philippe has applied for more than 10 patents and regularly speaks at international conferences in the mobile industry. His main interests reside in the development of services and the Internet environment as applied to the mobile domain, ensuring GSM/3G operators get maximum revenue from evolutions.

Jarnail Malra

Jarnail is a Chartered Engineer with a BEng Honours degree in telecommunications and has over fifteen years experience in the telecommunications industry. Currently he is engaged in activities associated with the design of the 3G IP network architecture for BT Cellnet, now O2. He has worked in both fixed and mobile telecommunication networks, which experience has been extended to include the IP world over recent years. As a member of GSMA, Jarnail and coauthor Kim Fullbrook have led the way on IP addressing for all GPRS operators by jointly producing the GSM-A IR.40 document on GPRS IP addressing policy for network infrastructure and mobile terminals.

Yves Martin

Yves Martin (Orange France) chairs the GSMA M-Services Interest Group. After some years as Head of the Handsets and SIM technical group of Orange France, Yves is following the development and validation of handsets closely as a key issue in the roll-out of networks and services.

R. Clark Misul

Clark Misul has diverse telecommunication experience. He started his career managing advanced distribution network projects in the research and development division of Telecom Italia, where he worked with SDH, ATM, PON, WLL, and MAN technologies.

He then held management and executive positions in the wireless industry, at Telecom Italia Mobile, Globalstar LP, and Iridium LLC.

Clark is currently Standards and Services Director at Detecon Inc. There he oversees, on top of the technical access and core network planning and operations, all the elements of the actual services business, such as marketing, billing and provisioning.

The above activities convinced him to make an active contribution to international standardization fora and in multinational environments.

Lauro Ortigoza Guerrero

Lauro Ortigoza Guerrero received his BSc. degree in Electronics and Communications Engineering from ESIME-UPC and his M.Sc. degree in Electrical Engineering from CINVESTAV, both from the Instituto Politécnico Nacional in Mexico City, Mexico. He received his Ph.D. in Electric Engineering from King's College London, University of London. He is now a technical consultant with Wireless Facilities, Inc. (WFI) in San Diego, California, where he provides technical advice to wireless network operators selecting network solutions. He has participated in the writing of two books, co-authoring "Resource Allocation in Hierarchical Cellular Systems, " (Artech House, 1999) and contributing to "Sistemas Inalámbricos de Comunicación Personal" (Alfaomega Grupo Editor, Mexico, 2001).

Carsten Otto

Carsten Otto is currently product manager for data service at T-Mobile Germany, with the primary responsibility for next generation data services. During his career at the Deutsche Telekom group he has also worked on projects for VoiceStream Wireless in Seattle, Deutsche Telekom's Asian Pacific headquarter in Singapore and T-Systems International in Bonn.

He has a Diploma in Communications Engineering (Dip.-Ing.) and an MBA in International Business (University of Birmingham, UK).

Stella Penso

Stella Penso has been in the telecom industry for five years. She is currently in charge of the Pricing Division of Marketing at Turkcell, where she has focused on inter-operator pricing and regulatory issues since 1999. Prior to Turkcell, between 1996 and 1999, she conducted analyses concerning the state of competition in domestic U.S. telecommunication markets and drafted expert reports for FCC and state regulatory proceedings as a senior analyst at NERA (National Economic Research Associates), an economic-litigation consulting firm based in New York. Stella Penso received her B.A. from Washington University in St. Louis in Economic Anthropology in 1993 and her Master of International Affairs from Columbia University in Economic Development in 1996.

Carol Politi

Carol Politi is cofounder and vice president of marketing for Megisto Systems, a technology leader deploying innovative IP-based mobile infrastructure equipment. Ms. Politi has 15 years of experience in data networking and telecommunications systems with responsibilities spanning product management, marketing, sales, and engineering. Prior to founding Megisto, Ms. Politi served as Vice President of Product Management for Ericsson's IP Infrastructure division where she was responsible for establishing and managing product development priorities for Ericsson's portfolio of carrier-based IP core networking products. Ms. Politi was previously Assistant Vice President of Marketing and Business Development at Torrent Networking Technologies. In this capacity she was responsible for Torrent's business development activities, taking a lead role in establishing strategic sales and investment partnerships. Ms. Politi also held senior marketing, product management, and engineering positions with Newbridge Networks and Hughes Network Systems. She has an MSEE from Johns Hopkins University, a BSEE from University of Maryland and an MBA from the University of Maryland.

Tage Rasmussen

In October of 2000 Tage Rasmussen joined the Deutsche Bank Venture Partner (DBVP) financed start-up Pre-Tel Wireless Ltd. as Chief Operating Officer. Now branded as End2End, it is one of the world's first Wireless Applications Infrastructure Providers (WAIP), focused on exploring the potential of wireless Internet access and the demand for intelligent and high-value data services.

Before joining End2End, Tage spent many years with SONOFON, the winner of the first Danish GSM license. As SONOFON's first Technical Director, he represented the company in the GSM MoU Association. When he became Chief Operating Officer and Department Managing Director in 1997, his purview shifted to leading the charge on product, IT, and business development. Rasmussen developed SONOFON' strategy for Wireless Internet, including launch of a nationwide GPRS network and a successful bid for Fixed Wireless Access licenses awarded after he left the company.

Jessica Roberts

Jessica Roberts currently works as a Solution Manager for Location Based Services at Nokia Networks. Jessica has worked in telecommunications since 1990 and has managed programs in software development and implementation programs to provide CLASS features and advanced intelligent network features. Prior to moving to Nokia, she was involved in developing IP telephony and fixed wireless solutions.

Jack Rowley

For ten years, Dr. Jack Rowley worked at the Telstra Research Laboratories (Australia) in the area of mobile communications with specific accountability for radio frequency safety and interference issues. He joined the GSM Association as Director of Environmental Affairs in January 2000 and his current responsibilities include overseeing the external EMF research programme, driving member initiatives and developing information resources on environmental issues. He has a degree in electronic engineering and a Ph.D. in the design of handset antennas.

Rafael Ruiz de Valbuena Bueno

Rafael has a six-year Degree in Telecommunications Engineering. His professional carrier began in 1996 There he was responsible for customer MAN/WAN-specific solutions associated with ATM and SDH technologies. In 2000 he joined Telefónica Móviles España as a marketing consultant in the Department of Value Added Services, where he is today in charge of wireless data corporate services, based on GSM, GPRS and TETRA technologies.

Telefónica Móviles was created in 2000 as a new business line to manage Telefónica S.A. mobile services worldwide. It now has offices in 14 countries worldwide and takes part in two joint ventures: Terra Mobile (a global mobile portal provider) and Mobipay International (a mobile payment solution supported by all the main banks and mobile operators in Spain).

Richard Schwartz

Richard Schwartz, President and CEO of SoloMio, has repeatedly created and led new organizations to innovate software that has had significant impact on the market. Before founding SoloMio, he was senior vice president and general manager of Convergence Applications at Vignette Corporation, after Vignette acquired his company Diffusion.

At Borland International, where Schwartz served as chief technology officer and senior vice president of technology, he created Borland Interactive, a joint venture with MCI focused on creation and delivery of a consumer online portal service. From 1990-93 Richard Schwartz created and developed the pan-European business and the European Development & Localization Center, headquartered in Paris. Schwartz joined Borland via the acquisition of his company Ansa

Software. He co-founded Ansa Software and was the co-creator of Paradox, the first widely used relational database software for business professionals that sold over 10 million copies.

Mark Smith

Mark Smith is Communications Director for the GSM Association (GSMA) the world's leading wireless industry representative body. The GSMA today consists of more than 639 second and third generation wireless network operators working collaboratively to define, prioritise and communicate requirements, as well as key manufacturers and suppliers to the wireless industry.

With more than seven years experience in the telecommunications sector, Smith is responsible for the Association's external media relations globally in addition to communications with the GSM Association's membership

Darren Thompson

Darren has worked in the wireless industry now for over 23 years. He is presently affiliated with VoiceStream Wireless where for the past 3 years he has managed Subscriber Equipment Standards development.

Earlier in his career Darren worked at Microcell Connexions as Manager of the terminal acceptance and smart cards group, and for major paging carriers in Canada as system manager for Eastern Canada. He is presently the chairman of the PTCRB (PCS Type Certification Review Board)and the GSMNA TWG (GSM North America Terminal Working Group).

Colin Watts

Colin Watts is Product Marketing Manager with Lucent Technologies, based in Lucent's GSM/UMTS Headquarters in Swindon, UK.

Joining Lucent in July 2000, Colin initially specialized in GSM-GPRS-EDGE-UMTS migration and TDMA-CDMA-UMTS convergence issues. Colin currently champions the UMTS UTRAN offer segment, in addition to his broader role supporting Lucent's UMTS and generic 3G solutions, and controlling Lucent's Mobility Web presence.

Colin has extensive experience in the Telecoms marketplace, gained in both large multinationals such as Philips and Marconi, and in smaller start-up companies. Initially from a design engineering and management background, with an Honours degree in Electronics, he then moved into the commercial world, where his roles have included Commercial Manager, Business Development Manager and Marketing Director.

Colin has represented Lucent on many occasions, and has spoken at a number of global events.

Randy Wohlert

Randolph Wohlert is a Principal Member of Technical Staff with SBC Communications. His charter includes management of next generation standards for both wireless and wireline initiatives. He is currently serving as the chair of the Technical Standards Committee within Committee T1 that is responsible for performance and reliability standards. Additionally he is vice-chair of the 3GPP group for service requirements. Mr. Wohlert has been involved in the telecommunications business for approximately 25 years, and holds degrees in biochemistry, education, and computer science.

Graham Wright
Graham Wright is the EMEA Marketing Director for Lucent Technologies' Mobile Division. With a strong background in Product Marketing, he has experienced first hand the evolution from GSM through GPRS and on to UMTS. Indeed if you have been a regular visitor to the GSM World Congress in Cannes you will have seen his team's presentation of the transition of Lucent's portfolio from GSM to one that encompasses the ever widening spectrum of products associated with GPRS and UMTS.

Prior to Lucent he worked for IBM in their PC Division, Cadbury/Schweppes as a Trade Marketing Manager, and Intel. He is a B.Sc. (Hons) Graduate in Management Science from UMIST and holds a postgraduate Diploma in Marketing from the Chartered Institute of Marketing.

INDEX

Note: Boldface numbers indicate illustrations; contributing authors names are italicized.

ABOUT THE AUTHOR

John Hoffman has worked with the GSM Association since early 1997 as a Senior Consulting Director responsible for numerous GSMA initiatives, including the M-Services Requirements Program, the GPRS Roaming Initiative, the M-Commerce Task Force, the GSM Certification Forum, and the initial 3G Program Plan. He has held various executive positions in the wireless industry: Regional Director of Central and North Florida for BellSouth Mobility; Chief Operating Officer of SONOFON in Denmark; Director of BellSouth International in Brussels, Belgium; General Manager of BellSouth Mobility DCS in Charlotte, North Carolina; and Chief Operating Officer of Pocket Communications in Washington, D.C. Mr. Hoffman has management experience in all phases of an operator's wireless business, including sales, marketing, technical engineering and operations, customer care, distribution, and general profit and loss management. He has managed the implementation, commercial launch, and operations of wireless networks which include GSM 900, GSM 1900, US AMPS, and US DAMPS technologies in the United States and Europe.

John holds a Bachelor of Science Degree in Architecture and a Master of Architecture Degree from the University of Michigan, as well as a Master of Business Administration Degree from the University of Phoenix. He is a licensed Architect in the state of Colorado and is a keynote speaker at numerous conferences and seminars around the world. John writes and publishes an electronic newsletter, *The Wireless Evolution Insider*, which provides information about wireless data business sector. John resides in Charlotte, North Carolina and travels extensively on business related activities.